FUNDAMENTAL ELECTRONIC DEVICES

PRENTICE HALL'S FUNDAMENTALS SERIES FOR ELECTRONIC TECHNOLOGY

Fundamental DC/AC Circuits: Concepts and Experimentation

Fundamental Electronic Devices: Concepts and Experimentation, Second Edition

Digital Electronics: Theory and Experimentation

Microprocessor Technology: Theory, Experimentation and Troubleshooting

FUNDAMENTAL ELECTRONIC DEVICES
Concepts and Experimentation

Second Edition

Fredrick W. Hughes

Electronics Training Consultant

Prentice Hall, Englewood Cliffs, New Jersey 07632

Library of Congress Cataloging-in-Publication Data

Hughes, Fredrick W., 1936–
Fundamental electronic devices : concepts and experimentation / Fredrick W. Hughes.—2nd ed.
p. cm.—(The Fundamentals series for electronic technology)
Rev. ed. of: Basic electronics. c1984.
ISBN 0-13-332883-X
1. Solid state electronics. 2. Solid state electronics—Experiments. I. Hughes, Fredrick W., 1936– Basic electronics. II. Title. III. Series.
TK7871.85.H76 1990 89-26438
621.381—dc20 CIP

Editorial/production supervision and
interior design: Tom Aloisi
Cover design: Karen Stephens
Manufacturing buyer: Dave Dickey

This book is dedicated to Tom Aloisi
for his work as production editor
and designing and developing such a good layout.

Previously published as Basic Electronics: Theory and Experimentation

ISBN 0-13-332883-X

Printed in the United States of America
10 9 8 7 6 5 4 3 2 1

Prentice-Hall International (UK) Limited, *London*
Prentice-Hall of Australia Pty. Limited, *Sydney*
Prentice-Hall Canada Inc., *Toronto*
Prentice-Hall Hispanoamericana, S.A., *Mexico*
Prentice-Hall of India Private Limited, *New Delhi*
Prentice-Hall of Japan, Inc., *Tokyo*
Simon & Schuster Asia Pte. Ltd., *Singapore*
Editora Prentice-Hall do Brasil, Ltda., *Rio de Janeiro*

Contents

Preface

This book has proved to be a practical approach for successfully training persons interested in becoming electronic technicians. It will provide the reader with the necessary skills for a job-entry position in the electronics industry.

The second edition of this book is enhanced with the addition of:

1. Unit objectives given for each unit, thereby showing the reader what will be learned.
2. A more in depth study of the theory of semiconductor materials.
3. A special experiment pertaining to voltage drops across diodes and the shorting effect of a forward-biased diode.
4. A section on miscellaneous diodes, including the tunnel diode, VCC diode, backward diode, *p-i-n* diode, Schottky diode, and snap diode.
5. A section on miscellaneous thyristors, including the Shockley diode, silicon-controlled switch, silicon unilateral switch, silicon bilateral switch, and gate-controlled switch.
6. Four new op-amp filter experiments that are easy to perform and show the effects of low-pass, high-pass, bandpass, and bandreject.
7. A more comprehensive discussion on the internal structure and operation of the 555 timer IC.
8. A 100-question final examination with illustrations on solid-state devices and circuits, with answers given in the Instructor's guidebook.

The learning of any skill requires thorough understanding of the basic theory of operation and proficient use of its tools or equipment. Acquiring such a skill is usually accomplished by repetitive efforts performed by the person desiring the skill. Learning electronics is no different and in many cases may seem more difficult. This manual teaches solid-state devices in a repetitious manner, but with varying methods that are easy, challenging, and often fun to perform. A person using this book should have a working knowledge of fundamental algebra and basic dc/ac circuit theory.

The manual is designed as an individualized learning package and involves the student in the activities of learning. Many line drawings are used to familiarize the student with circuit recognition and analysis, since this is an important part of being a good electronics technician. It provides the instructor with student-centered instructional material that does not require preparation.

Preliminary units are given to remind the reader of correct safety habits and to review the operation and use of multimeters, the oscilloscope, the signal generator, and the procedures for connecting a test circuit.

Each unit follows the same format so that the person using the book can become accustomed to the learning procedure. The first section introduces the theory and operation of the devices in a straightforward, practical manner. Section 2 provides a definition exercise for learning and remembering major terms and nomenclature given in the unit. In Section 3 there are basic drawing exercises and problems to familiarize the reader with schematic symbols, characteristics curves, proper methods of circuit connections, voltage and current nomenclature, and some basic calculations of the devices. Section 4 has basic experiments showing how to test the devices and verify their theory of operation. Fill-in questions are given at the end of each experiment to emphasize the important points gained from performing the experiment. A basic troubleshooting application is given in Section 5 in which the reader builds a simple circuit for the device and then introduces problems. Voltage measurements and oscilloscope waveforms are taken and entered into a table for analysis and future reference. More often than not, understanding is gained by finding a correct answer and solving specific problems. More exercises are presented in Section 6 to provide practice on the information covered in the experiments and in analyzing circuits. Section 7 is an instant review with simple remembering statements to reinforce the reader's understanding of the operation of the devices. Immediate feedback of the learning process is given in Section 8 with two self-checking quizzes.

A SPECIAL MESSAGE TO STUDENTS

You are studying electronics to obtain employment in a lucrative, exciting, and challenging industry. It is up to you, not your instructor (or anyone else, for that matter), how well you master the knowledge and skills required for such a fascinating mainstream career. To gain full value from this manual:

1. Perform all the experiments, exercises, and self-checking quizzes.
2. Review each section of the manual from time to time to refresh your memory on the basic concepts of the various semiconductor devices.
3. Keep this manual in your possession always, and review it before you take a job entrance examination.
4. Refer to this manual when you are on the job, to make you a more competent technician and prepare you for a higher-level position.
5. Have this manual handy when you are servicing equipment, to refer to the testing of individual devices and circuit troubleshooting applications. It can serve as a valuable aid to reducing time spent on repairing electronic devices.
6. Review the material in this manual when you are preparing to take professional license exams, such as that for the FCC Radio Telephone Operator's license and/or the CET (Certified Electronic Technician) exam.

Fredrick W. Hughes

Unit A

Preliminary Review—Safety

SECTION A-1 INTRODUCTION

The experiments in this manual do not use a voltage greater than 30 V (± 15 V); therefore, the chance of getting an electrical shock is greatly reduced. However, all voltages do have the potential to burn materials and start fires, destroy electronic components, and present hazards to the person performing the operations. Common sense and an awareness of electrical circuits is important whenever you are working on these experiments. An electronic technician or student may have to work with high voltages, power tools, and machinery. Before actual work is performed, sufficient instruction should be acquired in the proper use and safety requirements of all electronic devices.

SECTION A-2 CURRENT HAZARDS AND VOLTAGE SAFETY PRECAUTIONS

It takes a very small amount of current to pass through the human body from an electrical shock to injure a person severely or fatally. The 60-Hz current values affecting the human body are as follows:

Current Value	*Effects*
1 mA (0.001 A)	Tingling or mild sensation.
10 mA (0.01 A)	A shock of sufficient intensity to cause involuntary control of muscles, so that a person cannot let go of an electrical conductor.
100 mA (0.1 A)	A shock of this type lasting for 1 second is sufficient to cause a crippling effect or even death.
Over 100 mA	An extremely severe shock that may cause ventricular fibrillation, where a change in the rhythm of the heartbeat causes death almost instantaneously.

The resistance of the human body varies from about 500,000 Ω when dry to about 300 Ω when wet (including the effects of perspiration). In this case, voltages as low as 30 V can cause sufficient current to be fatal (I = voltage/wet resistance = 30 V/300 Ω = 100 mA).

Even though the actual voltage of a circuit being worked on is low enough not to present a very hazardous situation, the equipment being used to power and test the circuit (that is, power supply, signal generator, meters, oscilloscopes) is usually operated on 120 V ac. This equipment should have (three-wire) polarized line cords that are not cracked or brittle. An even better safety precaution is to have the equipment operate from an isolation transformer, which is usually connected to a workbench. To minimize the chance of getting shocked, a person should use only one hand while making voltage measurements, keeping the other hand at the side of the body, in the lap, or behind the body. Do not defeat the safety feature (fuse, circuit breaker, interlock switch) of any electrical device by shorting across it or by using a higher amperage rating than that specified by the manufacturer. These safety devices are intended to protect both the user and the equipment.

SECTION A-3
NEAT WORKING AREA

A neat working area requires a careful and deliberate approach when setting it up. Test equipment and tools should be set out on the workbench in a neat and orderly manner. Connecting wires from the test equipment to the circuit under test should be placed so as not to interfere with testing procedures.

Before power is applied to a circuit, the area around the circuit should be cleared of extra wires, components, hand tools, and debris (cut wire and insulation).

SECTION A-4
HAND TOOL SAFETY PRECAUTIONS

Hand tools can be dangerous and cause severe injuries. Diagonal cutters, wire strippers, long-nose pliers, and crimping tools can pinch and cut. Use care in cutting wire since small pieces can become projectiles and hit another person in the face or eye.

Screwdrivers should be held properly so that they do not slip and puncture some part of the body. Do not use them as chisels or cutters.

A soldering iron should have a holder on which to place it. Care must be used not to burn the body or other materials. Be careful of hot solder, which can splash and cause severe burns, especially to the eyes and face.

SECTION A-5
IN CASE OF ELECTRICAL SHOCK

When a person comes in contact with an electrical circuit of sufficient voltage to cause shock, certain steps should be taken as outlined in the following procedure:

1. Quickly remove the victim from the source of electricity by means of a switch, circuit breaker, pulling the cord, or cutting the wires with a well-insulated tool.

2. It may be faster to separate the victim from the electrical circuit by using a dry stick, rope, leather belt, coat, blanket, or any other nonconducting material.
   CAUTION: Do not touch the victim or the electrical circuit unless the power is off.
3. Call for assistance, since other persons may be more knowledgeable in treating the victim or can call for professional medical help while first aid is being given.
4. Check the victim's breathing and heartbeat.
5. If breathing has stopped but the victim's pulse is detectable, give mouth-to-mouth resuscitation until medical help arrives.
6. If the heartbeat has stopped, use cardiopulmonary resuscitation, *but only if you are trained in the proper technique.*
7. If both breathing and heartbeat have stopped, alternate between mouth-to-mouth resuscitation and cardiopulmonary resuscitation (*but only if you are trained*).
8. Use blankets or coats to keep the victim warm and raise the legs slightly above head level to help prevent shock.
9. If the victim has burns, cover your mouth and nostrils with gauze or a clean handkerchief to avoid breathing germs on the victim and then wrap the burned areas of the victim firmly with sterile gauze or a clean cloth.
10. *In any case, do not just stand there*—do something within your ability to give the victim some first aid.

Unit B

Preliminary Review—Multimeters

SECTION B-1 UNDERSTANDING MULTIMETERS

A *multimeter* is a general-purpose meter capable of measuring dc and ac voltage, current, resistance, and in some cases decibels. There are two types of meters: *analog,* using a standard meter movement with a needle, and *digital,* with an electronic numerical display (Figure B-1). Both types of meters have a positive (+) jack and a common jack (−) for the test leads; a function switch to select dc voltage, ac voltage, dc current, ac current, or ohms; and a range switch for accurate readings. The meters may also have other jacks to measure extended ranges of

Figure B–1 Multimeters: (a) analog VOM; (b) digital VOM. (From F. Hughes, *Illustrated Guidebook to Electronic Devices and Circuits,* Prentice-Hall, Englewood Cliffs, N.J., ©1981, Fig. 1–38, p. 41. Reprinted by permission.)

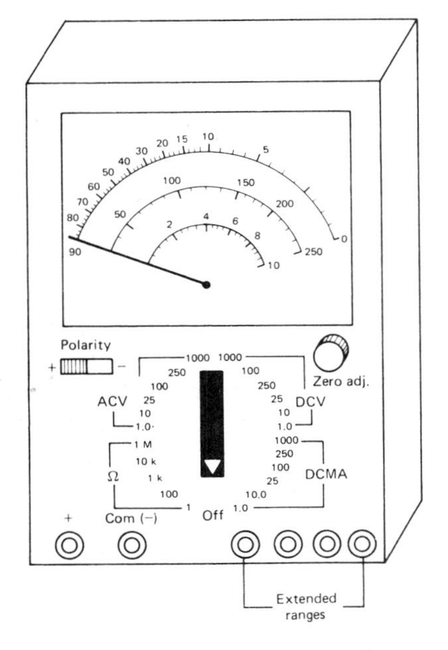

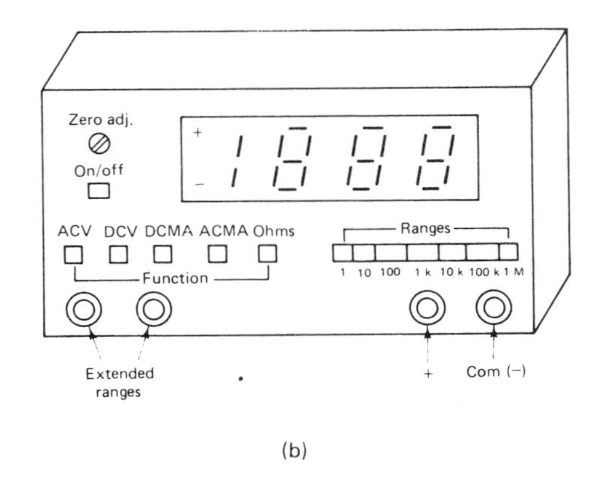

voltage (1 to 5 kV) and current (up to 10 A). There are some variations to the functions used for specific meters.

The analog meter usually includes the function and range switches in a single switch. It may also have a polarity switch to facilitate reversing the test leads. The needle will have a screw for mechanical adjust to set it to zero and also a zero adjust control to compensate for weakening batteries when measuring resistance. An analog meter can read positive and negative voltage by simply reversing the test leads or moving the polarity switch. A digital meter usually has an automatic indicator for polarity on its display.

A meter of reasonable quality will have an input resistance of 20 kΩ per volt or greater to prevent loading down a circuit, which causes an error in the reading. For example, if a dc voltmeter was set on the 10-V scale, its input resistance would be 200 kΩ. If it were placed across a 200-kΩ resistor in a circuit, the total effective resistance at that point would be 100 kΩ and would certainly cause an erroneous reading.

Meters must be properly connected to a circuit to ensure a correct reading (see Figure B-2). A voltmeter is always placed across (in parallel) the circuit or component to be measured. When measuring current, the circuit must be opened and the meter inserted in series with the circuit or component to be measured. When measuring the resistance of a component in a circuit, the voltage to the circuit must be removed and one end of the component opened from the circuit (to prevent any parallel paths from affecting the reading) and the meter placed in parallel with the component.

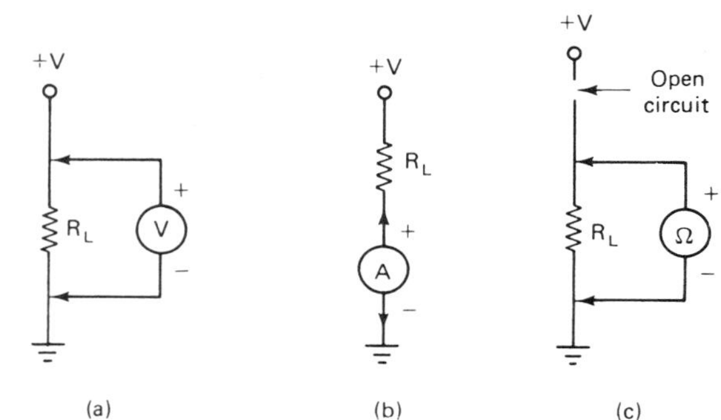

Figure B–2 Proper meter connections: (a) measuring voltage (parallel); (b) measuring current (series); (c) measuring ohms (open circuit). (From F. Hughes, *Illustrated Guidebook to Electronic Devices and Circuits*, Prentice-Hall, Englewood Cliffs, N.J., ©1981, Fig. 1–39, p. 42. Reprinted with permission.)

Special probes are used with meters for specific circuits. These include shielded cable, high-voltage, and capacitance types and radio-frequency (RF) detectors.

SECTION B-2 READING MULTIMETERS

On a standard analog meter there is a scale for ohms, dc, and ac (see Figures B-3 to B-5). When the function switch is set on 250ACV, a full-scale needle deflection indicates that the meter is measuring 250 V ac. If the needle is at 150, the meter is measuring 150 V ac. The various ranges of a specific function would use the same scale; therefore, the individual gradient values have to be determined. For example, if the 250ACV scale is used, there are 10 gradients between the numbers, and the value between numbers is 50 V. The value of each gradient can be found by dividing 50 by 10 ($50 \div 10 = 5$), which results in 5 V per gradient. The same scale would be used for 25ACV, except that the number $250 = 25$, $200 = 20$, $150 = 15$, and so on. Now there is a 5-V difference between numbers, so each gradient is worth 0.5 V. If the

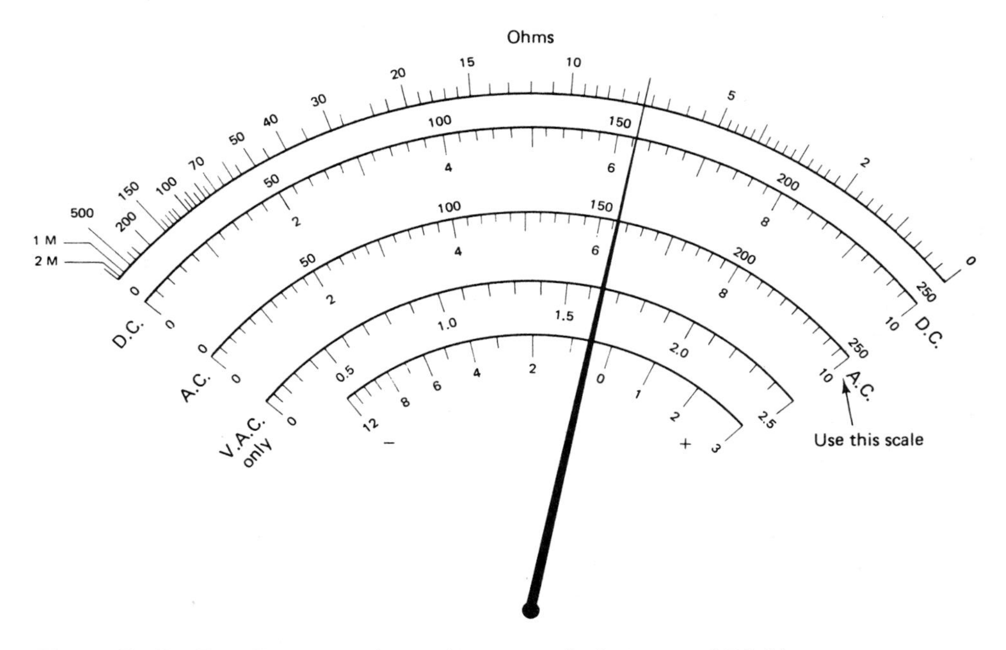

Figure B–3 Reading ac voltage (range switch set at 250 V ac, meter reads 155V). (From F. Hughes, *Illustrated Guidebook to Electronic Devices and Circuits,* Prentice-Hall, Englewood Cliffs, N.J., ©1981, Fig. 1–39, p. 42. Reprinted with permission.)

Figure B–4 Reading dc voltage (range switch set at 10 V dc, meter reads 9.2 V). (From F. Hughes, *Illustrated Guidebook to Electronic Devices and Circuits,* Prentice-Hall, Englewood Cliffs, N.J., ©1981, Fig. 1–41, p. 43. Reprinted with permission.)

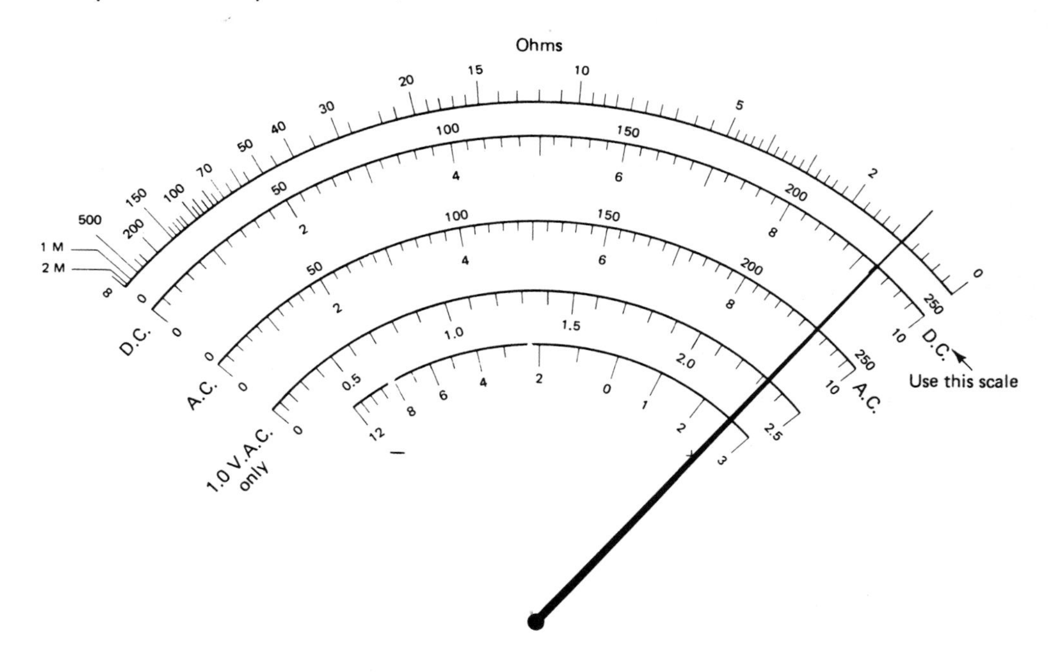

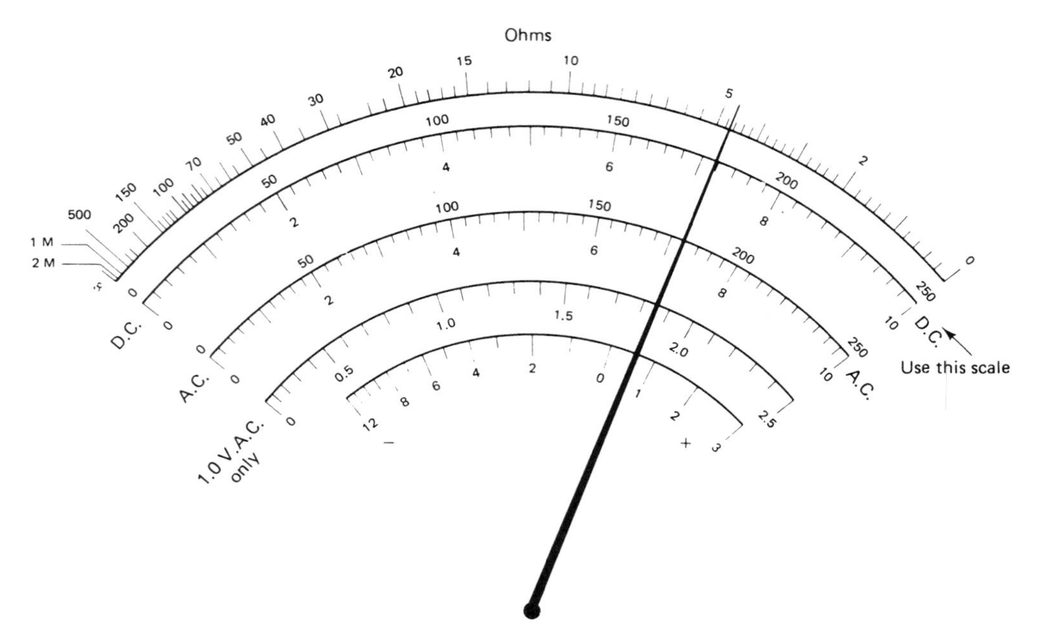

Figure B–5 Reading dc current range (range switch set at 25 mA dc, meter reads 18 mA. (From F. Hughes, *Illustrated Guidebook to Electronic Devices and Circuits,* Prentice-Hall, Englewood Cliffs, N.J., ©1981, Fig. 1–42, p. 44. Reprinted with permission.)

range switch is set to 10ACV, the same scale is used and each gradient is worth 0.2 V. When the range switch is set to 100ACV, the number 10 = 100, 8 = 80, 6 = 60, and so on, and each gradient is worth 2 V. All voltage and current scales are used the same way, remembering that the ac voltage is the effective or root-mean-square (rms) value.

The ohm scale is a nonlinear scale that may be indicated in reverse to the other scales (Figure B-6). The resistance function is used as a multiplier indicator. The function switch is placed to the desired range and the test leads are shorted together. The zero adj. (ohms adj.) control

Figure B–6 Reading ohms (range switch set a 1kΩ, meter reads 55kΩ). (From F. Hughes, *Illustrated Guidebook to Electronic Devices and Circuits,* Prentice-Hall, Englewood Cliffs, N.J., ©1981, Fig. 1–43, p. 45. Reprinted with permission.)

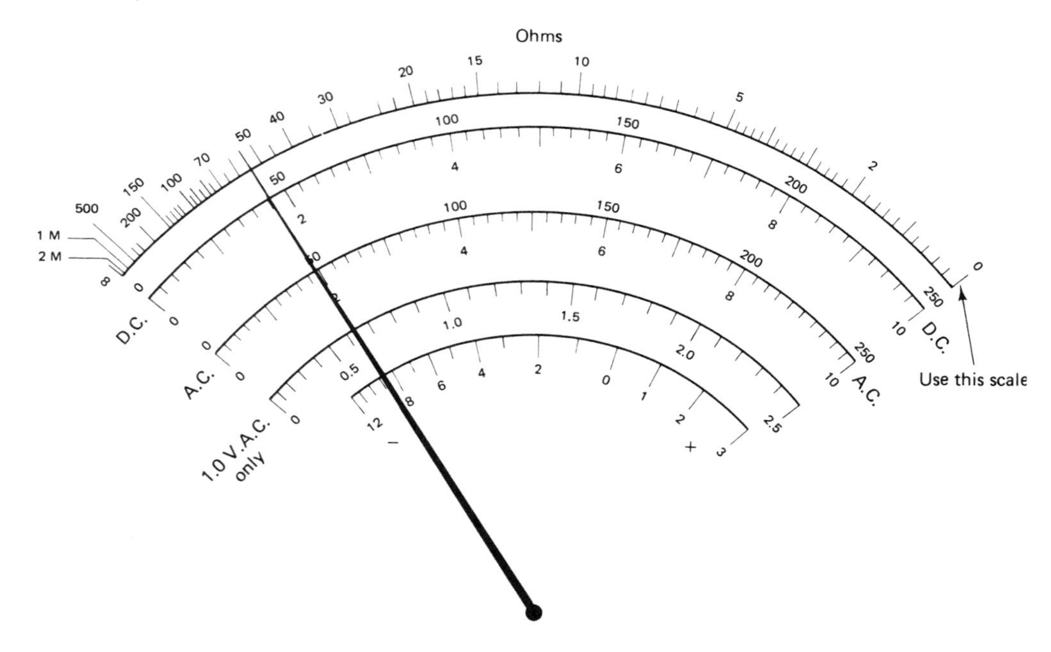

is then used to set the needle to zero on the scale. The leads are then opened and placed across the desired resistor to be read. If the function switch is set at 1 kΩ and the needle goes to 10, the value of the resistor being read is 10 kΩ. The meter may need to be zeroed each time a different range is selected.

With a digital meter, all values of dc, ac, and ohms measured will fall within the range selected. If the value being measured is greater than the range selected, an indication will be given, such as the display going blank or blinking, or perhaps only the most significant digit will light.

A user should spend some time getting oriented to meters and any test equipment being used. Equipment manuals will give detailed instructions as to their proper use.

Unit C

Preliminary Review—Oscilloscopes

SECTION C-1
UNDERSTANDING THE OSCILLOSCOPE

The oscilloscope presents an accurate electronic picture of changing voltages within a circuit. An electron beam is created, focused, accelerated, and properly deflected to display the voltage waveforms on the face of a cathode-ray tube (CRT). The basic circuits and controls of an oscilloscope (Figures C-1 and C-2) are:

Power supply: provides high dc voltage (up to a few thousand volts) for the CRT and lower dc voltages for other circuits

Intensity control: adjusts brightness of display

Focus control: adjusts sharpness of display

Time-base generator: provides the basic sawtooth voltage, which moves the trace on the face of the CRT from left to right horizontally

Time/CM selector: adjusts the frequency of the time-base generator

Horizontal amplifier circuits: amplifies the output of the time-base generator and applies it to the horizontal deflection plates

Horizontal gain control: adjusts full trace horizontally on the face of the CRT

Horizontal positioning control: centers trace horizontally on the face of the CRT

Vertical input: accepts voltage to be measured, either dc or ac

Vertical attenuator: reduces input voltage amplitude so as not to overdrive trace on face of the CRT

V/CM selector: selects desired input voltage attenuation

Vertical amplifier circuits: amplifies input voltage and applies it to the vertical deflection plates

Vertical gain control: manually adjusts the amplitude of input voltage displayed on the face of the CRT

Vertical positioning control: centers trace vertically on the face of the CRT

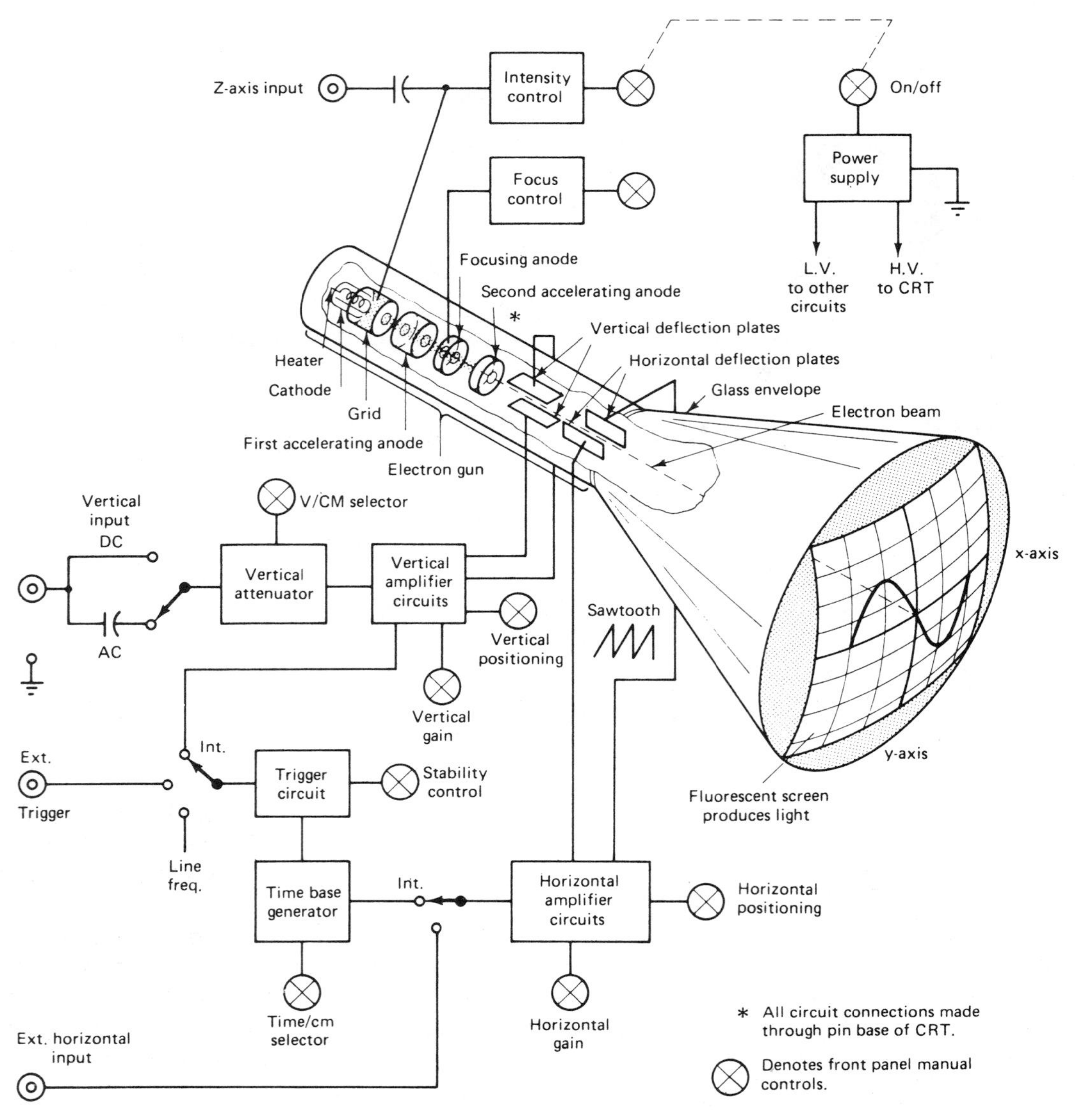

Figure C–1 Block diagram of basic oscilloscope. (From F. Hughes, *Illustrated Guidebook to Electronic Devices and Circuits,* Prentice-Hall, Englewood Cliffs, N.J., ©1981, Fig. 1–44, p. 46. Reprinted with permission.)

Trigger circuit: synchronizes time-base generator with input frequency, another external frequency, or 60-Hz line frequency; enables trace to be stopped for accurate measurements
Stability control: manual control for locking in display
External horizontal input: synchronizes horizontal trace for special measurements and displays as Lissajous patterns
Z input: used for intensity modulation of electron beam, perhaps for frequency measurements

SECTION C-2
READING THE OSCILLOSCOPE

The amplitude of a voltage waveform on an oscilloscope screen can be determined by counting the number of centimeters (cm) and/or fractions thereof, vertically, from one peak to the other peak of the waveform and then multiplying it by the setting of the volts/cm control. As an

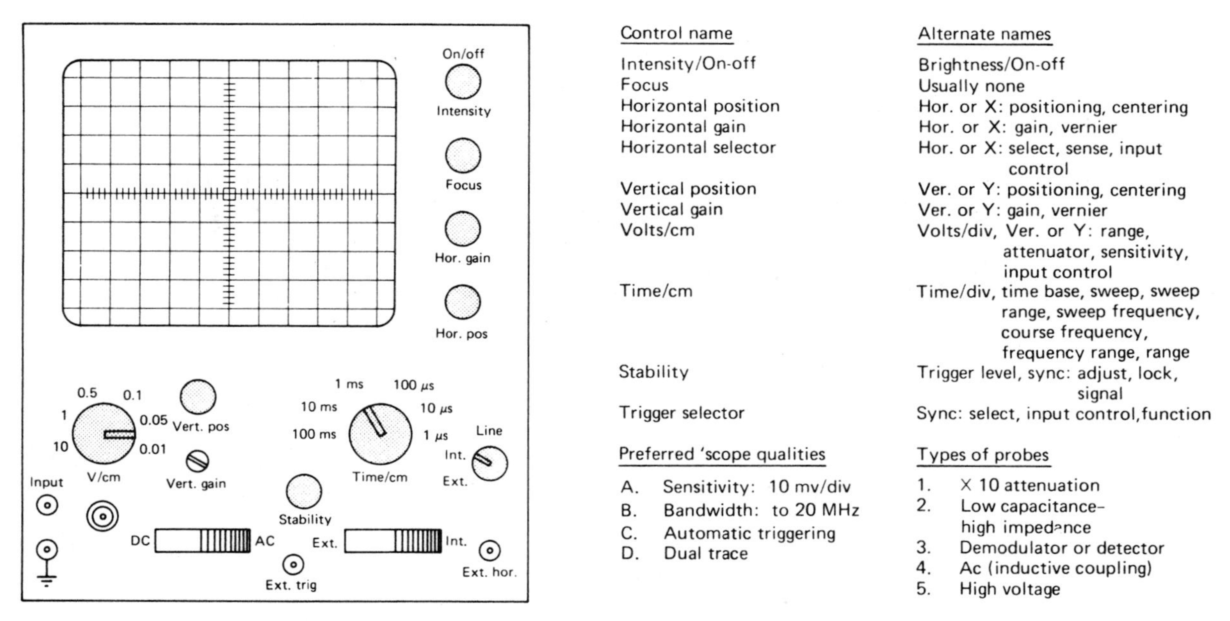

Typical oscilloscope control panel

Control name	Alternate names
Intensity/On-off	Brightness/On-off
Focus	Usually none
Horizontal position	Hor. or X: positioning, centering
Horizontal gain	Hor. or X: gain, vernier
Horizontal selector	Hor. or X: select, sense, input control
Vertical position	Ver. or Y: positioning, centering
Vertical gain	Ver. or Y: gain, vernier
Volts/cm	Volts/div, Ver. or Y: range, attenuator, sensitivity, input control
Time/cm	Time/div, time base, sweep, sweep range, sweep frequency, course frequency, frequency range, range
Stability	Trigger level, sync: adjust, lock, signal
Trigger selector	Sync: select, input control, function

Preferred 'scope qualities

A. Sensitivity: 10 mv/div
B. Bandwidth: to 20 MHz
C. Automatic triggering
D. Dual trace

Types of probes

1. × 10 attenuation
2. Low capacitance–high impedance
3. Demodulator or detector
4. Ac (inductive coupling)
5. High voltage

Figure C–2 Oscilloscope controls. (From F. Hughes, *Illustrated Guidebook to Electronic Devices and Circuits,* Prentice-Hall, Englewood Cliffs, N.J.,©1981, Fig. 1–45, p. 47. Reprinted with permission.)

example, referring to the sine wave in Figure C-3, if the amplitude is 4 cm and the control is set on 1 V/cm, the peak-to-peak voltage is 4 V (4 cm × 1 V/cm = 4 V). If the control is set on 0.5 V/cm, the voltage is 2 V peak to peak (4 cm × 0.5 V/cm = 2 V).

The frequency of a waveform can be determined by counting the number of centimeters and/or fractions thereof, horizontally, in one cycle or period of the waveform and then multiplying it by the setting of the time/cm control. For example, if the waveform is 4 cm long and the control is set at 1 ms, the period is 4 ms (4 cm × 1 ms = 4 ms). The frequency can now be found from the formula

$$f = \frac{1}{p} = \frac{1}{4\text{ ms}} = \frac{1}{4 \times 10^{-3}\text{s}} = 0.25 \times 10^{3} = 250\text{ Hz}$$

Figure C–3 Oscilloscope voltage waveforms: (a) sine wave; (b) square wave. (From F. Hughes, *Illustrated Guidebook to Electronic Devices and Circuits,* Prentice-Hall, Englewood Cliffs, N.J.,©1981, Fig. 1–46, p. 47. Reprinted with permission.)

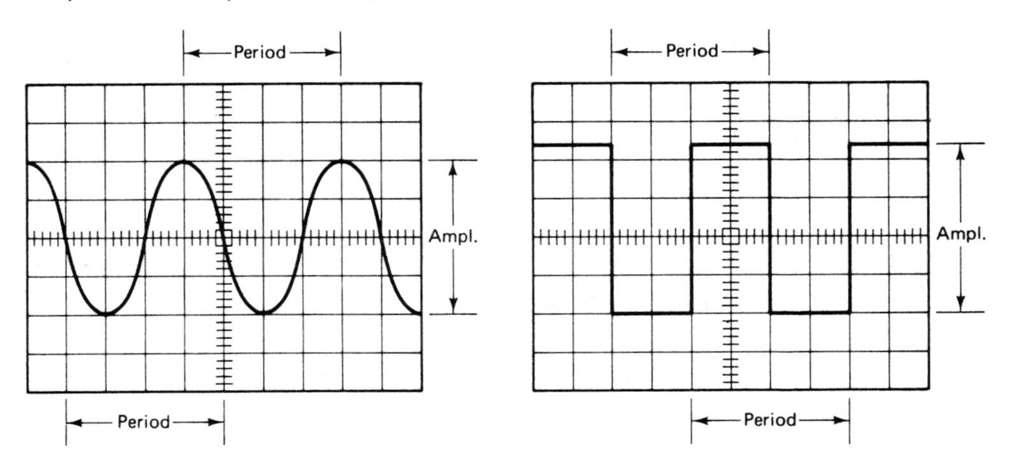

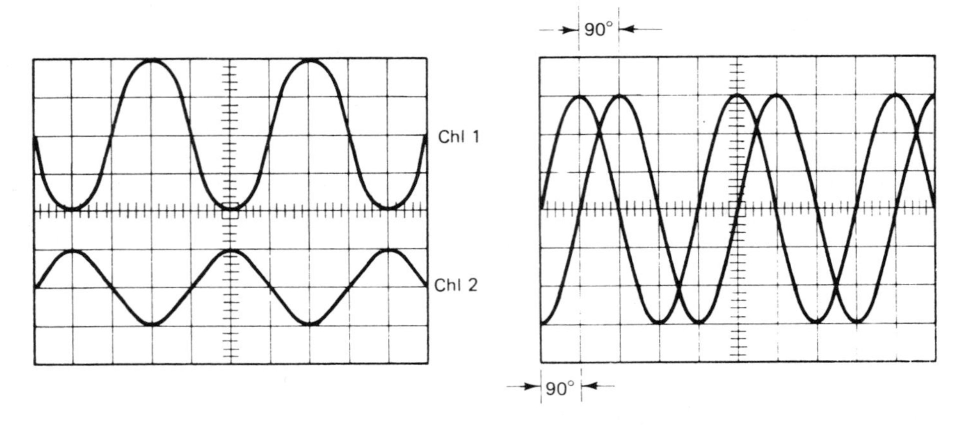

Figure C–4 Dual-trace oscilloscope voltage waveforms: (a) input/output signals of an amplifier; (b) both channels superimposed to show phase shift of two signals. (From F. Hughes, *Illustrated Guidebook to Electronic Devices and Circuits,* Prentice-Hall, Englewood Cliffs, N.J., ©1981, Fig. 1–47, p. 49. Reprinted with permission.)

If the control is set on 100 μs, the period is 400 μs (4 cm $\times$ 100 μs = 400 μs) and the frequency is 2.5 kHz:

$$f = \frac{1}{p} = \frac{1}{400\text{ ms}} = \frac{1}{4 \times 10^{-4}} = 0.25 \times 10^{4} = 2500\text{ Hz}$$

A dual-trace oscilloscope is advantageous to show the input signal and output signal simultaneously, to determine any defects, and to indicate phase relationships. The two traces may be placed over each other (superimposed) to indicate better the phase shift between two signals (Figure C-4).

Lissajous patterns can be used to show the phase relationship of two signals of the same frequency and to determine an unknown frequency from a known frequency (Figure C-5). One frequency is placed at the vertical input (f_v). The time-based generator is disengaged when the horizontal selector is set to external and the other signal is placed at the external horizontal input. This signal (f_H) now drives the horizontal sweep section. If both signals are the same frequency, a circle appears on the face of the oscilloscope. If f_v is twice f_H, a bow-tie type of pattern appears on the screen. The two peaks at the top (or bottom) and the single peak (side) indicate a ratio of 2 : 1. If the known frequency is at the horizontal input, say f_H = 1 kHz, then the frequency at the vertical input, f_v, is twice that at the horizontal input, or 2 kHz. If the frequencies are reversed at the inputs, the bow tie will turn on its side and indicate a 1 : 2 ratio. Other frequency ratios are also possible.

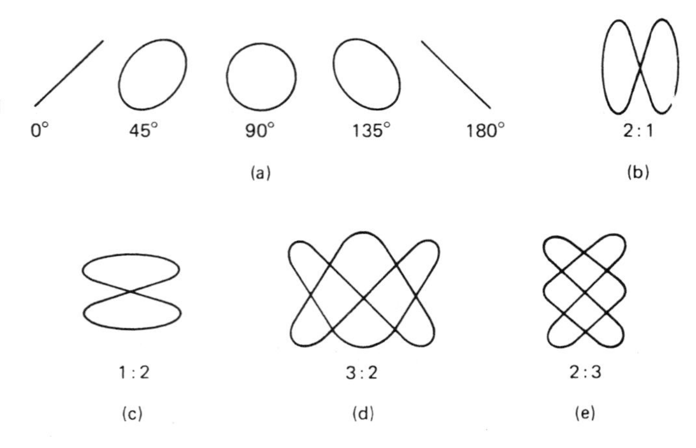

Figure C–5 Lissajous patterns: (a) two signals of the same frequency varying in phase; (b) $f_V = 2f_H$ (2:1); (c) $f_H = 2f_V$; (d) $f_V = 1.5f_H$; (e) $f_H = 1.5f_V$. (From F. Hughes, *Illustrated Guidebook to Electronic Devices and Circuits,* Prentice-Hall, Englewood Cliffs, N.J., ©1981, Fig. 1–48, p. 50. Reprinted with permission.)

Unit D

Preliminary Review—Basic Signal Generator

SECTION D-1
USING THE BASIC SIGNAL GENERATOR

A signal generator converts dc to ac or varying dc in the form of sine waves, square saves, triangle waves, or other types of voltage waveforms. The signal generator is used to inject a signal into a circuit or piece of equipment for troubleshooting or for calibration. Some generators may be used for audio, RF, or higher frequencies, whereas others have overlapping frequency ranges. A standard function generator usually has three types of waveforms. All generators will have a frequency range switch, a fine adjustmnet control for selecting a specific frequency, an amplitude control for varying the peak-to-peak output voltage, and output terminals (Figure D-1).

To select a sine wave of, say, 5 kHz, the user sets the function switch to the sine wave and the range switch to 1 k, and then adjusts the frequency fine adjust control to 5. The amplitude control is then adjusted to establish the desired peak-to-peak voltage output.

Some generators may have a dc component at the output terminals that could upset the circuit to which it is connected. In this case, a capacitor connected in series with the positive output terminal and the circuit will block the dc component.

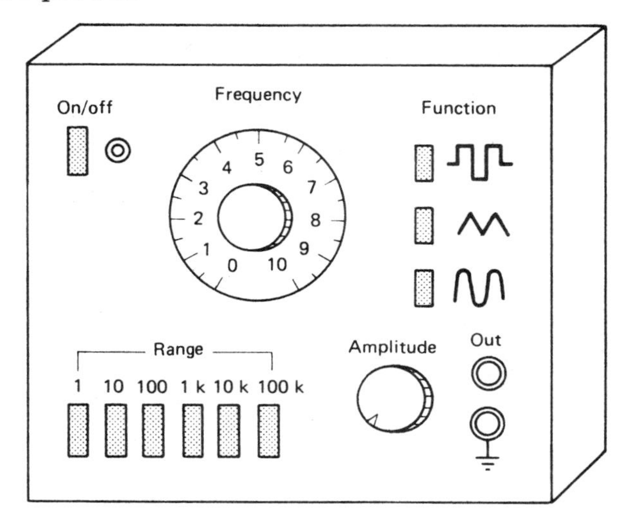

Figure D–1 Basic signal generator. (From F. Hughes, *Illustrated Guidebook to Electronic Devices and Circuits,* Prentice-Hall, Englewood Cliffs, N.J., ©1981, Fig. 1–49, p. 50. Reprinted with permission.)

In some cases a very small signal is required from the generator, but the noise at the output terminals may be too objectionable or the signal too large when the amplitude control is turned way down. To remedy this, the user can place a large-value resistor (100 kΩ to 1 MΩ) in series with the positive output terminal and the circuit. Sufficient voltage can be developed at the output terminals to overcome the problems mentioned, while the resistor drops some of the voltage, which permits the correct signal amplitude to be placed on the circuit.

Unit **E**

Preliminary Review—Procedure for Testing a Discrete Circuit

It may be necessary to disconnect a printed circuit (PC) board from a system to check it separately. The following procedure can be used as a guide for setting up the equipment to check a PC board or for experimenting with a new circuit on a breadboard (see Figure E-1)

1. Have the proper circuit schematic in front of you.
2. Have the proper equipment, parts, and test leads in front of you.
3. Construct the circuit if it is an experiment.
4. Connect the power supply to the circuit, which may be positive and ground, negative and ground, or both positive–negative and ground.
5. Connect all equipment grounds to the common circuit ground (as indicated by the dashed lines).
6. Turn on the power supply and set the proper dc voltages. Measure with a voltmeter.
7. Use a voltmeter to check dc voltages on the circuit.

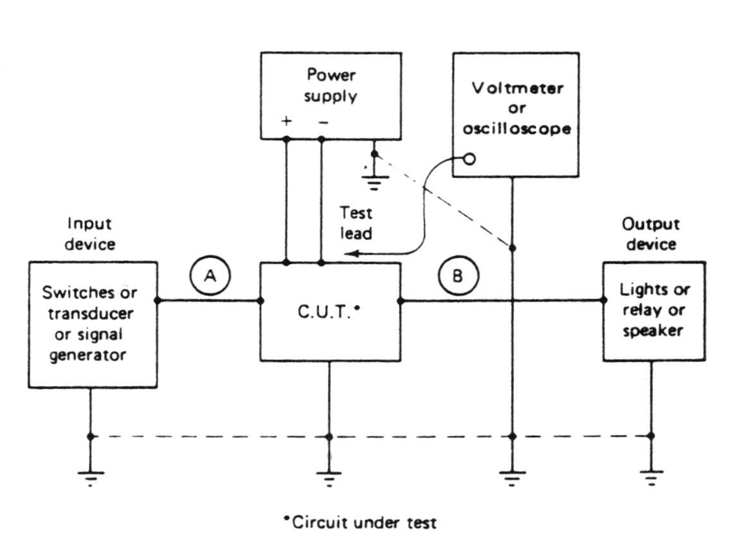

Figure E–1 General setup for testing a discrete circuit. (From F. Hughes, *Illustrated Guidebook to Electronic Devices and Circuits,* Prentice-Hall, Englewood Cliffs, N.J.,©1981, Fig. 1–50, p. 52. Reprinted with permission.)

8. Connect the output device to the circuit.
9. Connect the input device to the circuit.
10. Set the desired input signal to the circuit.
11. Use the oscilloscope or voltmeter to check the input signal at point *A*.
12. Use the oscilloscope or voltmeter to check the output signals at point *B*.
13. Observe the output device for the correct indication.

Unit 1

Semiconductor Diode

INTRODUCTION Diodes are semiconductor devices that are used quite extensively in the electronics industry. They are the basic unit of semiconductor devices.

UNIT OBJECTIVES Upon completion of this unit you will be able to:

1. Describe basic atomic theory.
2. Define the terms intrinsic, extrinsic, covalent bonding, crystal, doping, electron–hole pairs, donor and acceptor impurities, trivalent, tetravalent, pentavalent, majority carriers, minority carriers, depletion region, potential barrier, and hole flow.
3. Describe *N*- and *P*-type semiconductor material.
4. Explain the operation of a *PN* junction.
5. Show the conditions of forward bias and reverse bias of a *PN* junction.
6. Draw the schematic symbol for a diode.
7. Draw a current–voltage characteristic curve for a diode.
8. Test a diode with an ohmmeter.
9. Explain the operation of half-wave, full-wave, and bridge rectifiers.
10. Describe the function and use of a power supply.
11. Show the results of diode clippers.
12. Troubleshoot a power supply.

SECTION 1-1 THEORY OF STRUCTURE AND OPERATION

1-1a SEMICONDUCTOR MATERIALS

Understanding electrical circuits and devices depends on a fundamental knowledge of atomic theory and the laws of electrical charges. In the late 1940s, applications of these principles lead to the creation of the semiconductor device, which is the cornerstone of modern electronics and computer technology. A *semiconductor* is a material that will act as an insulator or a conductor depending on the type of bias voltage that is applied to it. Semiconductor devices are smaller in physical size, require less power to operate, are mechanically more rugged, have a longer life span, and produce products that are more efficient than any other previous methods. A basic knowledge of semiconductor theory will aid you in working with these types of devices.

1-1a.1 Review of Atomic Structure

The various natural elements have certain properties that make them adaptable to specific applications. Silicon and germanium have properties that are excellent for use in semiconductor material. Figure 1-1a shows a model of an actual silicon atom. Remember, an atom has a nucleus made up of protons (positively charged particles) and neutrons (neutral particles). Smaller particles called electrons (negatively charged particles) revolve or orbit around the nucleus.

The electrons revolve at different levels depending on the energy that was imparted to them at the time matter was created. The various energy levels or orbits are referred to as shells. There are also subshells. Those shells closer to the nucleus have a greater attraction between the electrons and nucleus, whereas the electrons in the outer shell have less attraction and can be pulled out of orbit to produce an electron flow. This outer shell of an atom is called the *valence shell,* and the electrons in this shell are called *valence electrons.* The ability to gain or lose an electron in this outer shell is called the *valence* of an atom, which determines the chemical and electrical combining properties of the atom.

Regardless of the number of shells an atom has with varying amounts of electrons in each shell, the valence or outer shell of each atom cannot contain more than eight electrons. The number of electrons or valence of an atom determines whether the element is a good conductor or insulator. Materials such as gold, silver, and copper have a valence of 1, which means they can accept 7 more electrons in the valence shell and are therefore excellent conductors. Elements such as boron, neon, and argon have a valence of 8 and are better insulators. Combinations of elements actually form practical insulators such as glass, plastic, and rubber.

Silicon and germanium have a valence of 4, which means they will donate their four existing valence electrons fairly easily, or they will

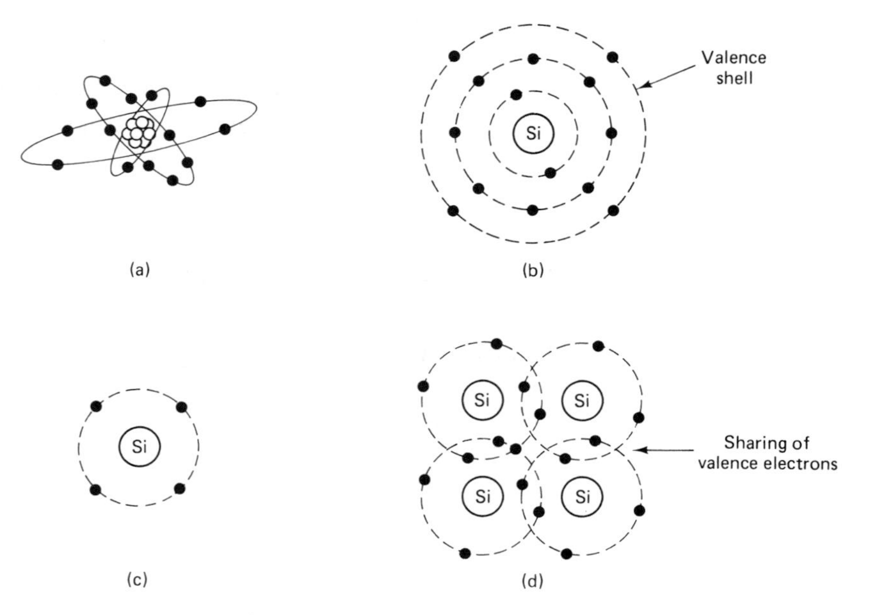

Figure 1–1 Silicon material: (a) actual silicon atom; (b) two-dimensional view of atom; (c) simplified silicon atom showing only valence band; (d) covalent bonding of silicon atoms.

accept four more electrons to their outer shell. Silicon is more stable under increasing temperature changes and is therefore most often used in semiconductor materials.

Figure 1-1b shows a two-dimensional drawing of a silicon atom, which is easier to understand. Notice the three different shells or layers and particularly the number of electrons in the valence shell. To better understand how semiconductor materials are produced, only the valence shell will be shown, as seen in the simplified silicon atom of Figure 1-1c.

Matter becomes apparent when large numbers of atoms are held together. Figure 1-1d shows how the atoms are held together for silicon. The atoms are close together, and the valence electrons travel from one atom to another, which forms an attraction between all the atoms. Each atom shares its electrons with the other atoms, which is referred to as *covalent bonding.* All the atoms are the same in the silicon material, which is thus referred to as *intrinsic,* or being of the same type.

1-1a.2 Producing *N*-type Semiconductor Material

The silicon atom has four valence electrons referred to as *tetravalent,* which is indicated by +4, as shown in Figure 1-2a. An arsenic atom has five valence electrons referred to as *pentavalent* and indicated by +5, also shown in Figure 1-2a. The two types of materials are brought together and combined in a fabricated manner or growing action called doping. This combining produces a material that is different from the original material. In other words, *doping* is the creation of a different material by combining two or more elements. The doping elements are called *impurities.*

When the silicon is doped with arsenic, the atoms combine as shown in Figure 1-2b. Since the arsenic atom has five valence electrons, only three more electrons from the silicon atoms are needed to fill the valence shell. Because there is a total of nine electrons, one electron is set free to roam about the other atoms. In effect, the arsenic atom has donated an extra electron to the material and is referred to as a *donor atom.* This process is accomplished with millions of atoms; therefore, millions of free electrons are in the new material. The material has an excess of electrons and is referred to as *n-type semiconductor material.*

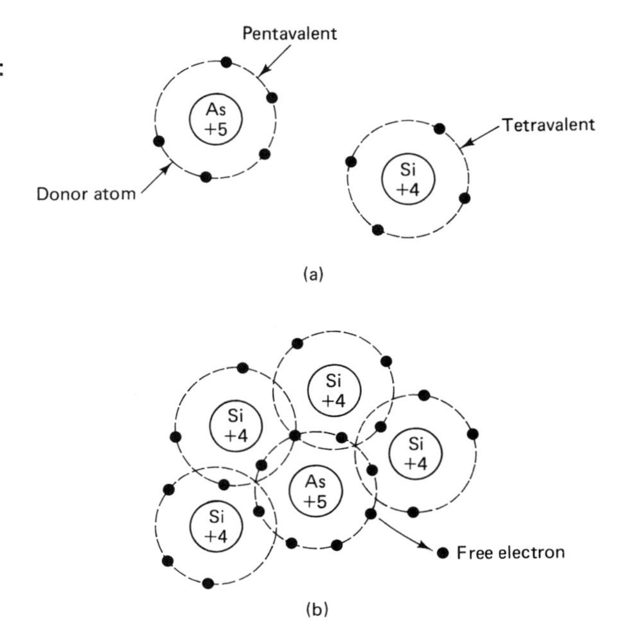

Figure 1–2 Donor atom producing *N*-type material: (a) separate atoms; (b) doped material.

In summary, a donor impurity produces *n*-type semiconductor material. A semiconductor material with impurities is referred to as an *extrinsic* semiconductor.

1-1a.3 Producing *P*-type Semiconductor Material

Indium has three valence electrons referred to as trivalent, as shown in Figure 1-3a. Unlike the arsenic atom with five valence electrons, which donates one electron to the silicon material, the indium atom accepts an electron from the silicon material and is called an *acceptor atom.*

During the doping process, the indium atom accepts an electron from the silicon atoms to fill the valence shell with eight electrons, which leaves a *hole* or absence of an electron. This hole has a positive charge, and since millions of holes are created, the result is *p*-type semiconductor material. A *p*-type semiconductor material has an excess of holes, or positive charges. In summary, an acceptor atom produces *p*-type semiconductor material.

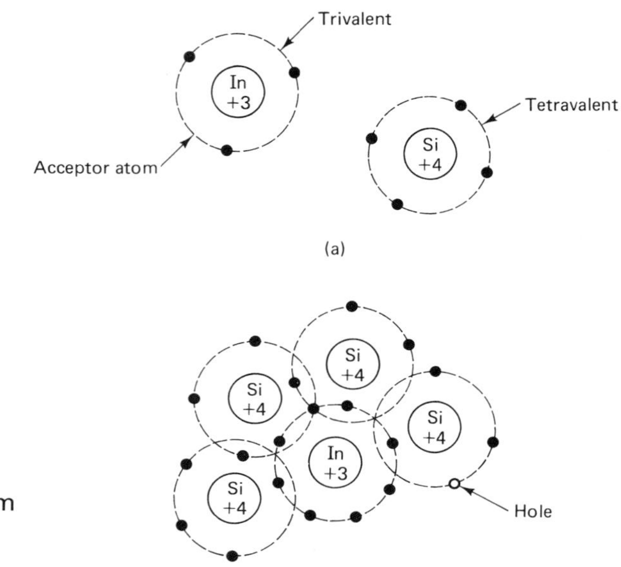

Figure 1–3 Acceptor atom producing *P*-type material: (a) separate atoms; (b) doped material.

Remember, the electrons are orbiting about the nucleus, and the atom is three dimensional, which results in a lattice structure called a *crystal.* Figure 1-4 shows a model of a silicon crystal. The electrons are shared by the atoms and form definite fields of influence, which take on the appearance of a specific lattice structure. Many of these crystals, when combined, create a solid.

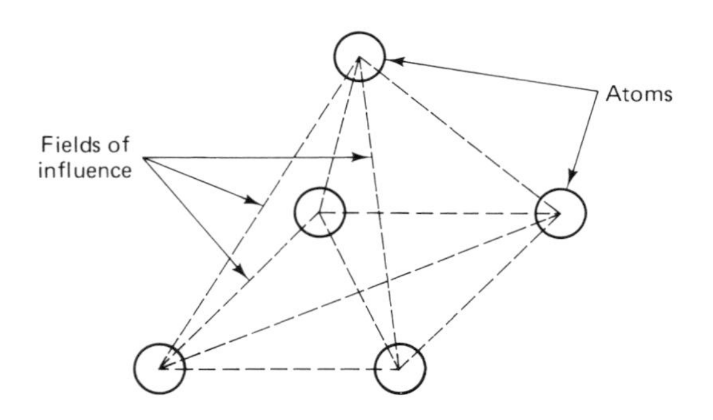

Figure 1–4 Single crystal for a silicon or germanium material.

1-1a.4 Energy Levels of Conduction

The orbits of the electrons of an atom are established by the energy levels and forces within the atom. When atoms are brought closer together, as in a solid, the electrons come under the influence of forces from the other atoms. These fields of influence cause the individual energy level of the electrons to merge into bands of energy levels. There are two distinct energy bands in which electrons may exist, the *valence band* and the *conduction band.* Separating these two bands is an *energy gap* in which no electrons may exist; it is called the *forbidden region.* Electrons in the conduction band can easily move from atom to atom, whereas electrons in the valence band are usually in normal orbit with their nucleus. Diagrams can be constructed to illustrate the relationship of the energy levels to insulators, semiconductor materials, and conductors, as shown in Figure 1-5.

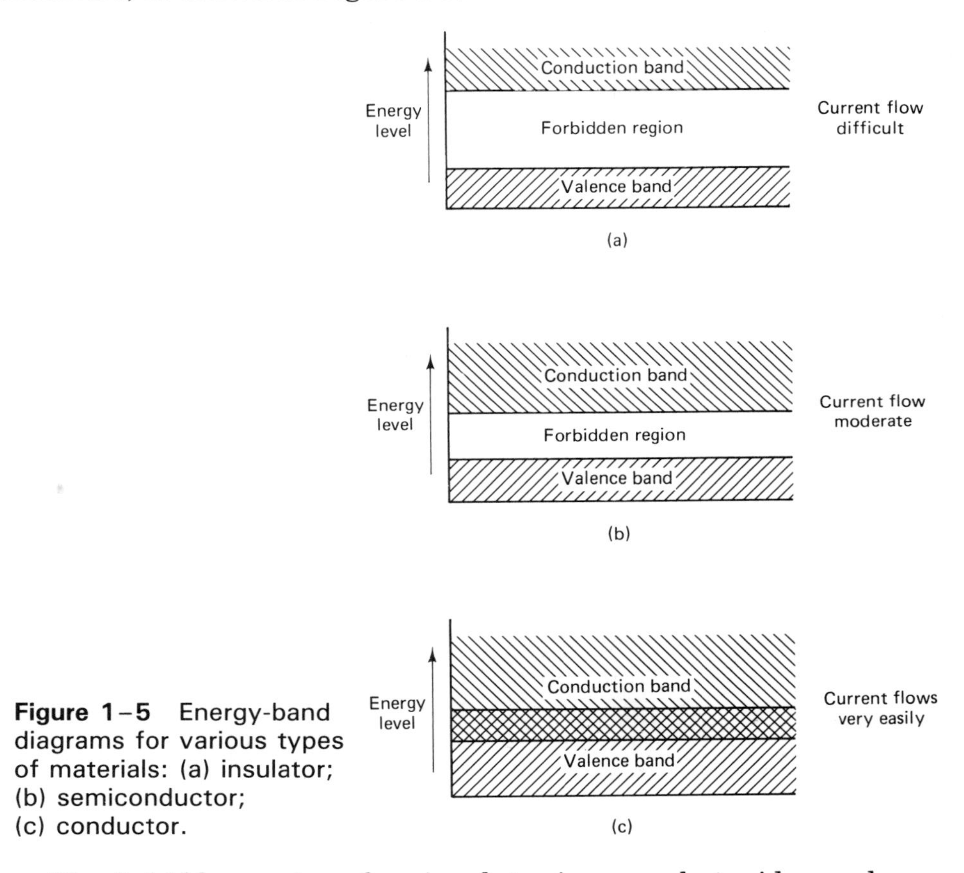

Figure 1–5 Energy-band diagrams for various types of materials: (a) insulator; (b) semiconductor; (c) conductor.

The forbidden region of an insulator is somewhat wide, as shown in Figure 1-5a. This indicates that a large amount of energy must be applied before current will flow. The forbidden region of a semiconductor material is much narrower, as shown in Figure 1-5b, which indicates that less energy is required to cause current flow. Little or no forbidden region exists for a conductor, and the conduction band and the valence band may even overlap, as shown in Figure 1-5c. Very little energy is required to cause a large current to flow.

1-1a.5 Conduction in Semiconductor Materials

Pure intrinsic silicon is an insulator and very little if any current will flow when a voltage is applied to it as shown in Figure 1-6a.

An n-type semiconductor material has an excess of electrons which is referred to as its *majority current carrier.* When a voltage is placed

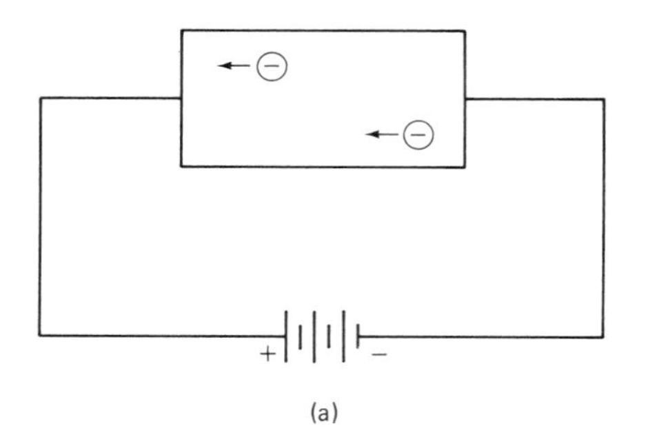

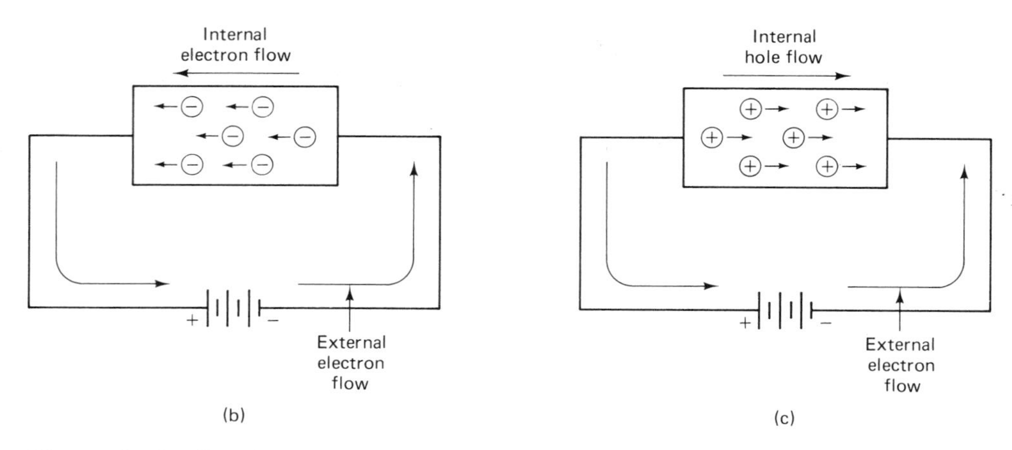

Figure 1–6 Conduction in semiconductor materials: (a) very little current flow in pure silicon; (b) electron flow in *n*-type material; (c) hole flow in *p*-type material.

across an n-type semiconductor material, the electrons will flow internally from the negative potential toward the positive potential as shown in Figure 1-6b. Holes are created by the traveling electrons and a general drift of holes is in the opposite direction through the material. The overall amount of holes moving is much less and this is referred to as the *minority current carriers.*

A p-type semiconductor material has an excess of holes which is referred to as its majority current carriers. When a voltage is placed across it the holes will flow internally from the positive potential toward the negative potential as shown in Figure 1-6c. Electron flow remains in the external connecting wires. A closer analogy of hole flow is shown in Figure 1-7.

We can analyze hole flow on a step-by-step basis. At point 1, an electron jumps left to the wire, leaving a hole or positive charge in the

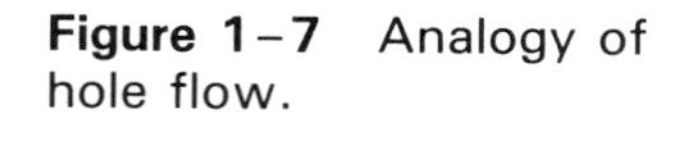

Figure 1–7 Analogy of hole flow.

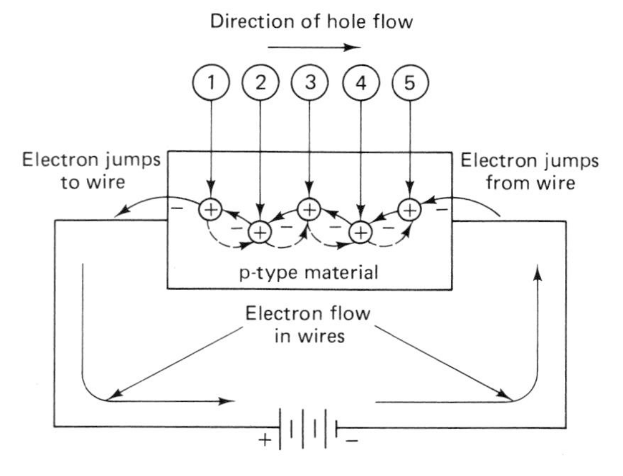

material. At point 2, an electron leaves an atom and jumps left to fill the hole at point 1. Now a hole exists at point 2. In effect, the hole has moved to the right. This same procedure is continued for points 3, 4, and 5 as the hole moves right. At point 5, an electron jumps left from the wire to fill the hole. A complete circuit is provided, and this action will continue until the voltage is removed from the material. There are of course, a few electrons moving through the material, which are referred to as its *minority current carriers.*

In summary, an *n*-type semiconductor material has electrons as its majority current carriers and holes as the minority current carriers. A *p*-type semiconductor has holes as its majority current carriers and electrons as the minority current carriers.

1-1b THE *PN* JUNCTION

During the process of making a diode, donor and acceptor impurities are added to the same basic semiconductor material, producing a *PN* junction, as shown in Figure 1-8. The *P*-type region is called the anode (A) and the *N*-type region is called the cathode (K). The diode schematic symbol is usually painted on larger diodes to identify the anode and cathode terminals, but on smaller diodes a band painted around one end identifies the cathode lead. Holes and electrons combine at the junction to form a neutral or depletion region, which exhibits a small internal voltage referred to as the potential barrier or space-charge region.

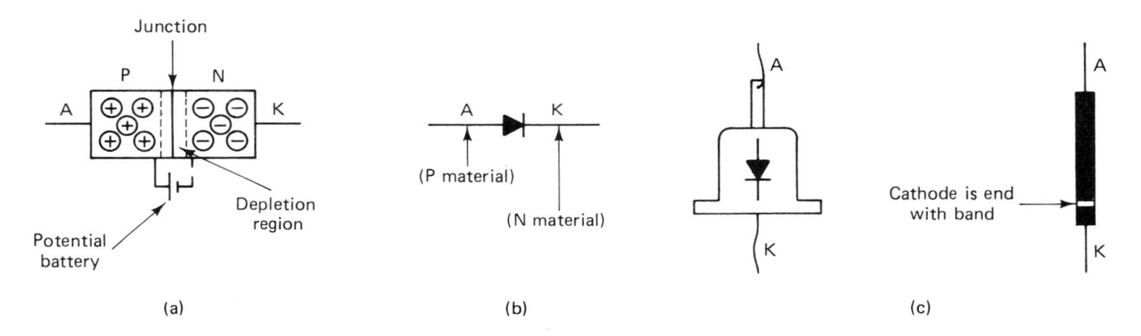

Figure 1–8 The diode: (a) *PN* junction structure; (b) schematic symbol; (c) lead identification.

1-1c BIASING THE DIODE

A diode is forward biased when the anode is more positive than the cathode, as shown in Figure 1-9a. The holes in the *P* region are repelled by the positive terminal of the battery toward the junction, while the electrons in the *N* region are repelled by the negative terminal of the battery toward the junction. The potential of the battery easily overcomes the potential barrier of the junction, and the electrons and holes easily combine at the junction, resulting in a large electron current flow in the circuit. In this forward-biased condition, there is a slight voltage drop across the diode (+0.3 V for germanium and +0.7 V for silicon).

In the reverse-biased condition, the anode is less positive than the cathode, as shown in Figure 1-9b. The holes in the *P* region are attracted away from the junction toward the negative terminal of the battery, while the electrons in the *N* region are attracted away from the junction toward the positive terminal of the battery. The depletion region at the junction increases, and there is very little electron–hole

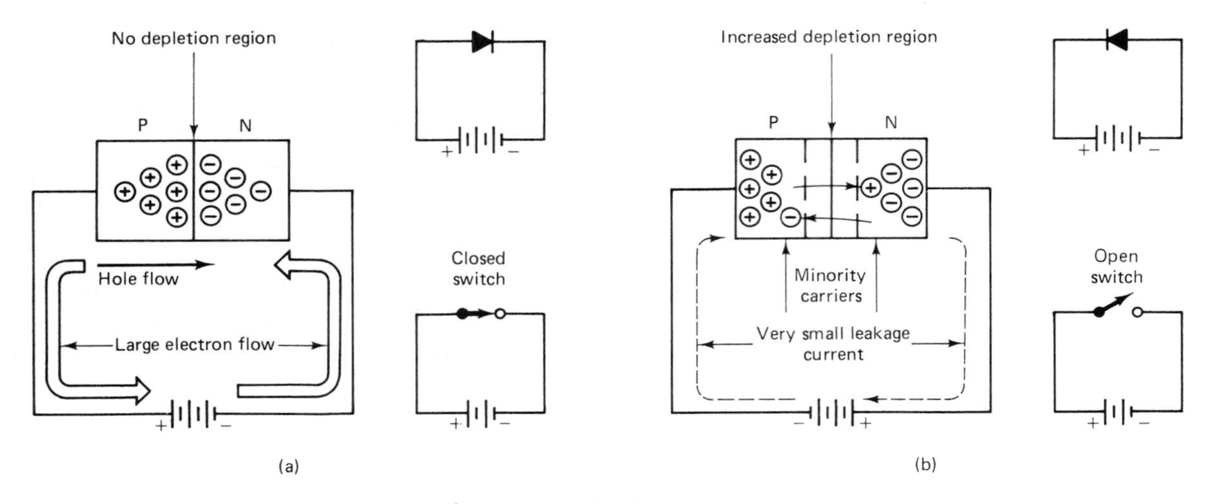

Figure 1–9 Biasing the diode: (a) forward bias; (b) reverse bias.

combination. Current in the circuit is practically zero; however, a few minority electron–hole carriers combine to produce a small leakage current of perhaps a few microamperes in the circuit.

In a practical application, the diode is equivalent to a switch. With forward bias, the diode conducts (closed switch) and electrons flow from the − terminal of a battery through the diode to the + terminal of a battery. With reverse bias (diode turned around), no current flows (switch open).

1-1d DIODE CURRENT–VOLTAGE CHARACTERISTICS

Figure 1-10 shows the current–voltage characteristics of a diode plotted on a graph. When a diode is forward biased with a voltage, often re-

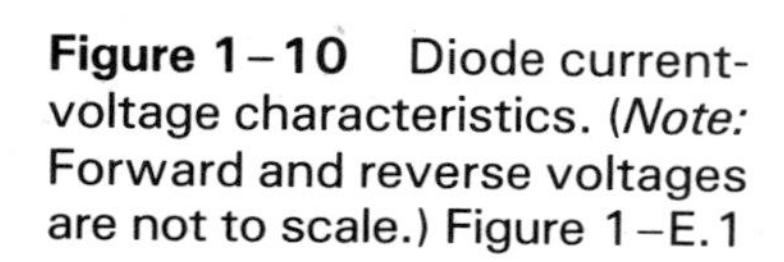

Figure 1–10 Diode current-voltage characteristics. (*Note:* Forward and reverse voltages are not to scale.) Figure 1–E.1

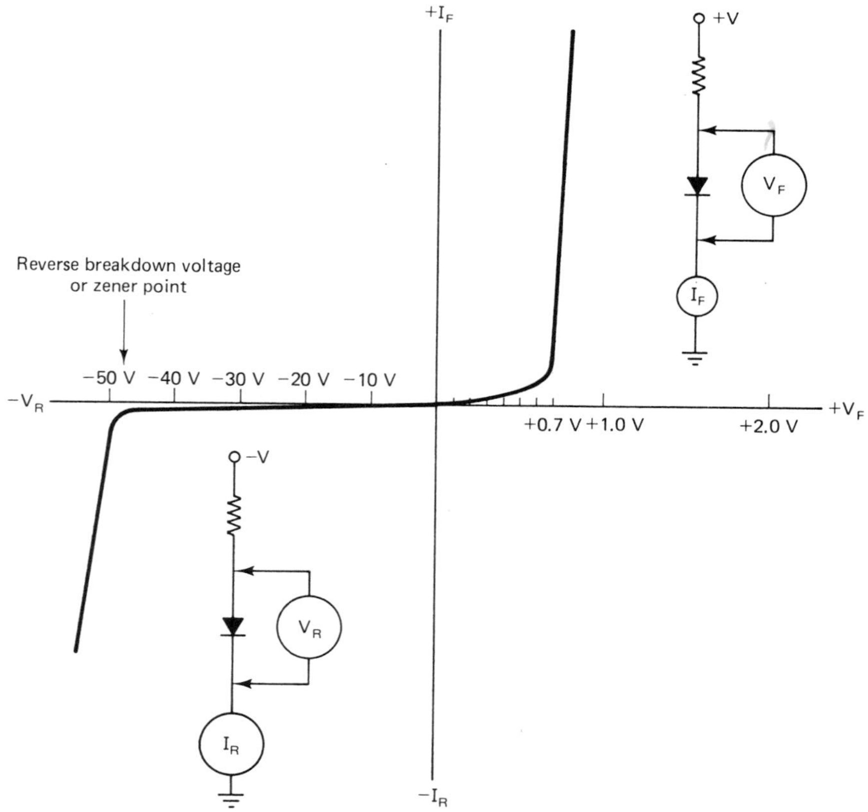

ferred to as forward voltage ($+V_F$), little or no forward current ($+I_F$) flows until the potential barrier of the junction is overcome (0.3 V for Ge and 0.7 V for Si). At this point, the current increases rapidly and is controlled primarily by the external resistance in the circuit.

When the diode is reverse biased with a voltage referred to as reverse voltage ($-V_R$), little or no reverse current ($-I_R$) will flow. However, if sufficient reverse voltage is applied to the diode, it will not withstand the pressure and breakdown occurs, resulting in an avalanche of current that flows in the direction opposite to normal conduction. This avalanche of current generates considerable heat, which can destroy or at least alter the semiconductor materials. If the diode does not burn open, it will usually be shorted and resemble the characteristics of a low-value resistor. The point at which the avalanche current occurs is called the *reverse voltage breakdown* or *zener point* of the diode.

1-1e DIODE DEFINITIONS

Generally, diodes are classified as rectifiers, switching diodes, and special-purpose diodes. It is important to know some of the characteristics of a diode for a particular application.

V_F (forward voltage) the voltage level needed to produce the desired forward current level

I_F (forward current) the amount of current in the forward-biased condition for a given V_F

$I_{F\max}$ (maximum forward current) the maximum forward current the diode can handle without being destroyed

V_{BR} (reverse breakdown voltage) the maximum reverse bias voltage that should not be exceeded to prevent the destruction of the diode

PIV (peak inverse voltage) the same as V_{BR}

I_R (reverse current or leakage current) the amount of current that will leak through the diode for various reverse bias voltages

T_{rr} (reverse recovery time) the time it takes a diode to recover from forward conduction and begin to block reverse current; usually more critical with switching diodes used in high-frequency applications

SECTION 1-2 DEFINITION EXERCISE

Write a brief definition for the following terms.

1. Valence ______________________________

____

____

2. Covalent bonding ______________________________

____

____

3. Intrinsic

4. Extrinsic

5. Trivalent

6. Tetravalent

7. Pentavalent

8. Doping

9. Acceptor impurities

10. Donor impurities

11. "Free electrons"

12. *N*-type semiconductor material

13. *P*-type semiconductor material

14. Majority carriers

15. Minority carriers

16. Energy-band diagram

17. Depletion region

18. Potential barrier

19. Forward bias

20. Reverse bias

21. Forward resistance

22. Reverse resistance

23. Electron current flow

24. Conventional current flow (hole flow)

25. Leakage current

26. Zener breakdown voltage

27. Avalanche current

28. V_F

29. I_F

30. $I_{F\max}$

31. V_R ______________________________

____

____

32. I_R ______________________________

____

____

33. V_{BR} ______________________________

____

____

34. PIV ______________________________

____

____

35. T_{rr} ______________________________

____

____

SECTION 1-3
EXERCISES AND PROBLEMS

Perform the following exercises and problems before beginning the next section.

1. Refer to Figure 1-1b and draw a two-dimensional view of a silicon atom.

2. Refer to Figure 1-5b and draw an energy level diagram for semiconductor material.

3. Place a T for true or an F for false to right of each statement.

_____ **a.** The sharing of electrons by atoms is called covalent bonding.

____ b. Silicon doped with a donor impurity creates *p*-type semiconductor material.

____ c. Hole flow is from positive to negative.

____ d. The minority carrier for *n*-type semicondcutor material is holes.

____ e. "Free electrons" are found in *n*-type material.

____ f. An *n*-type semiconductor material is intrinsic.

____ g. The majority carrier for *p*-type semiconductor material is electrons.

4. Draw the structure of a *PN* junction. (Label the anode and cathode.)

5. Draw a diode across a battery so that it is reverse biased.

6. Draw the schematic symbol of a diode. (Label the anode and cathode.)

7. Draw a *PN* junction connected to a battery in the forward-biased condition and indicate the relationship of holes and electrons in the *PN* junction.

8. Draw a diode across a battery so that it is forward biased.

9. Draw a *PN* junction connected to a battery in the reverse-biased condition and indicate the relationship of holes and electrons in the *PN* junction.

10. Draw the characteristic curve for a diode. (Show the zener breakdown point.)

11. Match circuits a-d with those below by writing a (1) or a (2) in the blanks provided. (Consider the diode a closed switch when conducting and an open switch when not conducting.)

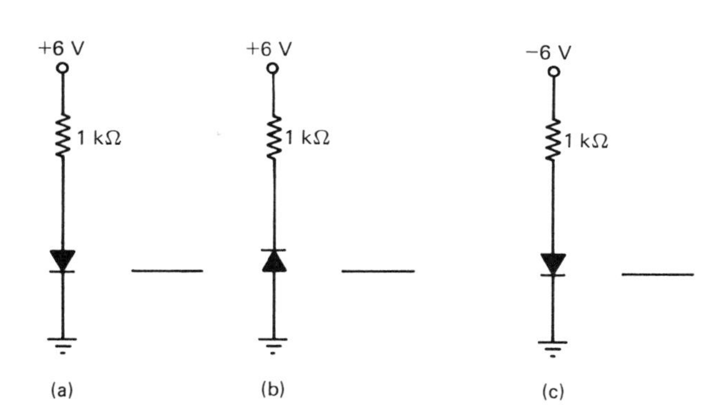

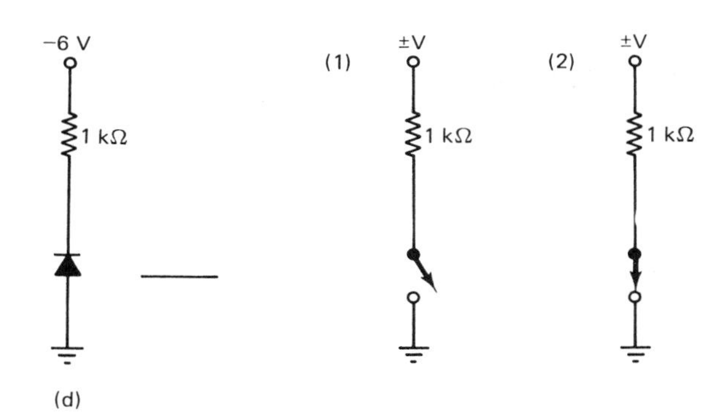

Figure 1–E.–1

12. Draw the schematic symbol of a diode and indicate the electron flow through the diode with an arrow.

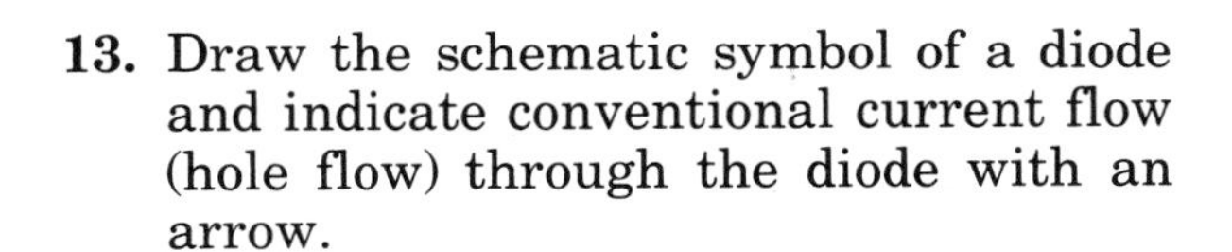

13. Draw the schematic symbol of a diode and indicate conventional current flow (hole flow) through the diode with an arrow.

14. Indicate in the blanks provided whether the diodes are conducting (On) or not conducting (Off) by the voltages given on their anode (V_A) and cathode (V_K).

a. $V_A = +0.7\ V$
$V_K = 0\ V$

______

b. $V_A = +2.2\ V$
$V_K = +1.5\ V$

______

c. $V_A = -3.0\ V$
$V_K = -3.7\ V$

______

d. $V_A = -0.7\ V$
$V_K = 0\ V$

______

e. $V_A = -2.2\ V$
$V_K = +1.5\ V$

______

f. $V_A = 0\ V$
$V_K = -0.7\ V$

______

15. List the approximate voltage drops across the diodes when they are conducting.

a. germanium ______ b. silicon ______

SECTION 1-4 EXPERIMENTS

EXPERIMENT 1. TESTING SEMICONDUCTOR DIODES

Objective:

To demonstrate a practical method of testing diodes with an ohmmeter. This is called a *go/no go test.*

Introduction:

An ohmmeter has a low-voltage potential placed at its leads when measuring resistance. One lead is positive (usually red in color) and the other lead is negative (usually black in color). When the positive lead is placed on the anode of a diode and the negative lead on the cathode, this forward resistance (R_F) should be low, since the diode is forward biased. When the leads are reversed, the reverse resistance (R_R) should be high, since the diode is reverse biased. This simple go/no go test can determine if the diode is open or shorted.

Resistance measurements will vary with different types of diodes, but a high-to-low ratio of 10 : 1 for rectifier diodes is acceptable, while a 100 : 1 ratio is considered good for switching diodes. A shorted diode will show low-resistance readings in both directions, and an open diode will show high resistance (infinity) in both directions.

Materials Needed:

A standard or digital ohmmeter
One or several diodes

Procedure:

1. Refer to Figure 1-11a and place the ohmmeter leads accordingly on the diode leads.
2. Set the ohmmeter to the lowest scale and record the R_F reading.
3. Refer to Figure 1-11b and place the ohmmeter leads accordingly on the diode leads.
4. Set the ohmmeter to the highest scale and record the R_R reading.
5. Calculate the ratio of reverse to forward resistance from the formula

R_R/R_F = ______.

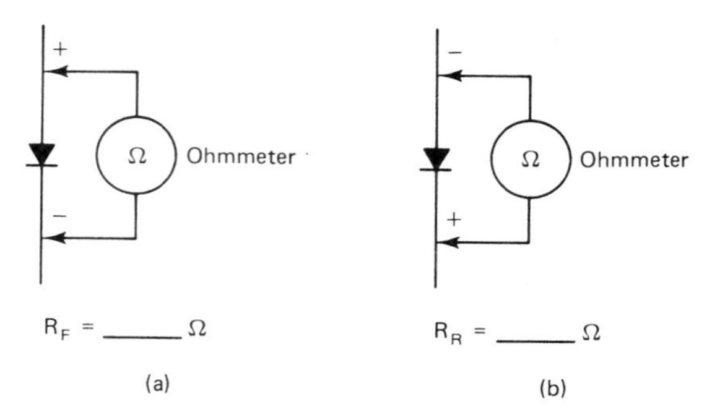

Figure 1–11 Testing a diode with an ohmmeter: (a) forward biased, minimum resistance (ideal = 0Ω); (b) reverse biased, maximum resistance (ideal = ∞Ω).

Fill-in Questions:

1. A forward-biased diode has __________ resistance.
2. A reverse-biased diode has __________ resistance.

EXPERIMENT 2. THE DIODE AS A SWITCH

Objective:

To show how to recognize a conducting and nonconducting diode by its circuit voltage drops and to determine the forward current.

Introduction:

Referring to Figure 1-12a, note that a forward-biased silicon diode has a voltage drop of 0.7 V across it with the remaining power supply voltage dropped across the load resistor (R_L). The voltage drop of R_L can be found by the formula $V_L = V_{DD} - V_F$. The forward current (I_F) through the circuit can be found by the formula $I_D = V_L/R_L$. Referring to Figure 1-12b, note that a reverse-biased silicon diode has the total power supply voltage dropped across it, while the voltage drop across the load resistor is zero, since no current is flowing in the circuit.

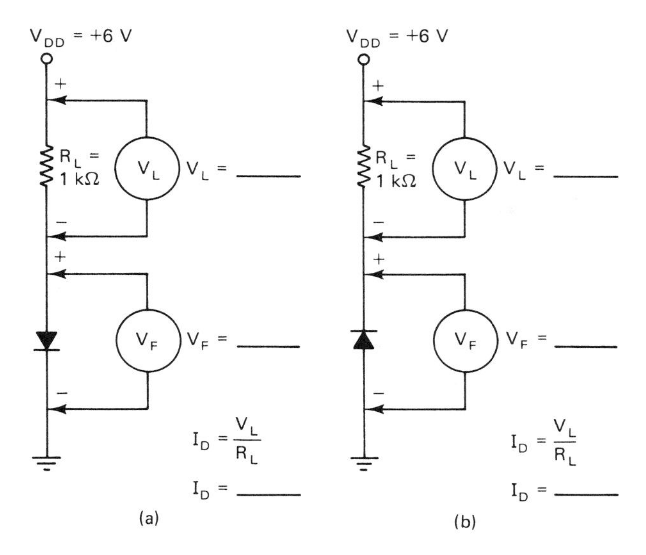

Figure 1–12 Measuring voltage drops in a diode circuit: (a) forward biased; (b) reverse biased.

Materials Needed:

1 Variable low-voltage power supply
1 Standard or digital voltmeter
1 1-kΩ resistor at 0.5 W
1 IN4001 silicon diode or similar type
1 Breadboard for constructing circuit

Procedure:

1. Construct the circuit shown in Figure 1-12a.
2. Set the power supply voltage at +6 V.
3. Measure and record V_F across the diode.
4. Measure and record V_L across R_L.
5. Calculate I_D and record.
6. Turn the diode around as shown in Figure 1-12b.
7. Measure and record V_F across the diode.
8. Measure and record V_L across R_L.
9. Calculate I_D and record.

Fill-in Questions:

1. The forward voltage across a silicon diode in a normally working circuit is ______ V.

2. A resistor in series with the diode of question 1 would have voltage drop equal to the ______ minus the voltage drop of the ______ .

3. If the diode in question 1 were to open, the voltage drop across it would be ____ V. (*Hint:* Refer to Figure 1-12b.)

EXPERIMENT 3. CURRENT–VOLTAGE CHARACTERISTICS OF A DIODE

Objective:

To demonstrate the relationships of forward voltage, current, and resistance of a diode and the reverse voltage, current, and resistance, also.

Introduction:

As the forward voltage across a diode increases, so does the forward current. Unlike a fixed resistor, a semiconductor junction's resistance decreases with an increase in current flow because of the increase in electron–hole combinations. The forward resistance of a conducting diode can be found by the Ohm's law calculation $R_F = V_F/I_F$. The voltage across the entire circuit can increase to several volts, while the external load resistor (R_L) mainly determines the current flowing in the circuit. In the reverse-biased condition, there is very little current (μA) flow, and hence the reverse resistance is high. The reverse resistance of a diode can be found by Ohm's law calculation: $R_R = V_R/I_R$.

Materials Needed:

1 Variable low-voltage power supply (up to 20 V)
1 Standard or digital voltmeter
1 Standard or digital dc current meter
1 IN4001 silicon diode or similar type
1 1-kΩ resistor at 0.5 W
1 Breadboard for constructing circuit

Procedure:

1. Construct the circuit shown in Figure 1-13a.
2. Adjust the variable power supply for a V_F reading of 0.1 V.
3. Record I_F in the second column of data table 1.
4. Calculate R_F from the two values given and record in the third column of data table 1.
5. Repeat steps 2, 3, and 4 for the V_F values given in data table 1 up to 0.8 V.
6. Set the variable power supply to zero and reverse its leads as shown in Figure 1-13b.
7. Reverse the leads of the meters.
8. Adjust the variable power supply for a V_R reading of -2 V.
9. Record I_R in the second column of data table 2.
10. Calculate R_R from the two values given and record in the third column of data table 2.
11. Repeat steps 8, 9, and 10 for the V_R values given in data table 2 up to -20 V.
12. From the values of V_F and I_F in data table 1 and the values of V_R and I_R in data table 2, plot a graph in Figure 1-13c.

Fill-in Questions:

1. When the forward current of a diode increases, its forward resistance ________ ________.

2. The current flowing in a diode circuit is determined primarily by the ________ ________ ________.

3. The current in a reverse-biased diode circuit is extremely ________, while the resistance of the diode is extremely ________.

EXPERIMENT 4. DIODE VOLTAGE DROPS

Objective:

To demonstrate how voltage drops across diodes in series are additive and to show the *shorting effect* of a forward-biased diode.

Introduction:

The voltage drop across a forward-biased diode is about 0.7 V. This voltage drop is somewhat stable, and in some instances diodes are placed in series to develop a reference voltage. In this experiment you will measure the voltage drops across diodes in series.

Materials Needed:

1 Variable power supply
1 Standard or digital voltmeter
1 Dual-trace oscilloscope
1 Sine-wave generator or function generator
1 100-Ω resistor at 0.5 W
1 1-kΩ resistor at 0.5 W
3 1N4001 silicon diode or similar type
1 Breadboard for constructing circuit

Procedure:

1. Construct the circuit shown in Figure 1-14.
2. Apply power to the circuit.
3. With the voltmeter, measure the voltage across each diode and record here:

   V_1 = ______; V_2 = ______; V_3 = ______.
4. Place the black lead of the voltmeter on ground.
5. With the red lead, measure the voltage across D_1, as shown by M_1. Record the value here: ______ . It should be the same as V_1.
6. Measure the voltage across D_1 and D_2 in series as shown by M_2. Record the value here: ______ . It should read the same as $V_1 + V_2$.
7. Measure the voltage across D_1, D_2, and D_3 as shown by M_3. Record the values here: ______ . It should read the same as $V_1 + V_2 + V_3$.

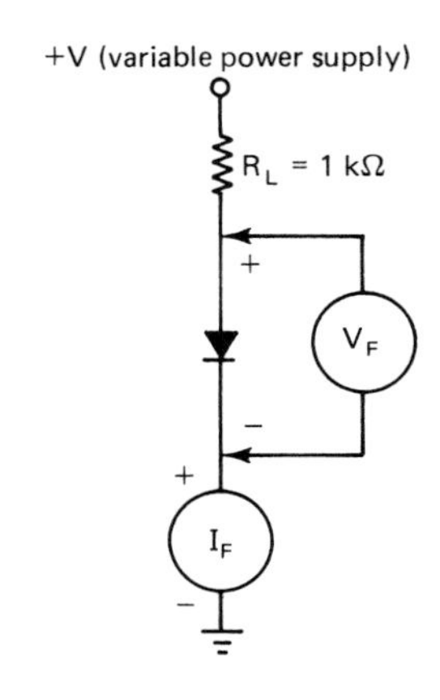

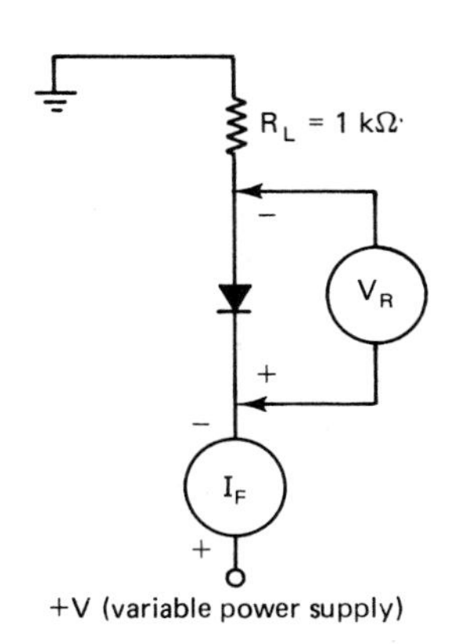

Data Table 1		
V_F	I_F (mA)	$R_F = V_F/I_F$
0	0	α
0.1		
0.2		
0.3		
0.4		
0.5		
0.6		
0.7		
0.8		

(a)

Data Table 2		
V_R	I_R (mA)	$R_R = V_R/I_R$
0	0	α
−2		
−5		
−7		
−10		
−12		
−15		
−18		
−20		

(b)

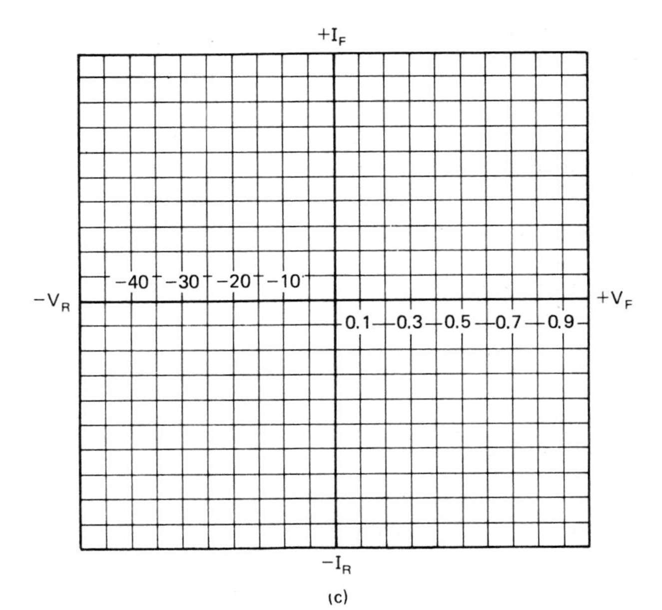

(c)

Figure 1–13 Measuring diode current–voltage characteristics: (a) forward biased; (b) reverse biased; (c)*I–V* graph.

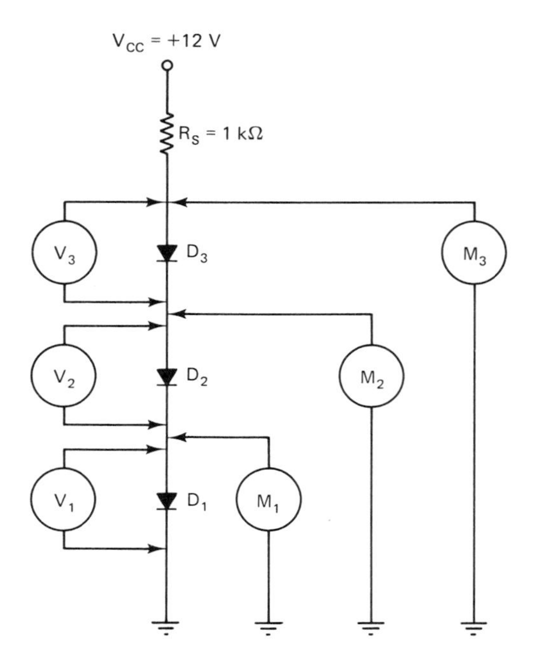

Figure 1-14 Diodes in series.

Procedure:

1. Construct the circuit shown in Figure 1-15.
2. Set the sine-wave generator for 1 kHz and an amplitude of 2 V.
3. Place one channel of the oscilloscope across v_{in} to the circuit.
4. Place the other channel of the oscilloscope across the v_{out} of the circuit.

If a forward-biased diode is placed in parallel with a resistor of some resistance, nearly if not all of the current will flow through the diode. In other words, in the forward-biased condition the diode "shorts out" the resistor. If the diode is reversed biased, the diode has an extremely high resistance and most of the current will flow through the resistor, developing a voltage drop.

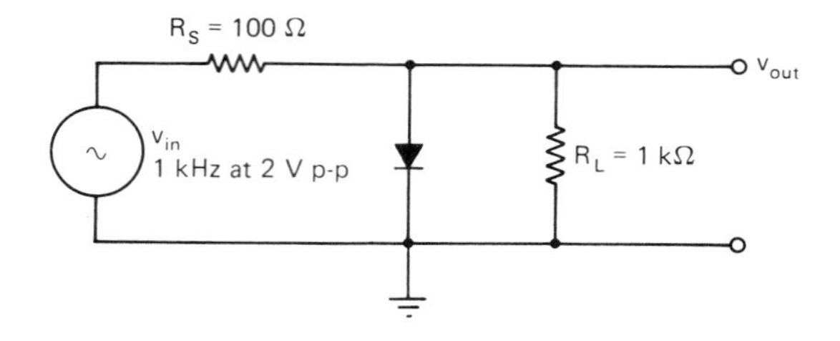

Figure 1-15 Diode shorting effect.

5. Notice that only the negative alternation is seen at the output of the circuit. When the positive alternation is present, the diode conducts and only a +0.7-V line appears across on the face of the CRT. When the negative alternation is present, the diode does not conduct, and the current flows through the resistor, developing the negative-going voltage.

Fill-in Questions:

1. The voltage drop of forward-biased diodes in series is ____________ .
2. If four silicon diodes are forward biased and connected in series, the total voltage drop across them will be ____ volts.
3. A forward-biased diode has ____________ ____________ resistance and will allow a ____________ current to flow through it.
4. A reverse-biased diode has ____________ or infinite resistance and allows little or ______ current to flow through it.

EXPERIMENT 5. HALF-WAVE RECTIFICATION

Objective:

To demonstrate how a diode will rectify (or pass) only one alternation of a sine wave.

Introduction:

Refer to Figure 1-16 and note that when point *A* of T_1 is positive point *B* will be negative. The diode will be forward biased, and current will flow from point *B*, up through R_L, through D_1 to point *A* and the transformer, developing a positive-going output voltage waveform across R_L. When point *A* is negative, point *B* will be positive. The diode is reverse biased, no current flows, and there is no voltage output across R_L. Therefore, the output voltage is positive pulsating dc, which can be filtered with capacitors in an attempt to produce pure dc.

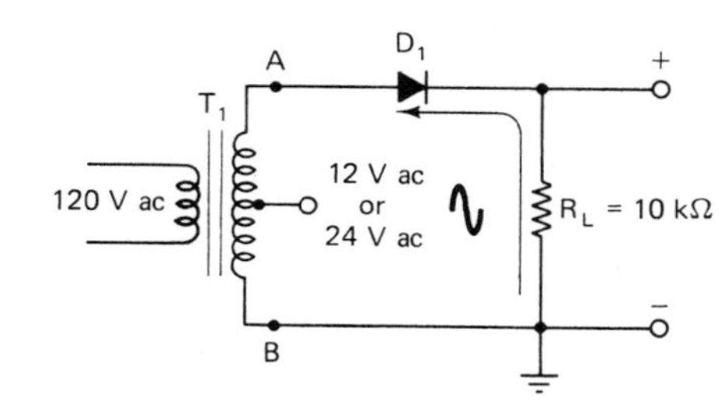

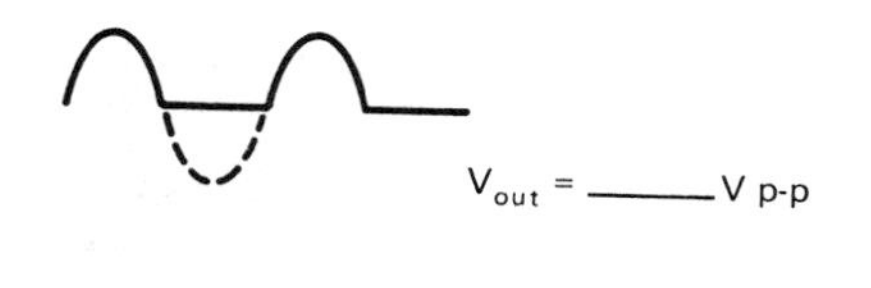

Figure 1–16 Half-wave rectification.

Materials Needed:

1 Oscilloscope

1 12- or 24-V center-tapped transformer (T_1)

1 1N4001 diode (D_1) or similar type

1 10-kΩ resistor at 0.5 W (R_L)

1 Breadboard for constructing circuit

Procedure:

1. Before applying power, construct the circuit shown in Figure 1-16.
2. Apply power to the circuit.
3. Place the oscilloscope across R_L, observing polarity.
4. Record the amplitude of the voltage seen on the oscilloscope in the location marked V_{out}.
5. Remove power from the circuit.
6. Turn the diode around in the circuit.
7. Apply power to the circuit.
8. Observe the oscilloscope and draw the output waveform, indicating its amplitude:

V_{out} =

The diode is now passing the negative alternation of the sine wave, and the polarity across R_L has been reversed. This voltage can be used to develop a negative dc supply voltage.

Fill-in Questions:

1. A diode will pass only ____________ alternation of a sine wave.

2. A half-wave rectifier produces only one ____________ of a sine wave.

3. A half-wave rectifier can produce positive or negative pulsating dc, depending on the ____________ that the diode is placed in the circuit.

EXPERIMENT 6. FULL-WAVE RECTIFICATION

Objective:

To show how two diodes can pass both alternations of a sine wave to produce a better pulsating dc voltage.

Introduction:

Refer to Figure 1-17 and note that when point A of T_1 is positive point B is negative. Diode D_1 is forward biased, and current flows from the center tap of T_1, up through R_L, through D_1 to point A and the transformer. This develops a positive-going voltage across R_L. Diode D_2 is reverse biased and no current flows through it. When point A is negative, point B will be positive. Diode D_2 is now forward biased, and current flows from the center tap, up through R_L, through D_2 to point B and the transformer. Since the current is in the same direction through R_L for this alternation, a positive-going voltage is again developed across R_L. Diode D_1 is reverse biased and no current flows through it. This voltage output is much easier to filter with capacitors in an attempt to produce pure dc.

Materials Needed:

1 12- or 24-V center-tapped transformer (T_1)

1 Oscilloscope

2 1N4001 diodes (D_1, D_2) or similar type

1 10-kΩ resistor at 0.5 W (R_L)

1 Breadboard for constructing circuit

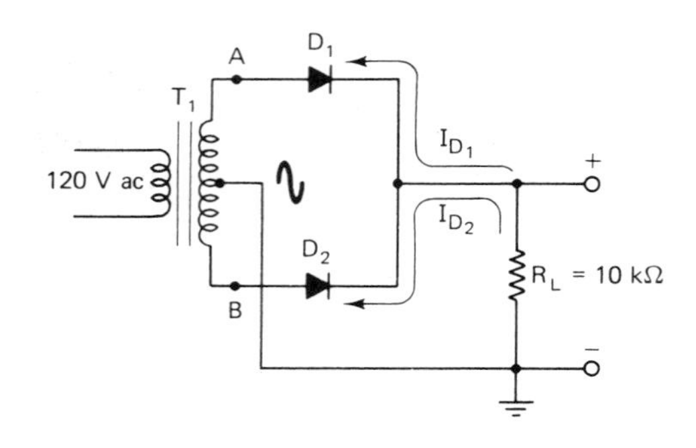

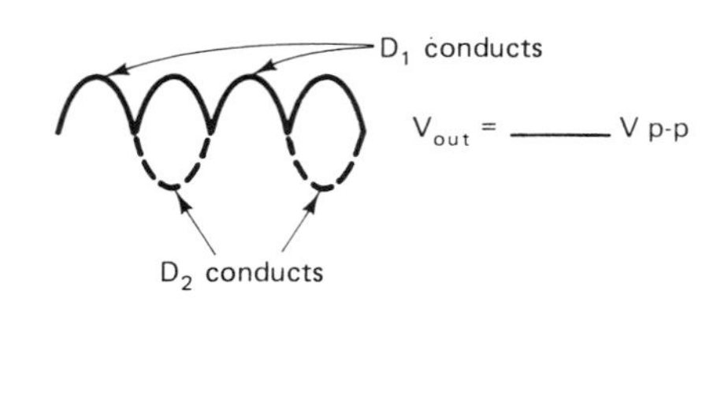

Figure 1–17 Full-wave rectification.

Procedure:

1. Before applying power, construct the circuit shown in Figure 1-17.
2. Apply power to the circuit.
3. Place the oscilloscope across R_L, observing polarity.
4. Record the amplitude of the voltage seen on the oscilloscope in the location marked V_{out}.
5. Remove power from the circuit.
6. Turn both diodes around in the circuit.
7. Apply power to the circuit.
8. Observe the oscilloscope and draw the output voltage waveform, indicating its amplitude:

$V_{out} =$

This is similar to the procedure in Experiment 5. The current is now reversed in R_L, and a negative-going pulsating dc voltage is developed.

Fill-in Questions:

1. A full-wave rectifier passes both ______________ of a sine wave.
2. In terms of polarity, the output of a full-wave rectifier has both alternations of the sine-wave input voltage going in the ______________ direction.
3. A full-wave rectifier can produce positive or negative pulsating dc, depending on the ______________ the diodes are placed in the circuit.

EXPERIMENT 7. BRIDGE (FULL-WAVE) RECTIFICATION

Objective:

To demonstrate how four diodes in a bridge arrangement can produce full-wave rectification without the use of a center-tapped transformer.

Introduction:

Refer to Figure 1-18 and notice how the diodes are connected and that they all point in one general direction, to the right. When point A is positive, D_2 is forward biased, while point B is negative and D_3 is forward biased. Diodes D_1 and D_4 are reverse biased at this time. Current flows from point B through D_3, up through R_L, through D_2 to point A and the transformer. A positive-going voltage waveform is developed across R_L. When point A is negative, point B is positive. Diodes D_2 and D_3 are now reverse biased, but diodes D_1 and D_4 are forward biased. Current now flows from point A and D_1, up through R_L, through D_4 to point B and the transformer. Since the current through R_L is in the same direction, another positive voltage waveform is produced for this alternation of the sine wave across the transformer. Full-wave rectification is accomplished with this circuit without the use of the center tap of the transformer.

Materials Needed:

1 12- or 24-V transformer (T_1)
1 Oscilloscope
4 1N4001 diodes (D_1 to D_4) or similar type
1 10-kΩ resistor at 0.5 W (R_L)
1 Breadboard for constructing circuit

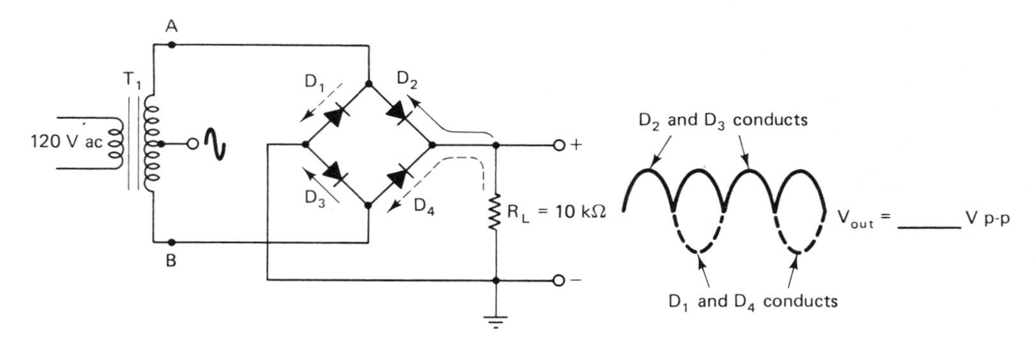

Figure 1–18 Bridge (full-wave) rectification.

Procedure:

1. Before applying power, construct the circuit shown in Figure 1-18.
2. Apply power to the circuit.
3. Place the oscilloscope across R_L, observing polarity.
4. Record the amplitude of the voltage seen on the oscilloscope in the location marked V_{out}.
5. Remove power from the circuit.
6. Remove D_1 from the circuit.
7. Apply power to the circuit.
8. Observe the oscilloscope and draw the output voltage waveform, indicating its amplitude:

   $V_{out} =$

   If one of the diodes in the bridge circuit is open, one of the alternations is lost at the output.
9. Compare the output voltage of all three rectifier experiments. Notice that the half-wave rectifier and the bridge rectifier have the same amplitude; however, the bridge rectifier will be easier to filter. The amplitude of the standard full-wave rectifier is one-half of the other two because of the center-tapped transformer needed.

Fill-in Questions:

1. The main advantage of the bridge rectifier over the standard full-wave rectifier is that it does not require a ____________ ____________ transformer.

2. When one of the diodes in the bridge circuit opens, one of the ____________ will be missing at the output.

EXPERIMENT 8. POWER SUPPLY FILTERING

Objective:

To show how capacitors attempt to filter pulsating dc and produce pure dc.

Introduction:

Refer to Figure 1-19a and note that when the positive-going pulse increases across R_L the capacitor C_1 charges up to the peak voltage. When the pulse begins to decrease, the electrons stores on the negative plate of C_1 discharge through R_L in an attempt to keep the voltage constant across R_L. Before the capacitor has a chance to discharge too far, another positive-going pulse arrives to charge it up to peak voltage. Pure dc is now present at the output, but a small ac component known as *ripple voltage* is riding along the top of it, due to the charging and discharging of the capacitor.

Refer to Figure 1-19b and note that the same action occurs with both capacitors. They attempt to keep the voltage across R_L constant; however, they are not allowed to discharge as long before another positive-going pulse is present due to the full-wave rectification. Resistor R_S separates the capacitors and provides better filtering action. The ripple voltage for this power supply will be much less.

Materials Needed:

1 12- or 24-V transformer (T_1)
1 Standard or digital voltmeter
1 Oscilloscope

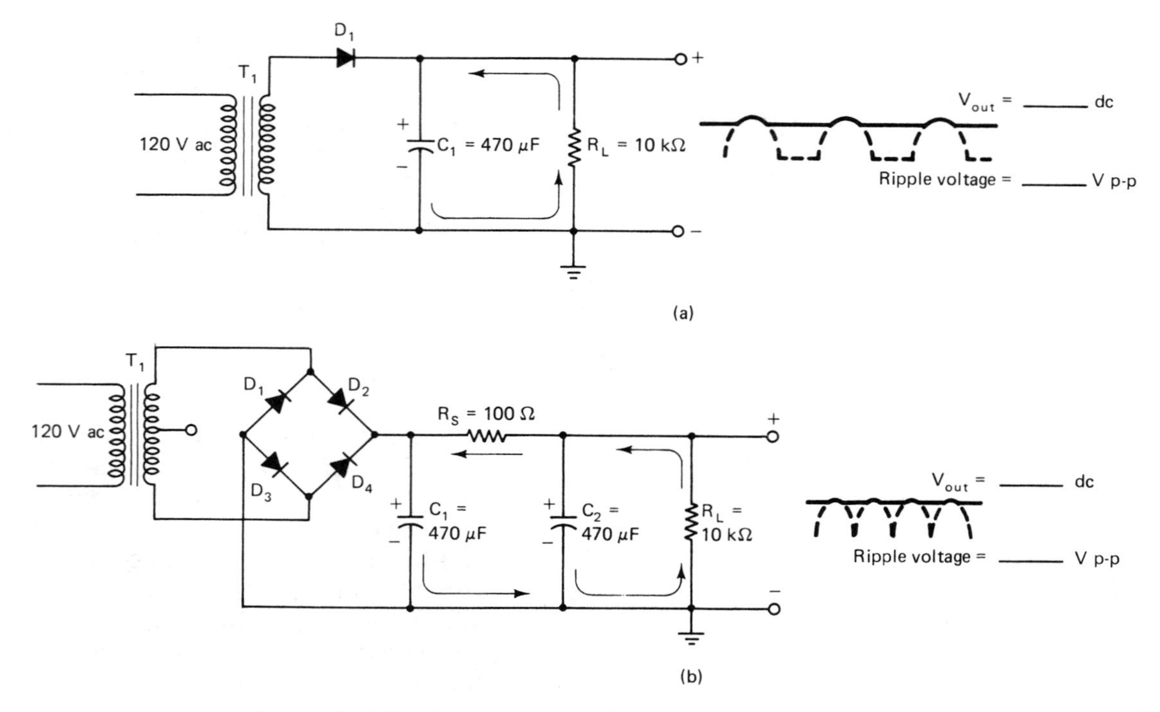

Figure 1–19 Power supply filtering: (a) simple capacitor; (b) pi-type *RC* filter.

4 1N4001 diodes (D_1 to D_4) or similar type
1 10-kΩ resistor at 0.5 W (R_L)
1 100-Ω resistor at 0.5 W (R_S)
2 470-µF electrolytic capacitors at 50 WV dc (C_1, C_2)

Procedure:

1. Before applying power, construct the circuit shown in Figure 1-19a.
2. Apply power to the circuit.
3. Place the voltmeter across R_L, observing polarity, and record the dc voltage in the location marked V_{out}.
4. Remove the voltmeter.
5. Set the input of the oscilloscope to ac and adjust its vertical input control to the lowest setting.
6. Place the oscilloscope across R_L, observing polarity, and record the voltage amplitude shown in the location marked ripple voltage.
7. Remove power from the circuit and disassemble the circuit.
8. Construct the circuit shown in Figure 1-19b.
9. Apply power to the circuit.
10. Place the voltmeter across R_L, observing polarity, and record the dc voltage in the location marked V_{out}.
11. Remove the voltmeter.
12. Place the oscilloscope across R_L, observing polarity, and record the voltage amplitude shown in the location marked ripple voltage.
13. Remove power from the circuit.
14. Remove C_1 from the circuit.
15. Apply power to the circuit.
16. Indicate if the ripple voltage remained the same, increased, or decreased. Ripple voltage ______________ .
17. Remove power from the circuit.
18. Replace C_1 and remove C_2 from the circuit.
19. Apply power to the circuit.
20. Indicate if the ripple voltage remained the same, increased, or decreased from the original value in step 12. Ripple voltage ______________ . Which capacitor had the most effect on the ripple voltage? ________ .

Fill-in Questions:

1. Capacitors in a power supply attempt to keep the dc output voltage constant by ______________ through the load resistor.

2. Full-wave rectification is easier to filter than ______________ rectification.

3. The ripple voltage for a full-wave power supply is ______________ than for a half-wave power supply.

SECTION 1-5
DIODE BASIC TROUBLESHOOTING APPLICATION: FULL-WAVE BRIDGE RECTIFIER POWER SUPPLY

Construct this circuit using your own values or those of a circuit from a previous section. Open or short the components as listed and record the voltages in the proper places in the table. The abbreviations will help indicate the voltage conditions associated with each problem. The voltage at piont A is greater than the output voltage $+V_{CC}$. All voltages are referenced to ground.

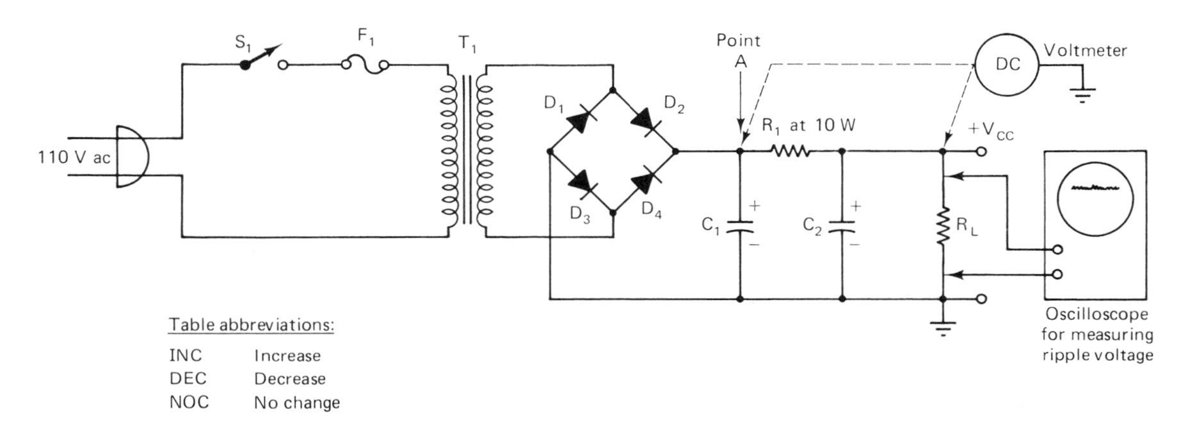

Condition	$+V_{CC}$	Voltage at point A	Ripple voltage at output	Remarks
Normal				All voltages are proper
Any diode open	NOC	NOC	INC	
C_1 open	DEC	DEC	INC	
C_2 open	NOC	NOC	INC	
C_2 shorted	DEC	DEC	0 V	
R_1 open	DEC	INV	0 V	
R_L open	INC	INC	0 V	
C_1 shorted	0 V	0 V	0 V	DANGER: Do not troubleshoot this problem; diodes draw too much current
Any diode shorted	0 V	0 V	0 V	DANGER: Do not troubleshoot this problem; other diodes draw too much current

Note (last two rows)

Figure 1–E.2

SECTION 1-6 MORE EXERCISES

Perform these exercises before beginning the next section.

1. Draw an ohmmeter across a diode with proper polarity for forward bias and reverse bias and indicate the resistance of the diode (low or high).

a. Forward bias

b. Reverse bias

2. What is V_{out} when S_1 is in position A and then in position B?

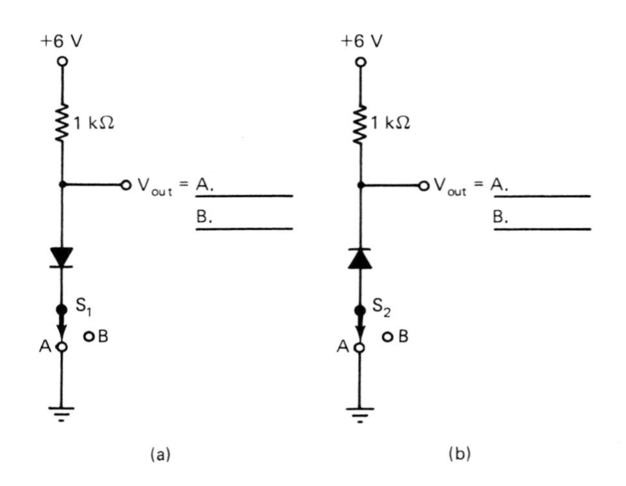

Figure 1–E.3

3. Draw the following listed circuits and show the output voltage waveforms for each.

a. Half-wave rectifier

b. Full-wave rectifier

c. Bridge rectifier with pi-type filter

4. Current flow through a diode depends on the power supply voltage (V_{CC}) and any other series resistance (R_L) in the circuit. Find the voltage drop across R_L and the current (I_L) in the circuit. (*Hint:* $V_L = V_{CC} - V_D$, and $I_L = V_L/R_L$.) Fig. 1-E4.

5. Find the indicated voltages. Formulas are given to aid in the computations. The voltage drop across each forward-biased diode is 0.7 V. Fig. 1-E5.

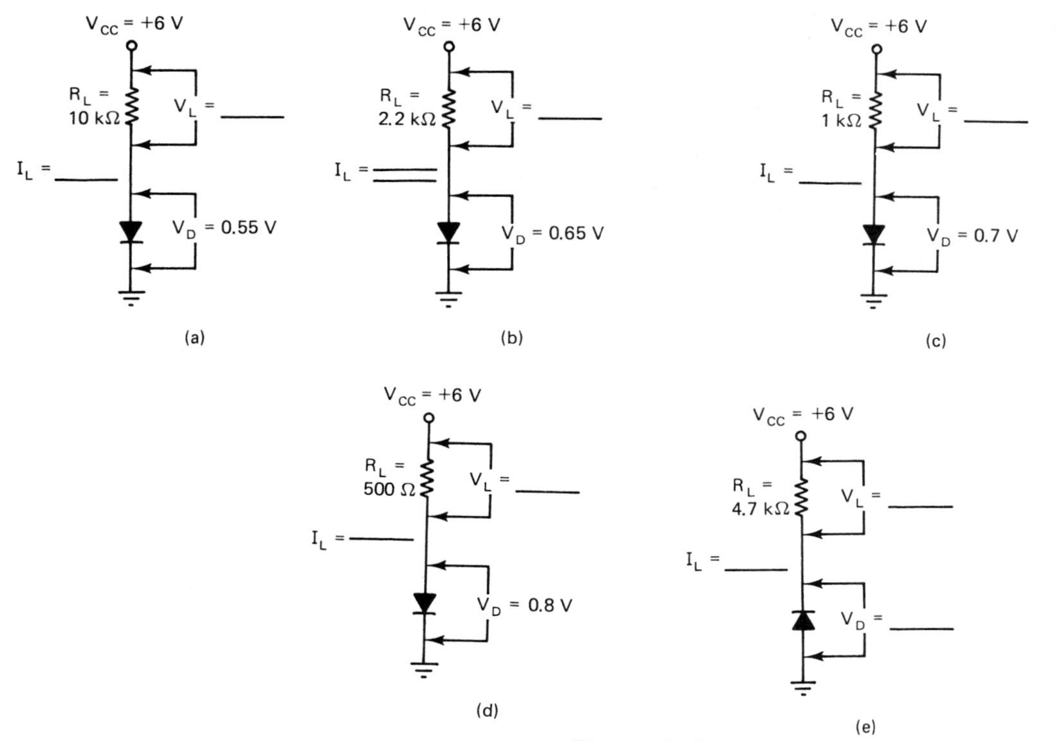

Figure 1–E.4

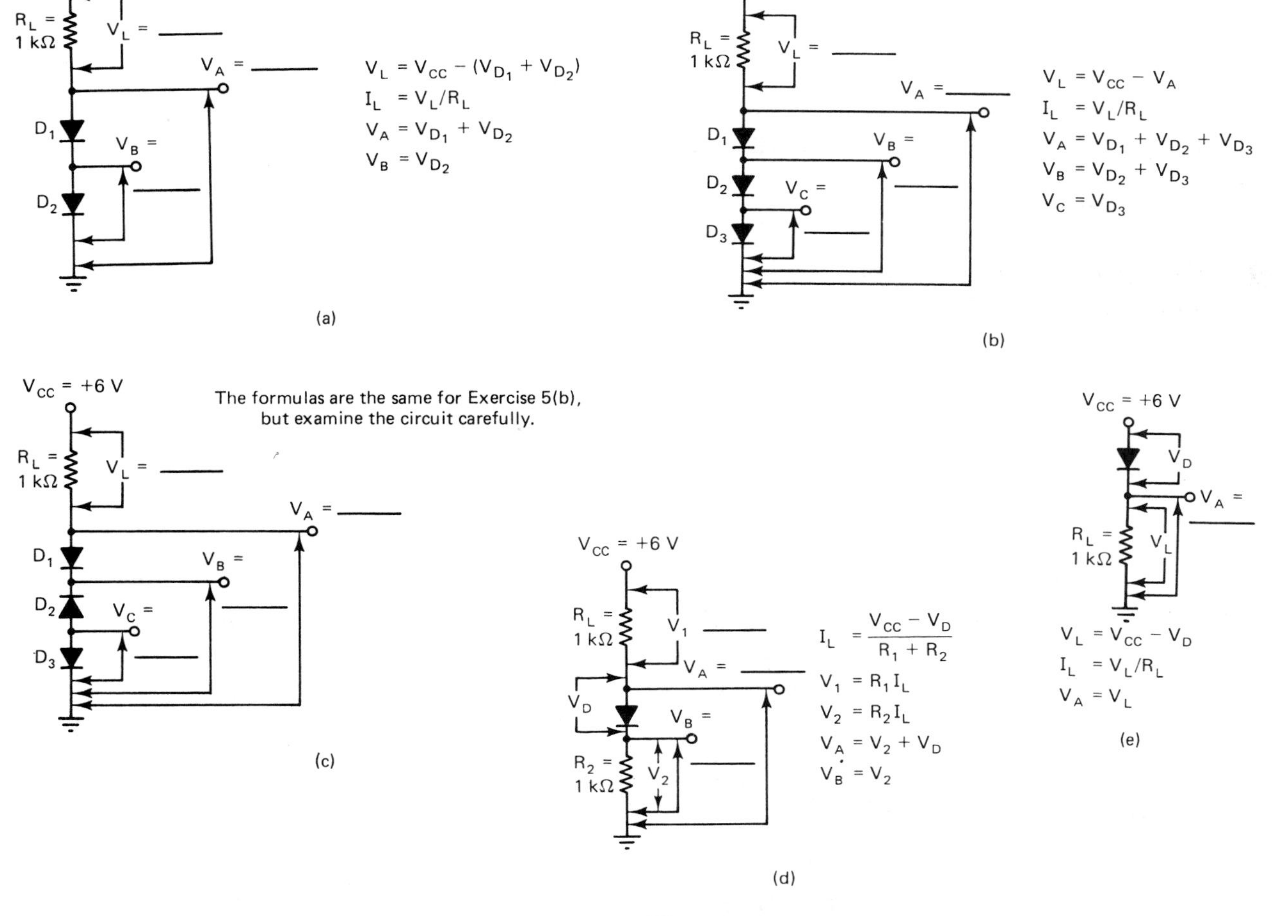

Figure 1–E.5

6. Diodes can be used to rectify, clip a portion of a sine wave, or clamp a pulsating voltage to a dc reference voltage. Remember that a diode will conduct only when its anode is more positive than its cathode. When the diode is conducting, its resistance is considered very low. Study the following examples:

Draw the output voltage waveform for each of the following circuits shown in Figure 1-E.7, indicating the proper voltage values. Consider the diodes to be ideal: either completely open or completely closed. The input voltage to each circuit is a 24-V p-p sine wave.

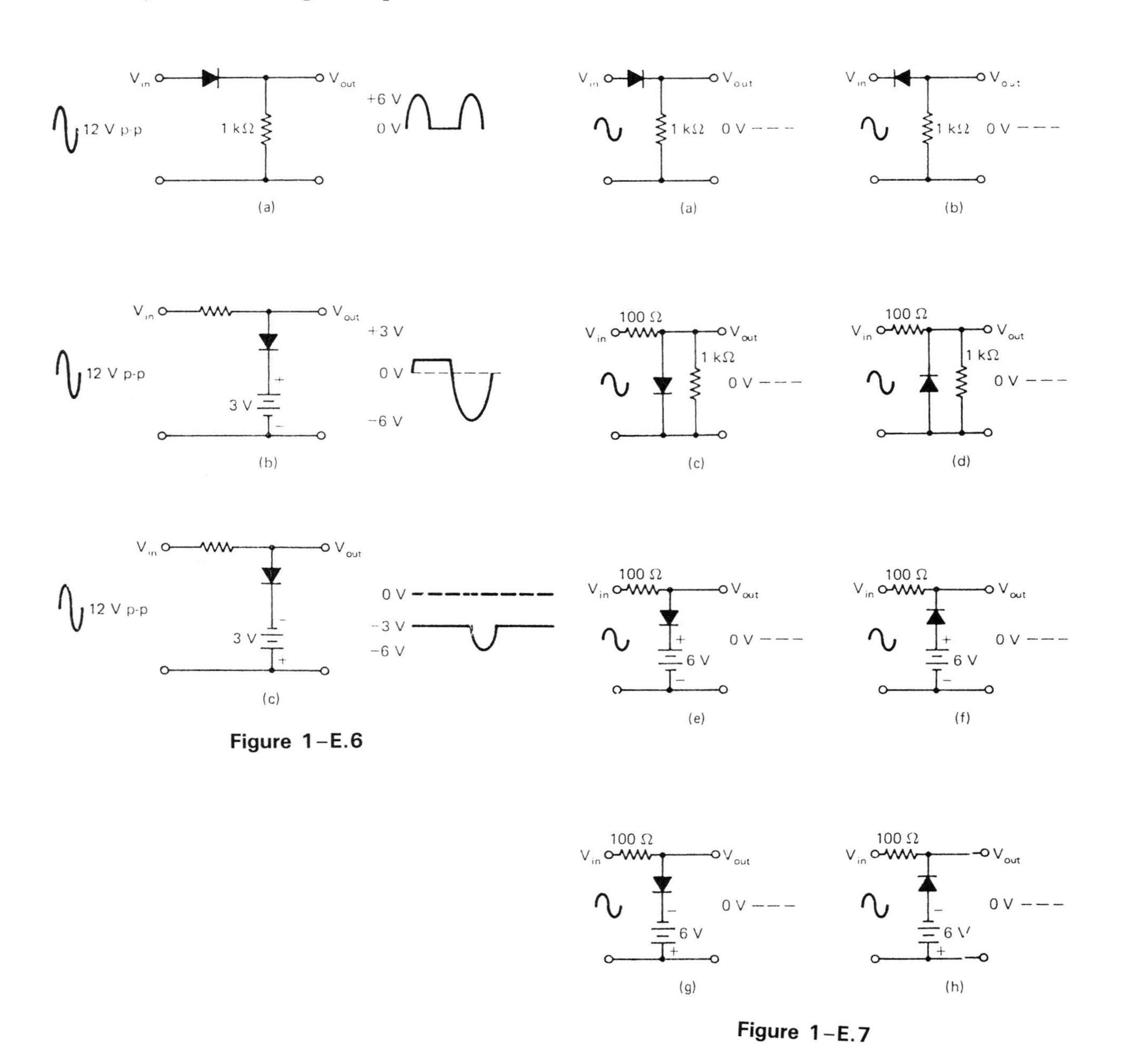

Figure 1–E.6

Figure 1–E.7

SECTION 1-7 DIODE INSTANT REVIEW

How a diode works: A diode conducts (turned on) when the anode is more positive than the cathode (greater than the potential barrier). For any other condition, the diode does not conduct (turned off).

SECTION 1-8 SELF-CHECKING QUIZZES

1-8a DIODE: TRUE–FALSE QUIZ

Place a T for true or an F for false to the left of each statement.

____ **1.** The majority carriers for *P*-type semiconductor material are electrons.

____ **2.** The diode is forward biased when the anode is more positive than the cathode.

____ **3.** In a forward-biased diode, both holes and electrons will flow toward the junction.

____ **4.** The resistance of a reverse-biased diode is low.

____ **5.** When a diode is reverse biased, its depletion region increases.

____ **6.** Leakage current occurs when a diode is reverse biased.

____ **7.** The arrow in the diode schematic symbol points in the direction of conventional current flow.

____ **8.** Current will flow in the reverse direction through a diode when the zener breakdown voltage point is exceeded.

____ **9.** The average voltage drop across a forward-biased silicon diode is 0.3 V.

____ **10.** Diodes cannot rectify ac voltage.

1-8b DIODE: MULTIPLE-CHOICE QUIZ

Circle the correct answer for each question.

1. The majority carriers in *N*-type semiconductor material are:

a. holes **b.** electrons

c. ions **d.** protons

2. With the positive end of a battery placed on the *p*-type material and the negative end placed on the *n*-type material, the *pn* junction is:

a. forward biased **b.** reverse biased

3. The forward resistance of a diode is:

a. low **b.** high

4. Leakage current in a diode occurs when it is:

a. forward biased **b.** reverse biased

5. If a diode has -1.8 V on its anode and -2.5 V on its cathode with reference to ground, it is:

a. conducting

b. not conducting

c. cannot be determined

6. With the positive lead connected to the anode and the negative lead connected to the cathode, an ohmmeter placed across a diode should read:

a. low ohms **b.** high ohms

c. no ohms **d.** zero ohms

7. The voltage of the meter shown in Figure 1-A would read about:

a. +0.3 V **b.** +0.7 V
c. +9 V **d.** 0 V

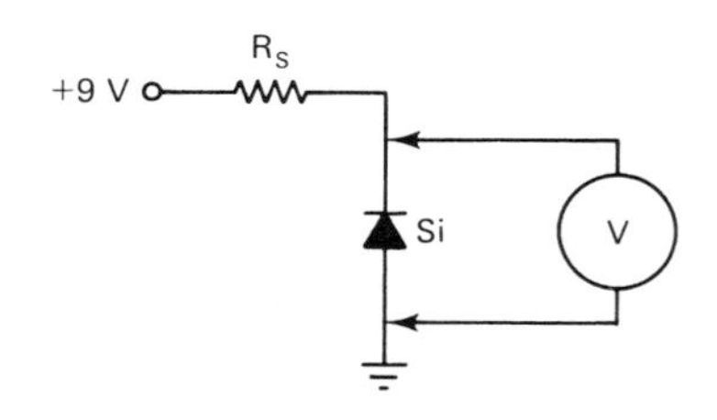

Figure 1–A

8. If three forward-biased silicon diodes are connected in series with a resistor to a +10-V source, the total voltage drop across the diodes is about:

a. +0.7 V **b.** +1.4 V
c. +2.1 V **d.** +10 V
e. 0 V

9. The output waveform for the circuit in Figure 1-B is:

a. ______________________
b. ______________________
c. ______________________
d. ______________________

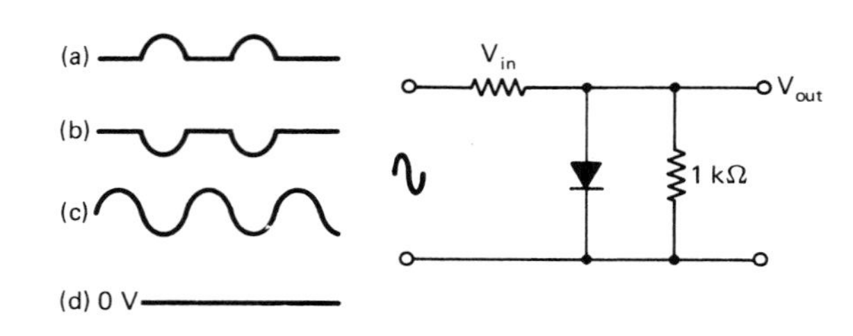

Figure 1–B

10. The light in Figure 1-C will turn on with pulses:

a. P_1 and P_2 **b.** P_1 and P_4
c. P_2 and P_3 **d.** P_2 and P_4

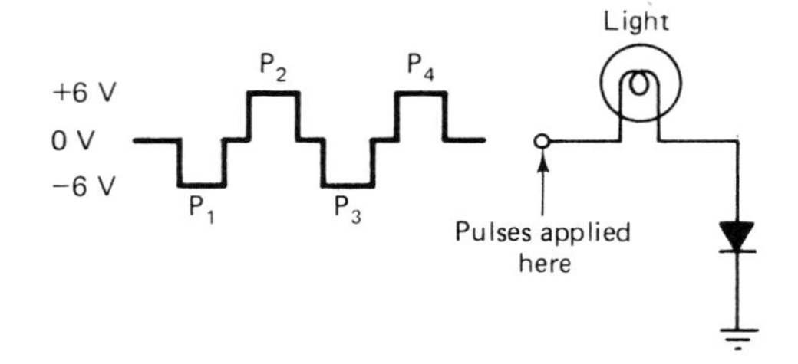

Figure 1–C

ANSWERS TO EXPERIMENTS AND QUIZZES

Experiment 1. **(1)** low **(2)** high

Experiment 2. **(1)** 0.7 **(2)** supply voltage (V_{DD}), diode forward voltage (V_F) **(3)** +6

Experiment 3. **(1)** decreases **(2)** external load resistor **(3)** low, high

Experiment 4. **(1)** additive **(2)** 2.8 **(3)** low, large **(4)** high, no

Experiment 5. **(1)** one **(2)** alternation **(3)** direction

Experiment 6. **(1)** alternations **(2)** same **(3)** direction

Experiment 7. **(1)** center tap **(2)** alternations

Experiment 8. **(1)** discharging **(2)** half-wave **(3)** less

True–False. **(1)** F **(2)** T **(3)** T **(4)** F **(5)** T **(6)** T **(7)** T **(8)** T **(9)** F **(10)** F

Multiple Choice. **(1)** b **(2)** a **(3)** a **(4)** b **(5)** a **(6)** a **(7)** c **(8)** c **(9)** b **(10)** d

Unit 2

Zener Diode

INTRODUCTION The zener diode is one of the most used devices in electronics.

UNIT OBJECTIVES Upon completion of this unit, you will be able to:

1. Draw the schematic symbol of a zener diode.
2. Explain how the zener diode operates in the forward-biased and reverse-biased condition.
3. Draw a current–voltage characteristic curve of a zener diode.
4. Calculate the value of the series resistor required in a zener diode circuit.
5. Describe how the zener diode regulates voltage by varying the current flowing through it.
6. Solve for the various currents in a zener diode circuit.
7. State briefly the operation of miscellaneous diodes, including the tunnel diode, VCC diode, backward diode, *p-i-n* diode, Schottky diode, and snap diode.
8. Test zener diodes.
9. Show how zener diodes are used as voltage reference devices, voltage limiters, and clippers.
10. Troubleshoot a zener diode regulator circuit.

SECTION 2-1 THEORY OF STRUCTURE AND OPERATION

2-1a COMPARISON OF ZENER DIODE TO REGULAR DIODE

A zener diode is similar to a regular diode in appearance, as shown in Figure 2-1. The zener diode is forward biased when its anode is more positive than the cathode. It is reversed biased when the cathode is more positive than the anode. In the forward-biased condition, the zener can conduct a large current (I_F) with a forward voltage drop (V_F) across it of about 0.7 V. In the reverse-biased condition, very little current flows until the reverse voltage (V_R) exceeds the zener voltage (V_Z)

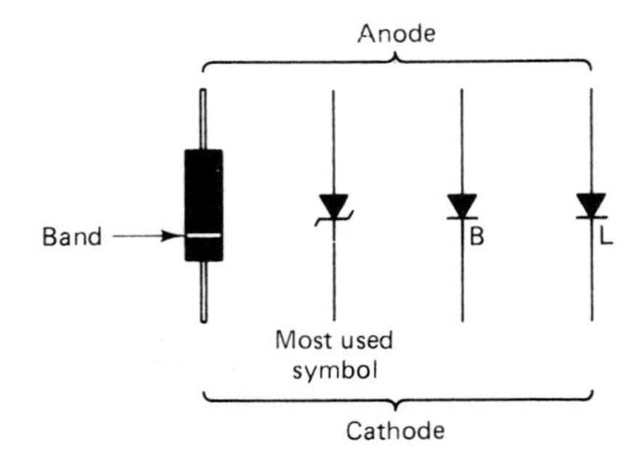

Figure 2–1 Zener diode and schematic symbols.

breakdown point. At this point a large reverse current called the zener current (I_Z) flows through the zener diode, while the zener voltage (V_Z) remains fairly constant across its terminals. A regular diode would most likely be altered or destroyed by this reverse current, but the zener diode is specially doped during manufacturing to operate safely in this zener breakdown region. Figure 2-2 shows these relationships. A zener diode is usually operated in the reverse-biased condition. In a circuit using a positive voltage, the cathode is placed toward the more positive side of the voltage, while the anode is placed toward ground.

Since the zener voltage (V_Z) is fairly constant when the diode is conducting, it is sometimes referred to as a voltage reference diode. These diodes are manufactured with zener voltages ranging from a few tenths of a volt to over 1000 V, capable of handling milliwatts to 100 W of power.

Figure 2–2 Zener diode current–voltage characteristics.

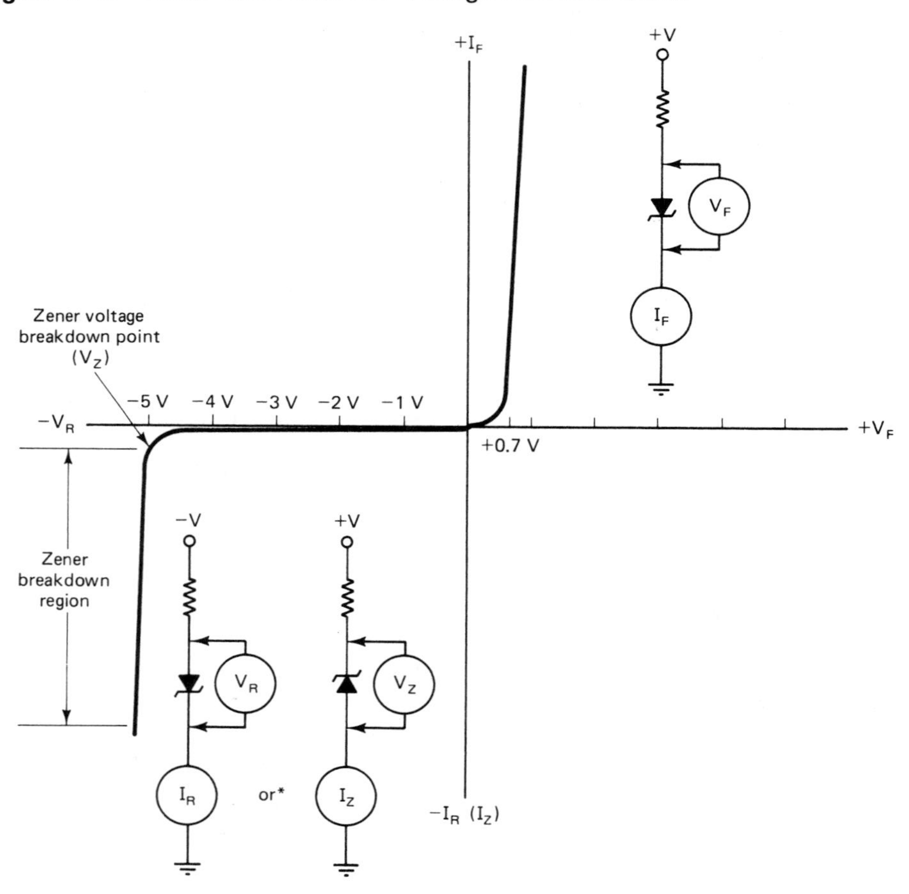

2-1b MAXIMUM ZENER CURRENT ($I_{Z\,max}$)

There is a limit to how much reverse or zener current (I_Z) can flow before the diode is destroyed. This maximum current, $I_{Z\,max}$, can be calculated for each zener diode. Manufacturers specify zener diodes by their power (in watts) and zener voltage (V_Z). For example, as shown in Figure 2-3, if a 1-W zener diode had a V_Z of 5.1 V, rearranging the basic power formula ($P = VI$), the $I_{Z\,max}$ would be equal to $P_{Z\,max}/V_Z$ = 1 W/5.1 V = 196 mA. This formula does not take into account the existing ambient temperatures under which the zener diode will be operating; therefore, it is advisable to operate the diode with much less current flowing through it if permitted by other circuit conditions. A safe practice is to operate the zener diode at one-half of $I_{Z\,max}$.

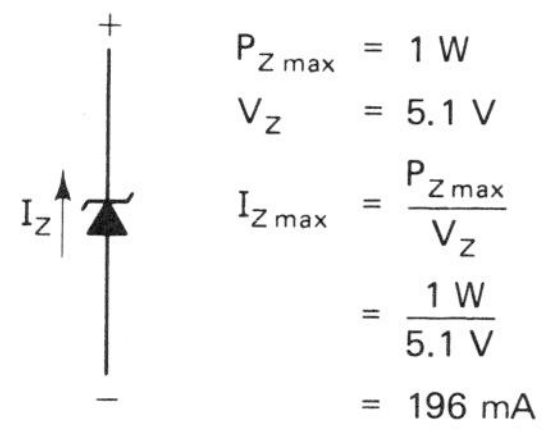

$$P_{Z\,max} = 1\text{ W}$$
$$V_Z = 5.1\text{ V}$$
$$I_{Z\,max} = \frac{P_{Z\,max}}{V_Z} = \frac{1\text{ W}}{5.1\text{ V}} = 196\text{ mA}$$

Figure 2–3 Finding the maximum zener current ($I_{Z\,max}$) for a zener diode.

2-1c SIMPLE ZENER DIODE REGULATOR

In a simple zener diode regulator, as shown in Figure 2-4, the supply voltage ($+V$) should be at least 1 V greater than the desired regulated voltage (V_Z). The series resistor (R_S) is used to set the desired I_Z through the diode with no load connected at the output. Ohm's law is used to find the value of R_S, where V_{R_S}, the voltage across the resistor, is divided by I_Z ($R_S = V_{R_S}/I_Z$). Voltage V_{R_S} is found by subtracting the zener voltage from the supply voltage ($V_{R_S} = +V - V_Z$). As an example, if $+V$ = 8 V and V_Z = 5.1 V and we decide on a I_Z of 90 mA, then

$$V_{R_S} = +V - V_Z = 8\text{ V} - 5.1\text{ V} = 2.9\text{ V}$$

$$R_S = \frac{V_{R_S}}{I_Z} = \frac{2.9\text{ V}}{90\text{ mA}} = 32\ \Omega$$

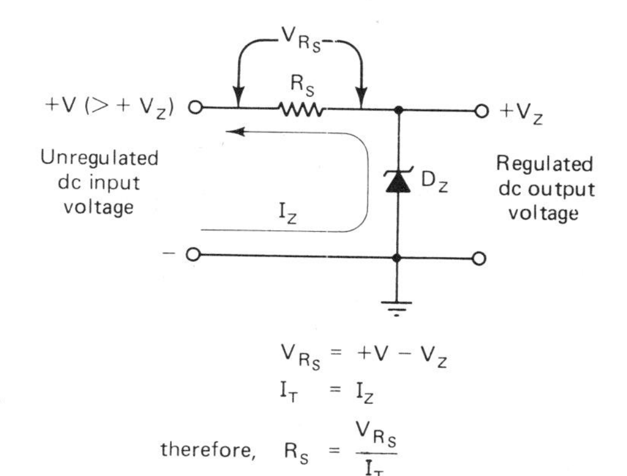

Figure 2–4 Finding the series resistor (R_S) value for a simple zener diode voltage regulator circuit.

2-1d HOW THE ZENER DIODE REGULATES VOLTAGE

When a load (R_L) is connected to the output of a zener regulator as shown in Figure 2-5, the load current (I_L) plus the zener current (I_Z) is equal to the total current (I_T) flowing through R_S. This I_T must remain

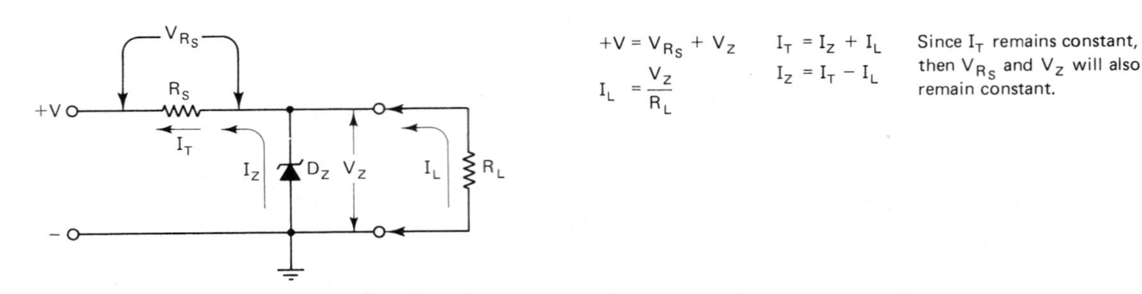

Figure 2–5 How a zener diode regulates voltage V_Z.

constant at all times to keep V_{R_S} constant, which in turn keeps V_Z constant. When R_L is connected to the circuit, the current flowing through the zener diode (I_Z) decreases by the amount that is flowing through R_L, so I_T remains constant.

For example, if $+V = 8$ V, $V_Z = 5.1$ V, and $R_S = 32\ \Omega$, then

$$V_{R_S} = {+V} - V_Z = 8\text{ V} - 5.1\text{ V} = 2.9\text{ V}$$

$$I_T = \frac{V_{R_S}}{R_S} = \frac{2.9\text{ V}}{32\ \Omega} = 90\text{ mA}$$

With no load, I_Z would equal I_T (90 mA). If R_L is a 1-kΩ resistor connected to the regulated output, its current is

$$I_L = \frac{V_Z}{R_L} = \frac{5.1\text{ V}}{1\text{ k}\Omega} = 5.1\text{ mA}$$

Now I_Z decreases to

$$I_Z = I_T - I_L = 90\text{ mA} - 5.1\text{ mA} = 84.9\text{ mA}$$

I_T remains constant and V_{R_S} remains constant, causing V_Z, the regulated output voltage, to remain constant.

If R_L is changed to 500 Ω, then

$$I_L = \frac{V_Z}{R_L} = \frac{5.1\text{ V}}{500\ \Omega} = 10.2\text{ mA}$$

$$I_Z = I_T - I_L = 90\text{ mA} - 10.2\text{ mA} = 79.8\text{ mA}$$

Again, since I_T is constant, so is the output voltage, V_Z.

2-1e ZENER DIODE DEFINITIONS

- V_Z — reverse bias voltage or zener voltage drop across the zener diode when it is operating in the reverse voltage breakdown region
- I_Z — current through the zener diode when it is operating in the reverse voltage breakdown region
- $I_{Z\,max}$ — maximum allowable value of I_Z
- $I_{Z\,min}$ — minimum holding current to keep the zener diode conducting in the reverse voltage breakdown region
- R_Z — resistance of the zener diode when it is operating in the reverse voltage breakdown region
- $P_{Z\,max}$ — maximum power dissipation of the zener diode according to manufacturers' specifications

2-1f MISCELLANEOUS DIODES

A few other specialized diodes are being used more in various other electronic circuits. Figure 2-6 shows the schematic symbol and voltage–current characteristic curve or structure of these devices.

2-1f.1 Tunnel Diode

The *tunnel diode* (Figure 2-6a) is a semiconductor diode in which the *p* and *n* materials are heavily doped, resulting in an extremely thin depletion region. Useful operation of the tunnel diode occurs below the turn-on voltage for a regular germanium diode. When forward bias is applied, a significant amount of forward current flows. As the forward bias is increased, amazingly, the current decreases nearly to zero. Because the current decreases when forward bias increases, this is referred to as the negative resistance region. It is in this region that the tunnel diode is very useful for switching circuits and in high-frequency amplifiers and oscillators. In effect, current has tunneled through the normal potential barrier region of the diode. When the forward bias continues to increase, normal diode action results as the normal potential barrier is overcome. Because of the heavy doping, reverse bias also produces a significant current flow.

2-1f.2 Voltage-variable Capacitance (VVC) Diode

A reverse biased diode (Figure 2-6b) exhibits the properties of a *voltage-variable capacitor* (VVC). The *p* and *n* regions serve as the plates, and the depletion region acts as the dielectric. The least amount of reverse bias voltage produces the greatest capacitance: increasing the bias voltage reduces the amount of capacitance. The range of capacitance is in picofarads. If the diode becomes forward biased, the capacitance is destroyed and the diode performs as a normal diode. VVC diodes are used in tuning circuits.

2-1f.3 Backward Diode

A *backward diode* (Figure 2-6c) is similar to a zener diode, except that it is especially doped and breaks down sooner in the reverse-biased condition than in the forward-biased condition. This diode can be used to rectify small-level signals that are difficult to amplify before rectification.

2-1f.4 The *p-i-n* Diode

The *p-i-n diode* (Figure 2-6d) has a thin layer of intrinsic (undoped) silicon between the *p* and *n* regions. With no bias applied, the intrinsic (*i*) region is empty of free charges. In the forward-biased condition, electrons from the *n* region flow into the *i* region and lower its ac resistance. The greater the dc current, the lower the ac resistance is. This particular phenomenon finds application in RF modulator circuits.

2-1f.5 Schottky Diode

The *Schottky diode* (Figure 2-6e) is a unipolar device in which the anode is made of metal and the cathode is made of *n*-type material. The metal has no holes, so there is no depletion region or potential barrier present, enabling the diode to switch off and on faster than a bipolar diode. A Schottky diode performs well as a high-frequency rectifier.

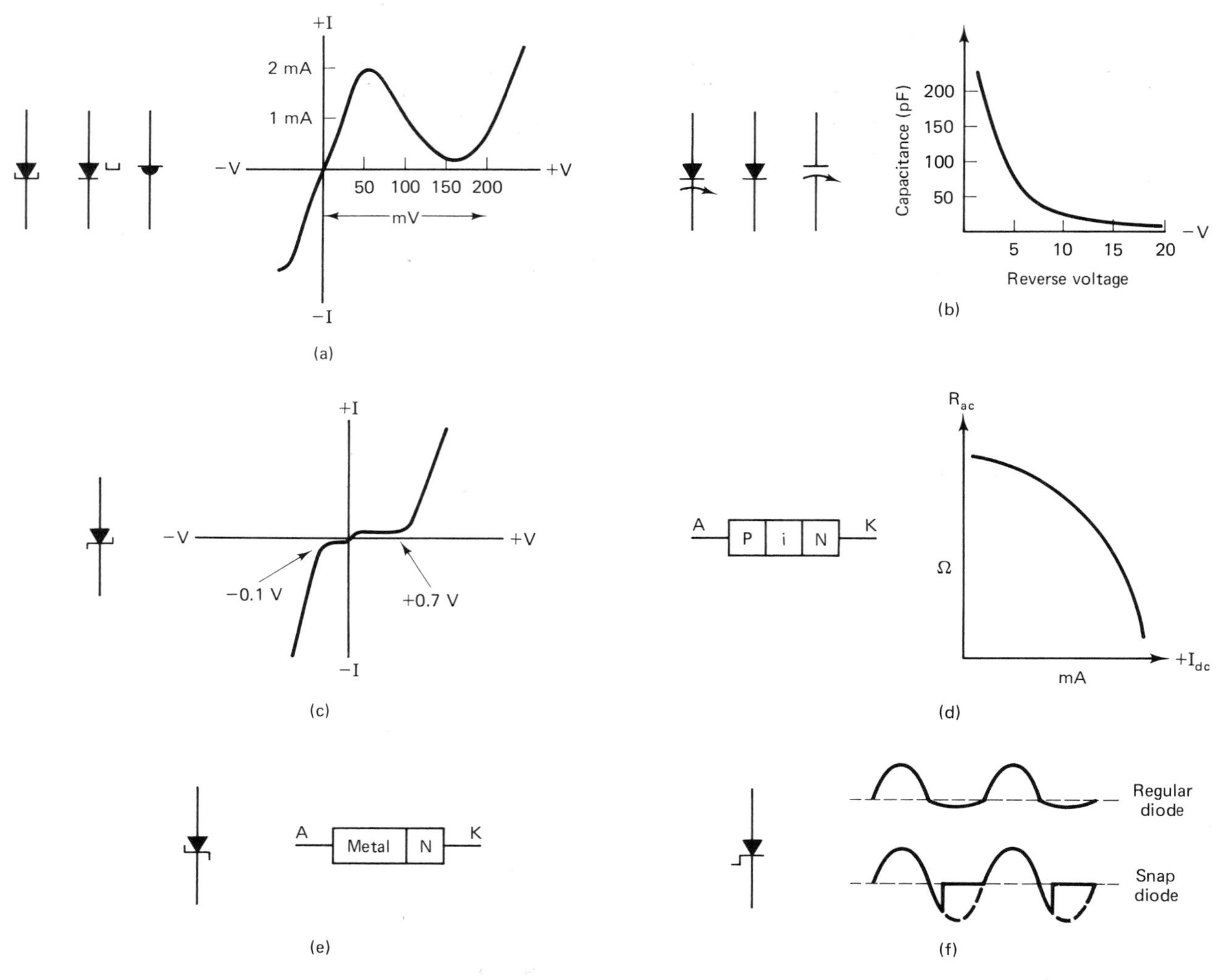

Figure 2–6 Miscellaneous diodes: (a) tunnel diode; (b) voltage-variable capacitance (VVC) diode; (c) backward diode; (d) *p-i-n* diode; (e) Schottky diode; (f) step-recovery (snap) diode.

2-1f.6 Step-recovery or Snap Diode

A *step-recovery diode* (sometimes called a *snap diode*) is especially doped so that the concentration of impurities drops off near the junction (Figure 2-6f). A regular diode rectifies well at low frequencies, but at high frequencies the depletion region does not have time to switch back and forth, resulting in a rather large reverse current. This action also occurs in a snap diode, except that at a certain point in the reverse-biased condition the diode appears to "snap" open and reverse current stops. This diode finds applications in RF amplifiers and high-frequency multipliers.

SECTION 2-2
ZENER DIODE DEFINITION EXERCISE

Refer to the previous sections and write a brief definition of each term:

1. V_Z ____

____

2. I_Z

3. $I_{Z\,max}$

4. $I_{Z\,min}$

5. R_Z

6. $P_{Z\,max}$

7. Tunnel diode

8. VVC diode

9. Backward diode

10. *p-i-n* diode

11. Schottky diode ____________________

____

____

12. Snap diode ____________________

____

____

SECTION 2-3 EXERCISES AND PROBLEMS

Perform the following exercises before beginning Section 2-4.

1. Draw the schematic symbol of a zener diode. (Label the anode and cathode.)

2. Draw the characteristic curve for a zener diode.

3. Draw the circuit in Figure 2-5 and enter the data given: $+V = +12$ V, $R_S = 50$ Ω, $V_Z = 9$ V and $R_L = 1$ kΩ.

4. Circle the correct statement.

a. A zener diode is normally used in the forward-biased condition.

b. A zener diode is normally used in the reverse-biased condition.

5. Find $I_{Z\,max}$ for the following zener diodes. Use the formula $I_{Z\,max} = P_Z/V_Z$.

a. $P_{Z\,max} = 0.5$ W
$V_Z = 9$ V

$I_{Z\,max} =$ ______

b. $P_{Z\,max} = 1$ W
$V_Z = 9$ V

$I_{Z\,max} =$ ______

c. $P_{Z\,max} = 1$ W
$V_Z = 12$ V

$I_{Z\,max} =$ ______

d. $P_{Z\,max} = 1$ W
$V_Z = 15$ V

$I_{Z\,max} =$ ______

e. $P_{Z\,max} = 0.5$ W
$V_Z = 15$ V

$I_{Z\,max} =$ ______

f. $P_{Z\,max} = 0.5$ W
$V_Z = 6$ V

$I_{Z\,max} =$ ______

6. Find I_T, I_L, and I_Z for the following circuits. Use the formula $I_L = V_Z/R_L$; therefore, $I_T = V_{R_S}/R_S$ (no load) and $I_Z = I_T - I_L$.

4. Calculate V_{out} from the formula $V_{out} = V_{Z1} + V_{Z2}$.
5. Measure and record V_{out}.
6. Connect a 1-kΩ load resistor across V_{out} and ground.
7. Measure V_{out} and record here: ______ .

Fill-in Questions:

1. The regulated output voltage is equal to ______ plus ______ .

2. This circuit operates the same as a ____________ zener diode regulator.

3. A regulated output of +14.2 V may use a ______-V zener diode and a ______-V zener diode.

EXPERIMENT 4. USING ZENER DIODES AS VOLTAGE LIMITERS

Objective:

To demonstrate how a zener diode can limit an ac voltage.

Introduction:

Refer to Figure 2-10 and note that a zener diode can be used to limit peak voltages, thereby protecting other circuits that cannot withstand high voltages. In the forward-biased condition (when the positive alternation is present), the zener diode conducts and a +0.7-V drop is seen across the load resistor. In the reverse-biased condition (when the negative alternation is present), the zener diode does not conduct until the voltage exceeds its zener voltage breakdown point (V_Z). The diode then conducts, and output voltage will now be clipped or clamped at the V_Z level.

Materials Needed:

1 18-V p-p ac source or a 6.3-V rms transformer
1 Oscilloscope
1 1N5231 zener diode (5.1 V) or equivalent (Z_1)
1 100-Ω resistor at 0.5 W (R_S)
1 1-kΩ resistor at 1.0 W (R_L)
1 Breadboard for constructing circuit

Procedure:

1. Construct the circuit shown in Figure 2-10.
2. Connect the oscilloscope across V_{out} and ground.
3. Verify the peak-to-peak voltage output waveform as seen in the figure.
4. Turn off the 18-V p-p input voltage (V_{in}).
5. Turn the zener diode around in the circuit.
6. Apply the 18-V p-p input voltage (V_{in}).
7. Draw the output voltage waveform below and indicate its voltage levels.

Figure 2-10 Voltage limiting.

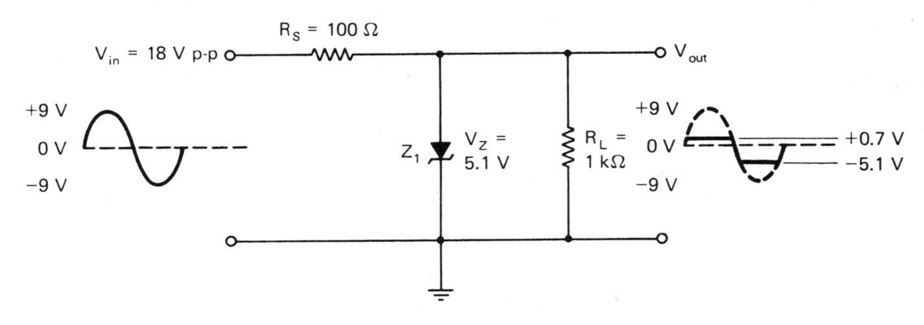

Fill-in Questions:

1. The zener diode can be used as a voltage ______ .

2. In the forward-biased condition, the voltage drop across the zener diode is ______ .

3. In the reverse-biased condition, the output voltage is clamped at ______ .

4. When the diode was turned around in the circuit, the highest output voltage was ______ .

EXPERIMENT 5. ZENER DIODE CLIPPING OF BOTH ALTERNATIONS OF A SINE WAVE

Objective:

To show how two zener diodes can limit or clamp the peaks of an ac sine wave.

Introduction:

Referring to Figure 2-11, when two zener diodes are connected face to face or back to back in an ac circuit, both the positive and negative peaks will be clipped. When the positive alternation is present, Z_1 is forward biased but no current flows initially, because Z_2 is reverse biased. When the input voltage rises to the level of V_Z of Z_2, current flows and the output voltage is clipped at a level equal to the V_Z of Z_2 plus the forward voltage drop (V_F) of Z_1. In this case $V_{out} = V_{Z2} + V_{F1} = 5.1\text{ V} + 0.7\text{ V} = +5.8\text{ V}$. When the negative alternation is present, Z_2 is forward biased and Z_1 is reverse biased. No current will flow until the input voltage reaches the V_Z of Z_1. At this time, the output voltage will be clipped at the level of $V_{Z1} + V_{F2}$, or -5.8 V.

Materials Needed:

1 18-V p-p ac source of 6.3-V rms transformer
1 Oscilloscope
2 1N5231 zener diode (5.1 V) or equivalent (Z_1, Z_2)
1 100-Ω resistor at 0.5 W (R_S)
1 1-kΩ resistor at 0.5 W (R_L)
1 Breadboard for constructing circuit

Procedure:

1. Construct the circuit shown in Figure 2-11.
2. Connect the oscilloscope across V_{out} and ground.
3. Verify the peak-to-peak voltage output waveform as seen in the figure.
4. Turn off the 18-V p-p input voltage (V_{in}).
5. Turn both zener diodes around in the circuit.
6. Apply the 18-V p-p input voltage (V_{in}).
7. Draw the output voltage waveform below and indicate its voltage levels.

Figure 2–11 Clipping both alternations.

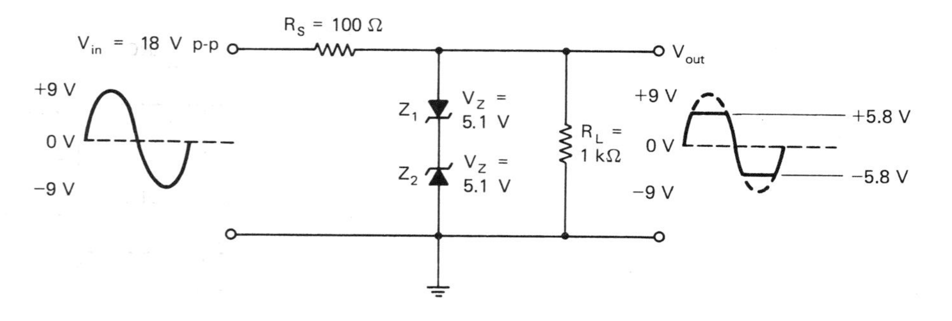

Fill-in Questions:

1. To clip both peaks of an ac sine wave, two zener diodes can be placed in

   series ____________ ______ __________

   or ____________ ______ __________ .

2. The output waveform will be clipped at a level corresponding to the __________ of one diode plus the ____________ of the other diode.

SECTION 2-5
ZENER DIODE BASIC TROUBLESHOOTING
APPLICATION: SIMPLE ZENER DIODE VOLTAGE REGULATOR

Construct this circuit using your own values or those of a circuit from a previous section. Open or short the components as listed and record the voltages in the proper place in the table. The abbreviations will help indicate the voltage conditions associated with each problem. The voltage at point *A* is greater than the output voltage $+V_{CC}$. All voltages are referenced to ground.

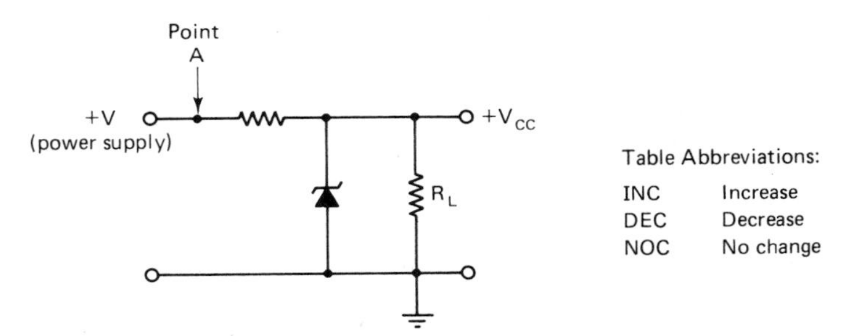

Condition	$+V_{CC}$	Voltage at point A(+V)	Remarks
Normal			All voltages are proper
R_S open	DEC	INC	No voltage applied to D_Z or R_L
R_L open	NOC	NOC	D_Z has full load current
D_Z open	INC	INC	No regulation of $+V_{CC}$
D_Z shorted	DEC	DEC	Short across $+V_{CC}$; maximum current through R_S
+V low (below V_Z)	DEC	DEC	D_Z not regulating $+V_{CC}$; trouble before point A
+V high (in limits)	NOC	INC	Normal operation within limits of $I_{Z\,max}$

Figure 2–E.3

SECTION 2-6
MORE EXERCISES

Perform the following exercises before beginning Section 2-8.

1. Indicate V_{out} of each circuit.

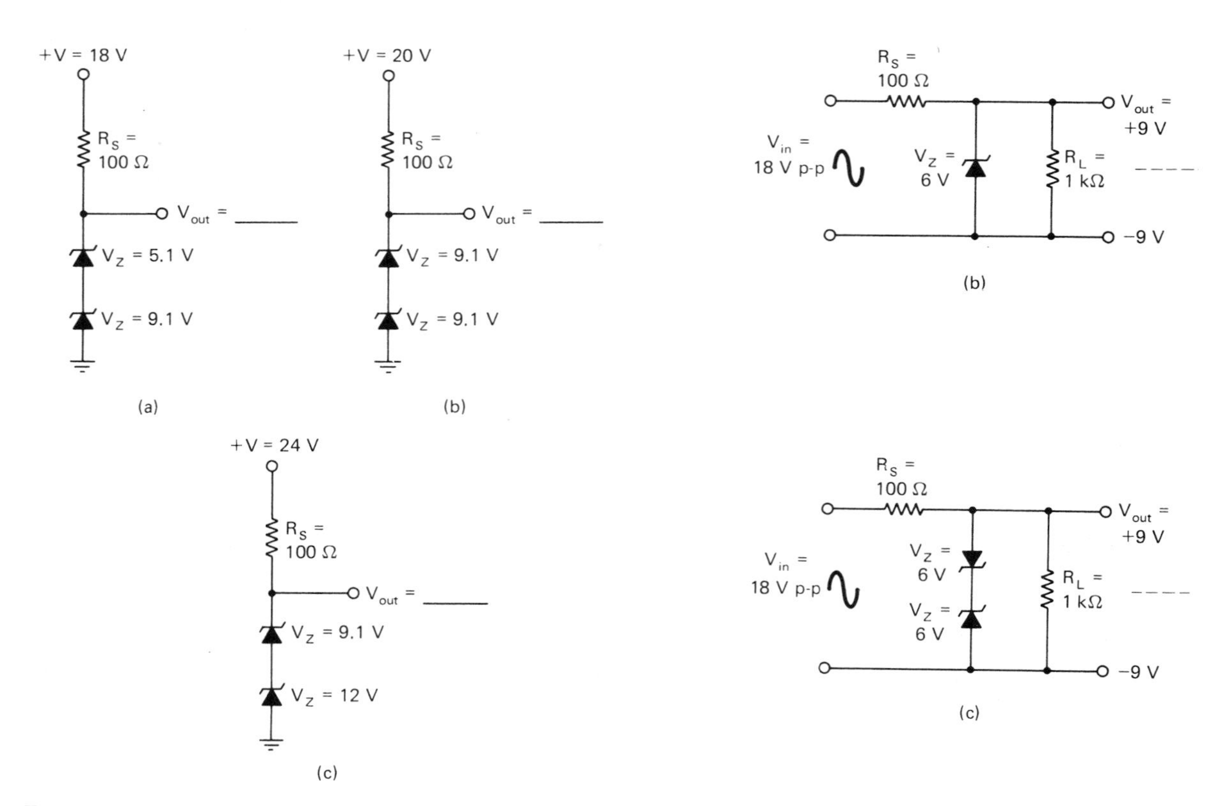

Figure 2–E.4

2. Draw the output voltage waveforms and indicate the proper voltage levels.

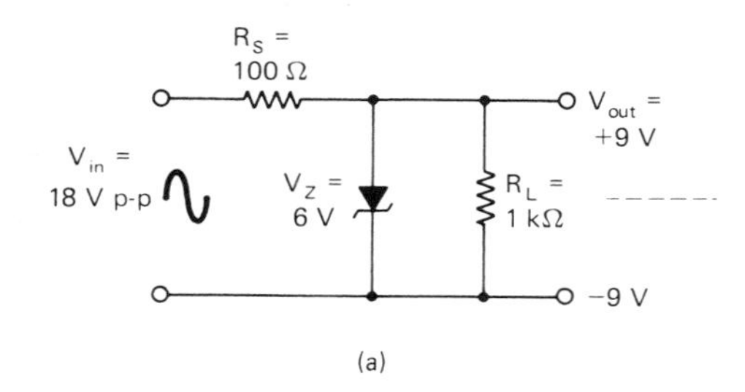

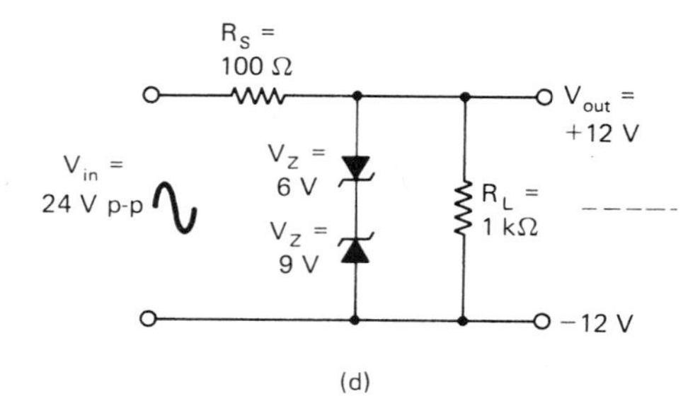

Figure 2–E.5

SECTION 2-7
ZENER DIODE INSTANT REVIEW

A zener diode operates the same as a normal diode, except that in the reverse-biased breakdown region it will attempt to maintain the zener voltage (V_Z) across its terminals by varying the amount of current flowing through it (I_Z), depending on the amount of load current (I_L) of the circuit.

SECTION 2-8 SELF-CHECKING QUIZZES

2-8a ZENER DIODE: TRUE–FALSE QUIZ

Place a T for true or an F for false to the left of each statement.

_____ **1.** A forward-biased zener diode has low resistance.

_____ **2.** A zener diode attempts to maintain or regulate the voltage across it when it is in the breakdown region.

_____ **3.** A zener diode will maintain a constant current when it is in the breakdown region.

_____ **4.** A zener diode is usually operated in the reverse-biased condition when it is used as a voltage regulator.

_____ **5.** If the load current in a zener diode regulator circuit increases, the current through the zener diode increases.

_____ **6.** Zener diodes are used as circuit-protection devices.

_____ **7.** If the input voltage to a zener diode regulator circuit increases, the zener current will increase while the load voltage and current will remain fairly constant.

_____ **8.** The maximum zener current ($I_{Z\ max}$) can be determined by dividing the zener voltage (V_Z) into the manufacturer's power rating for the zener diode ($I_{Z\ max} = P_Z/V_Z$).

_____ **9.** Zener diodes are ideal for rectifying ac voltage.

_____ **10.** If the current through a zener diode falls below the minimum holding current, the result is as if the zener diode were not in the circuit.

2-8b ZENER DIODE: MULTIPLE-CHOICE QUIZ

Circle the correct answer for each question.

1. If the load current decreases in a zener diode regulator circuit, the zener current:

a. increases **b.** decreases

c. remains the same

2. In Figure 2-A, R_S equals:

a. 600 Ω **b.** 500 Ω

c. 166 Ω **d.** 100 Ω

3. In Figure 2-A, I_L equals:

a. 12 mA **b.** 18 mA

c. 30 mA **d.** 42 mA

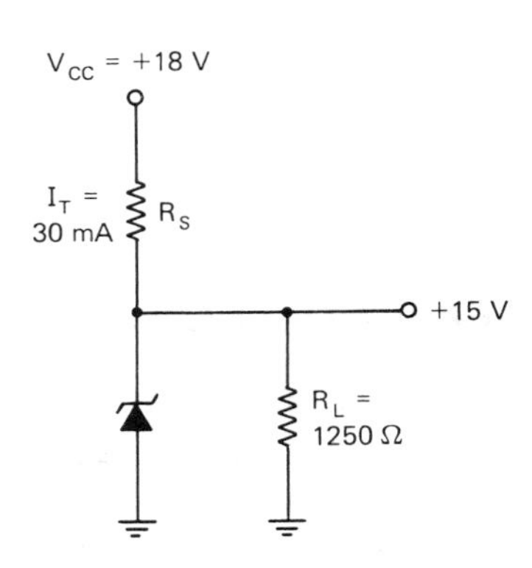

Figure 2–A

4. In Figure 2-A, I_Z equals:

a. 12 mA b. 18 mA
c. 30 mA d. 42 mA

5. In Figure 2-A, V_Z equals:

a. 12 V b. 15 V
c. 18 V d. 33 V

6. In Figure 2-B, V_{out} equals:

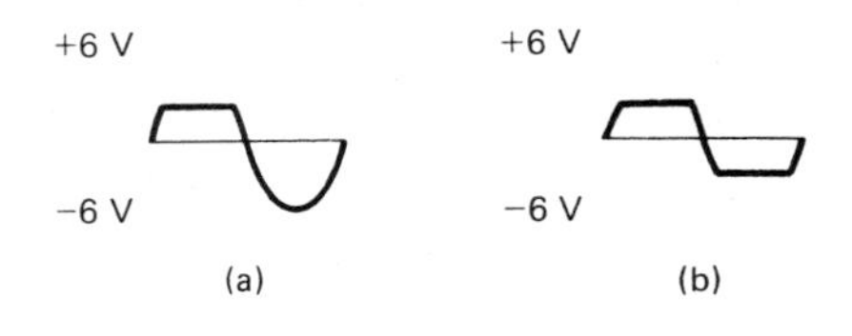

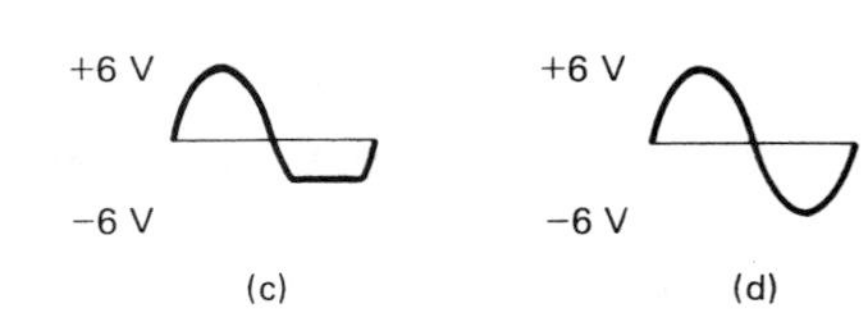

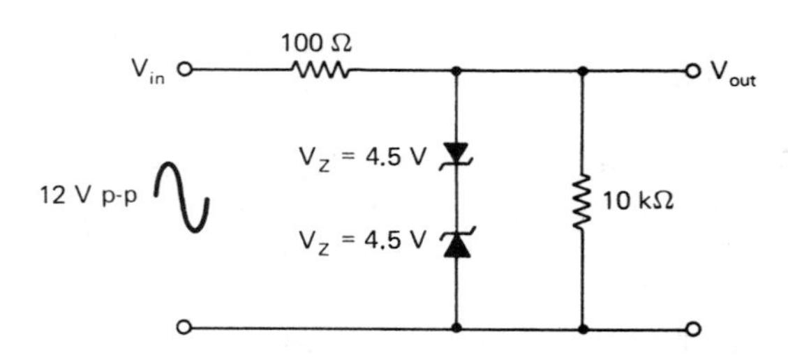

Figure 2–B

7. The maximum zener current for a 1-W zener diode with a V_Z of 5.1 V is about:

a. 5.1 mA b. 10.2 mA
c. 98 mA d. 196 mA

8. If two zener diodes are connected correctly (series aiding) in a circuit with V_{CC} = 14 V, and each of their V_Z = 6 V, then the total voltage across them is:

a. +6 V b. +8 V
c. +12 V d. +14 V

9. In a simple zener diode regulator circuit, V_Z = + 9 V, I_Z = 31 mA, and I_L = 9 mA. If R_L is reduced by one-half, I_Z equals:

a. 9 mA b. 18 mA
c. 22 mA d. 31 mA

10. If the input voltage (V_{CC}) to a zener diode regulator circuit increases, the output voltage (V_Z) will:

a. increase b. decrease
c. remain fairly constant

ANSWERS TO EXPERIMENTS AND QUIZZES

Experiment 1. (1) reverse biased (2) remain, varies (3) limit, safe

Experiment 2. (1) equals (2) decreases (3) increases (4) constant (5) constant

Experiment 3. (1) V_{Z_1}, V_{Z_2} (2) single (3) 5.1, 9.1 (or vice versa)

Experiment 4. (1) limiter (2) +0.7 V (3) −5.1 V (4) +5.1 V

Experiment 5. (1) face to face, back to back (2) V_Z, V_F (or vice versa)

True–False. (1) T (2) T (3) F (4) T (5) F (6) T (7) T (8) T (9) F (10) T

Multiple Choice. (1) a (2) d (3) a (4) b (5) b (6) b (7) d (8) c (9) c (10) c

Unit 3

Bipolar Transistor

INTRODUCTION

Bipolar transistors were the first attempt to produce a small, physically rugged, low-power consumption device to replace vacuum tubes.

UNIT OBJECTIVES

Upon completion of this unit you will be able to:

1. Draw the schematic symbol for a *PNP* and *NPN* transistor.
2. Explain the principles of electron flow and hole flow inside a bipolar transistor.
3. Describe how a bipolar transistor is a current-operated device.
4. Calculate the current gains of alpha and beta.
5. Show how a transistor operates as a voltage-controlled switch.
6. Draw a current–voltage characteristic curve for a bipolar transistor.
7. Plot a load line for a class A transistor amplifier.
8. Define leakage current and show how to measure it.
9. Test bipolar transistors with an ohmmeter.
10. List and define the three basic amplifier configurations.
11. Construct and test a two-stage amplifier circuit.
12. Troubleshoot a common-emitter amplifier.

SECTION 3-1 THEORY OF STRUCTURE AND OPERATION

3-1a STRUCTURE, SCHEMATIC SYMBOLS, AND PACKAGE CONFIGURATIONS

The bipolar transistor consists of three elements of semiconductor material as shown in Figure 3-1. An *NPN* transistor consists of an *N*-material emitter (*E*) and collector (*C*) separated by a very thin (0.001 in. or less) *P* material called the base (*B*). A *PNP* transistor has *P* material for its emitter and collector with an *N*-material base. The emitter is heavily doped to provide current carriers, which are controlled by the base element as they travel toward the less doped collector element. The *NPN* transistor schematic symbol has the emitter arrow pointing

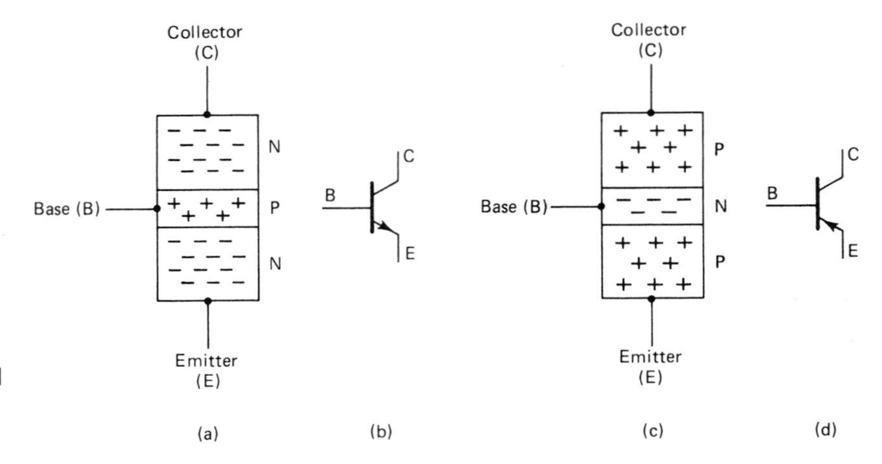

Figure 3–1 Bipolar transistor structures and schematic symbols: (a) and (b) *NPN* transistor; (c) and (d) *PNP* transistor.

away from the base, while the *PNP* transistor emitter arrow points toward the base.

Although the element leads that come out of the transistor case can be any arrangement, a general configuration for many transistors is shown in Figure 3-2. However, any transistor used should first be checked for its proper lead positions from a transistor manual or dependable source before placing it in a circuit.

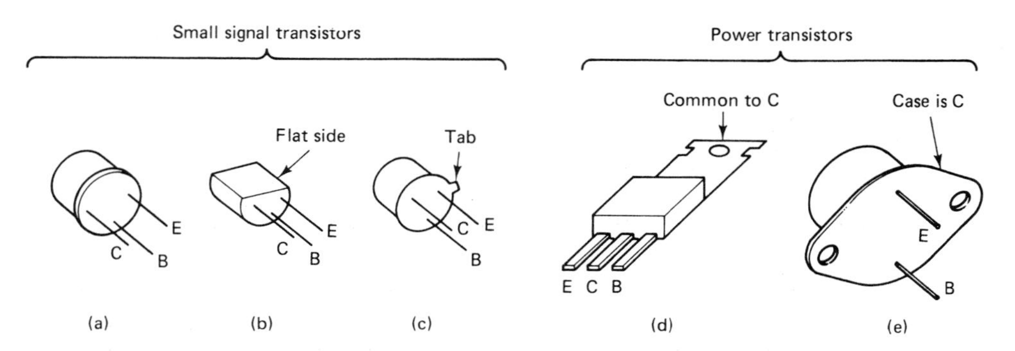

Figure 3–2 General transistor package configurations and lead identification: (a) TO-5 case; (b) TO-92 case; (c) TO-18 case; (d) TO-220 case; (e) TO-3 or TO-66 case.

3-1b BIASING AND CURRENT PATHS

Bipolar transistors are "normally off" devices (no current flowing through them), and voltage biasing methods are needed to cause current to flow. To accomplish this, the emitter–base junction is forward biased and the collector–base junction is reverse biased. With an *NPN* transistor, the base is made more positive than the emitter and the collector is made more positive than the base, as shown in Figure 3-3. The negative end of the small battery at the emitter repels electrons toward the base, and the positive end of the same battery repels holes in the base toward the emitter. Current is considered flowing in two directions, hence the term *bipolar*. There are some electron–hole combinations that produce a very small base current (I_B), but most of the emitter current (I_E) is attracted through the thin base region toward the higher battery voltage placed on the collector and is referred to as a collector current (I_C). Therefore, I_C is I_E less the small amount of I_B. These current relationships can be stated by the formulas

$$I_C = I_E - I_B$$
$$I_B = I_E - I_C$$

or

$$I_E = I_C + I_B$$

A small base current must flow in order for a large emitter–collector current to flow, and the bipolar transistor is called a *current-operated* device.

Since I_B is very small (in μA), I_C is approximately equal to I_E, and this current is usually referred to as just I_C. The base current directly controls the collector current; that is, more I_B produces more I_C.

A single battery can be used to bias the transistor properly, as shown in Figure 3-3c. Resistor R_A is made large enough to limit I_B and I_C so that the transistor does not overheat and possibly burn out. Resistor R_L also limits I_C.

Figure 3–3 *NPN* bipolar transistor biasing methods and current paths: (a) structure analysis; (b) schematic representation; (c) single voltage source biasing.

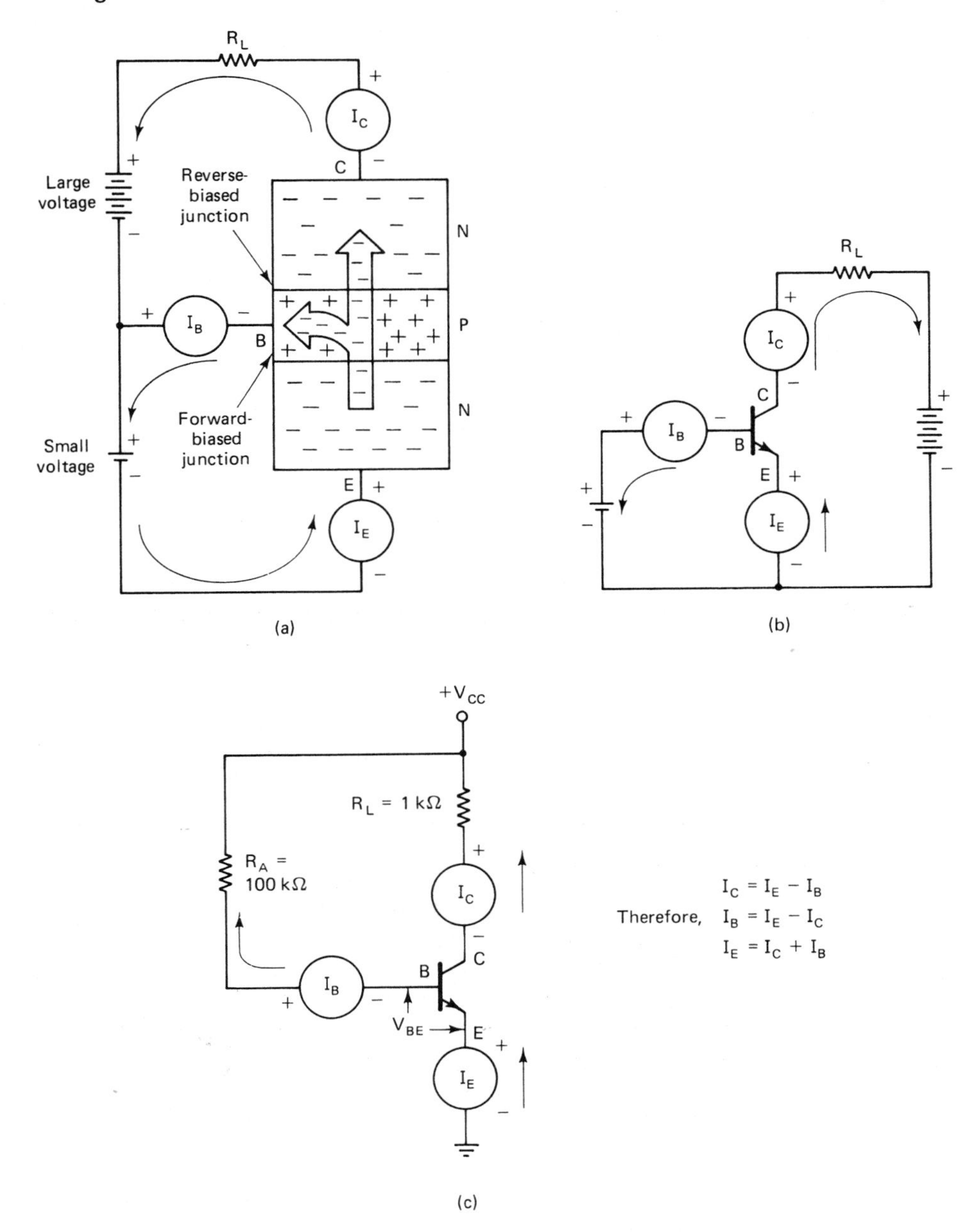

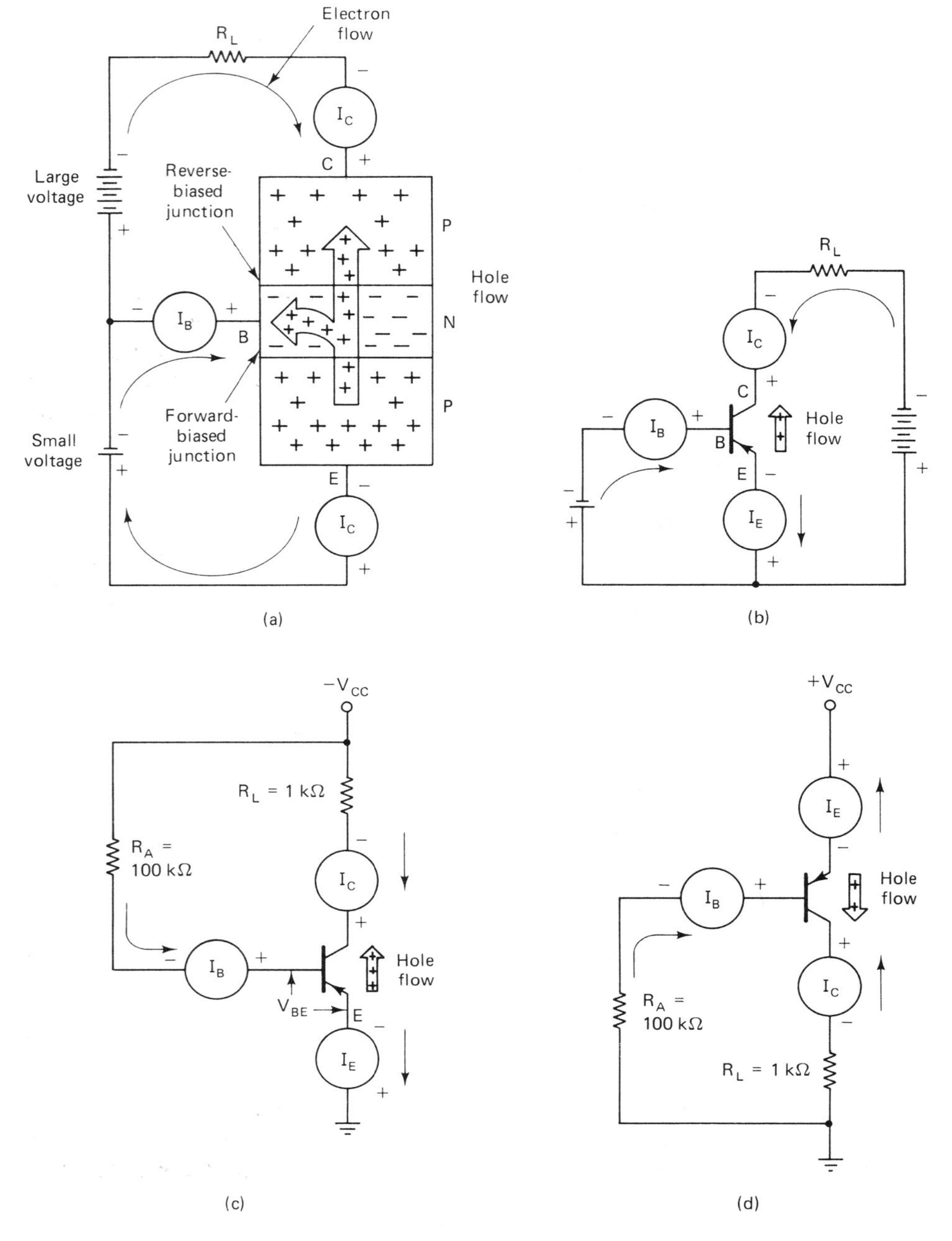

Figure 3–4 *PNP* bipolar transistor biasing methods and current paths: (a) structure analysis; (b) schematic representation; (c) single voltage source biasing; (d) using a positive supply voltage.

When a bipolar transistor is conducting, the emitter–base junction is forward biased, and there will always be a voltage drop (V_{BE}) of about 0.2 V for germanium material and about 0.6 V for silicon material.

The same biasing conditions are required for a *PNP* transistor to cause current to flow, except that the voltage polarities are reversed and the major current through the transistor is hole flow. These current relationships can be seen in Figure 3-4.

3-1c BIPOLAR TRANSISTOR AS A SWITCH

Control of I_B can cause the transistor to perform as a switch, which is shown in Figure 3-5. With the base shorted to ground (Figure 3-5a), there is no forward bias on the emitter–base junction and no current

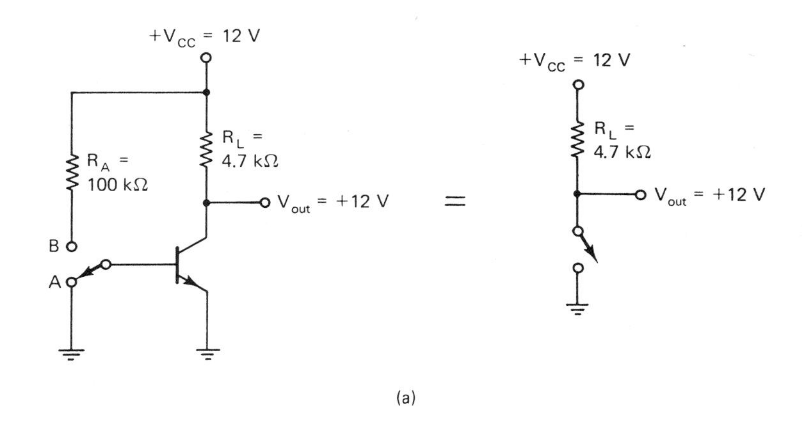

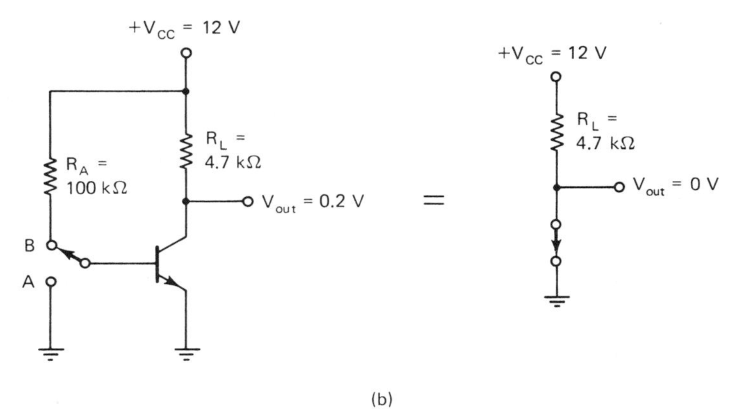

Figure 3–5 Bipolar transistor used as a switch: (a) open switch; (b) closed switch.

flow in the transistor. The transistor is cut off (switch open), and in effect both junctions are reverse biased. An indication that the transistor is not conducting is found by measuring the collector voltage, which will equal the supply voltage ($+V_{CC}$). An open switch will have the full voltage present across its terminals.

When the base is connected to R_A (Figure 3-5b), the emitter–base junction is forward biased and current flows through the transistor. The transistor is on (switch closed), and in effect both junctions are forward biased. An indication that the transistor is conducting is again found by measuring the collector voltage, which will be about $+0.2$ V (nearly zero). A closed switch will have 0 V present across its terminals. *The technique of measuring collector voltage to determine if the transistor is conducting or cut off is an important troubleshooting procedure for testing nearly all solid-state active-device circuits.*

3-1d BIPOLAR TRANSISTOR AS A VARIABLE RESISTOR

Since the amount of I_B directly controls I_C, the bipolar transistor acts as a variable resistor, as shown in Figure 3-6a. This transistor resistance (R_X) can be calculated by dividing I_C into the voltage across the collector and emitter (V_{CE}), in this case $+V_{CC}$. When the wiper of potentiometer R_A is moved up, I_B increases, causing I_C to increase, which in effect reduces R_X. When the wiper of R_B is moved down, I_B decreases, causing I_C to decrease and in effect R_X increases. A transistor operating as a variable resistor is said to be in the active region and is used for amplification purposes.

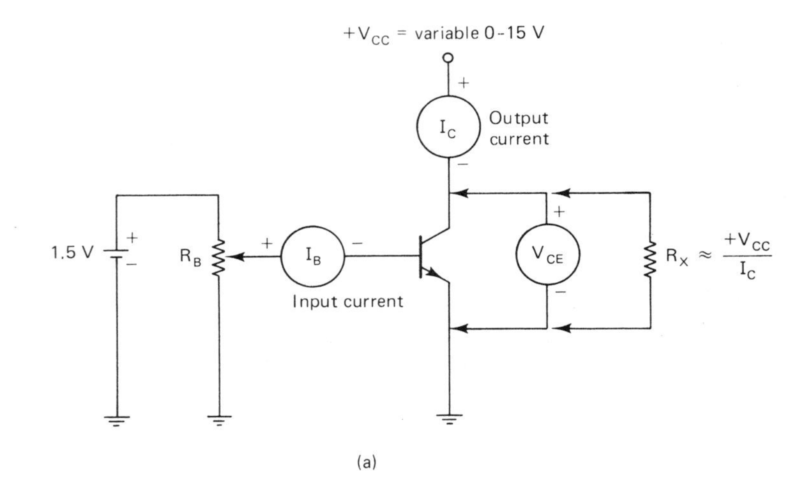

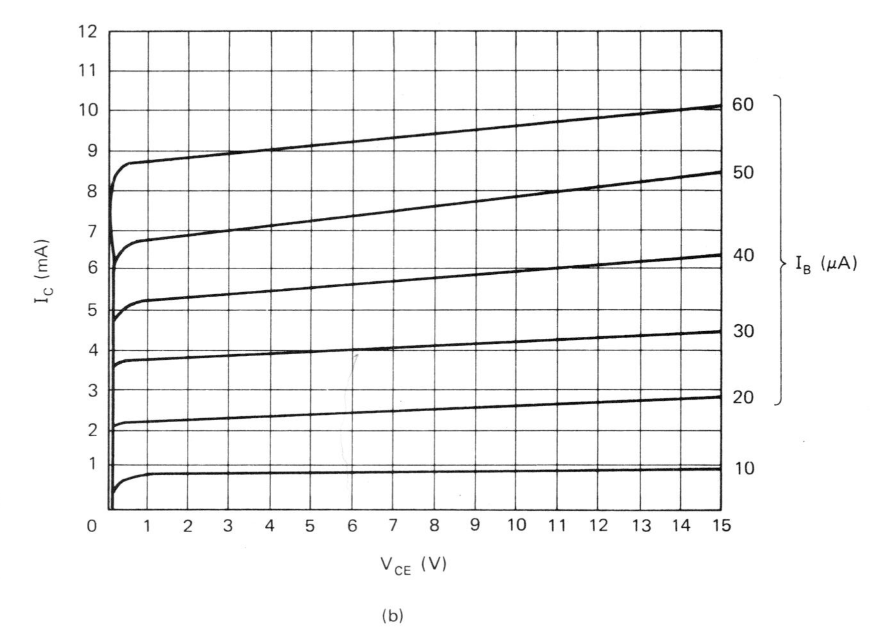

Figure 3–6 I_C–V_{CE} characteristic curves for an *NPN* transistor common-emitter circuit: (a) test circuit; (b) I_C versus V_{CE} for various values of I_B graph.

3-1e CURRENT–VOLTAGE CHARACTERISTICS AND AMPLIFICATION

A graph technique known as plotting collector characteristic curves can be used to show the amounts of I_B needed to control certain amounts of I_C and is helpful in understanding the operation of the bipolar transistor. A test circuit to accumulate data used in the graph is shown in Figure 3-6a. Potentiometer R_B is adjusted to set a desired I_B. A variable power supply is then set at various voltages and the I_C readings are recorded. Base current is then set for other values and the procedure is repeated. The data are then plotted on a graph, as shown in Figure 3-6b.

Inspecting the graph reveals that I_B controls I_C more than the voltage across the collector and emitter (V_{CE}). As an example, with I_B set at 10 μA, I_C is about 1 mA. Even though V_{CE} is increased, I_C remains fairly constant. When I_B is set at 20 μA, I_C increases to about 2.5 mA and remains fairly constant even when V_{CE} is increased.

Characteristic curves can be used to predict the operation of a standard common-emitter amplifier, as shown in Figure 3-7. Resistors R_A, R_B, and R_E are used to establish the dc operating bias point, called the quiescent (Q) point, when there is no signal present at the input of the base. Capacitor C_E helps to maintain this Q point, even though I_C varies

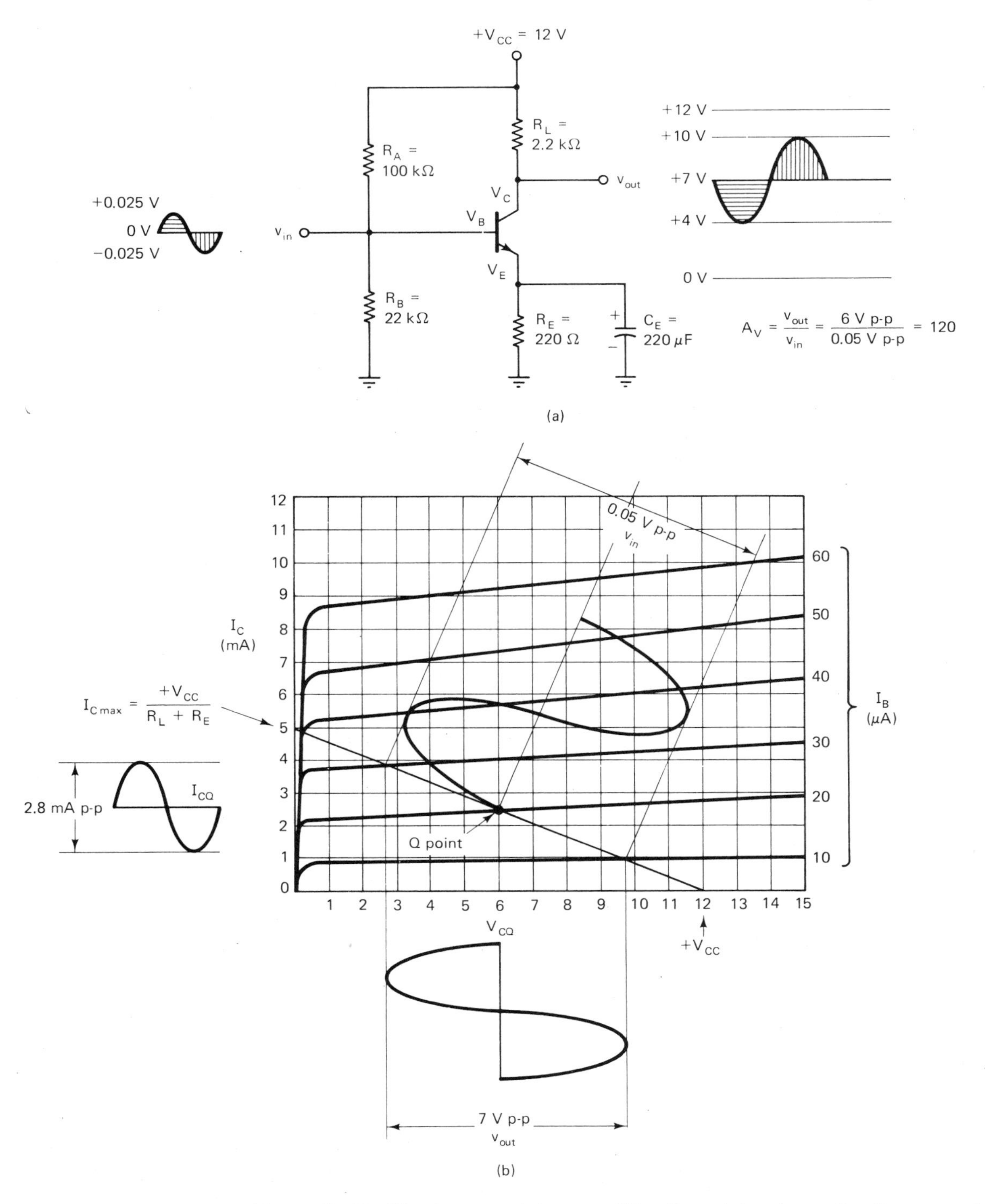

Figure 3–7 Bipolar transistor amplification: (a) schematic diagram of the most used common-emitter circuit; (b) graphic representation.

with an input signal to the base. A load line is drawn from the value of the power supply voltage to the maximum I_C that could flow if the transistor is considered an absolute "short." This $I_{C\,\max}$ is found by dividing $+V_{CC}$ by the value of $R_L + R_E$. The Q point is selected about halfway on the load line, where $I_B = 20\ \mu$A intersects it. A line projected downward shows that the quiescent voltage at the collector (V_{CQ}) will be about +6.0 V. A line projected from the Q point to the left shows a quiescent collector current (I_{CQ}) of about 2.5 mA. Lines are drawn at right angles to the load line at intersections I_B = 10, 20, and 30 μA.

These lines represent the amount of I_B variation when a signal of 0.05 V p-p is placed at the base.

When I_B increases to 30 μA, I_C increases to 4 mA, while V_C decreases to +2.7 V. When I_B decreases to 10 μA, I_C decreases to about 1 mA, while V_C increases to about +9.7 V. The remaining lines projected on the graph show the I_C swing and the V_C swing of the amplifier circuit. The voltage gain of the amplifier can be calculated from the voltage swings by the formula

$$A_V = \frac{v_{\text{out}}}{v_{\text{in}}} = \frac{7\text{ V p-p}}{0.05\text{ V p-p}} = 140$$

The main points to remember are as follows:

1. When I_B increases, I_C increases and V_C decreases.
2. When I_B decreases, I_C decreases and V_C increases.

This situation accounts for a signal phase inversion of 180° between the input and output voltage signals.

3-1f AMPLIFICATION FACTORS BETA AND ALPHA

The ability of a bipolar transistor to amplify is referred to as *current gain.* Beta (β) is the current gain of a common-emitter circuit, sometimes called the forward current-transfer ratio (H_{FE}). Beta is equal to the current gain of I_C divided by I_B. This rating can be given in static (dc) or dynamic (ac) terms expressed as

$$\beta_{\text{dc}} = \frac{I_C}{I_B} \quad \text{and} \quad \beta_{\text{ac}} = \frac{\Delta i_c}{\Delta i_b}$$

where the symbol Δ means "a small change." For example, if I_B = 20 μA and I_C = 2.5 mA, then

$$\beta_{\text{dc}} = \frac{I_C}{I_B} = \frac{2.5\text{ mA}}{20\ \mu\text{A}} = 125$$

Referring to the graph in Figure 3-7b, note that if I_B changes from 20 to 10 μA a corresponding change occurs in I_C of from 2.5 to 1.0 mA. The dynamic beta is then found:

$$\beta_{\text{ac}} = \frac{\Delta i_c}{\Delta i_b} = \frac{1.5\text{ mA}}{10\ \mu\text{A}} = 150$$

The bipolar transistor can be arranged in a common-base circuit where the input current is I_E and the output current is I_C. The current gain of this circuit is called alpha (α), where I_C is divided by I_E. It can also be expressed in static or dynamic terms:

$$\alpha_{\text{dc}} = \frac{I_C}{I_E} \quad \text{and} \quad \alpha_{\text{ac}} = \frac{\Delta i_c}{\Delta i_e}$$

Alpha is always less than 1, since some of I_E flows out of I_B, leaving less I_C.

Conversion formulas can be used to find either alpha or beta, if the other is known.

$$\beta = \frac{\alpha}{1 - \alpha} \quad \text{and} \quad \alpha = \frac{\beta}{\beta + 1}$$

For example, if $\beta = 100$, then

$$\alpha = \frac{\beta}{\beta + 1} = \frac{100}{101} = 0.99$$

or if $\alpha = 0.99$, then

$$\beta = \frac{\alpha}{1 - \alpha} = \frac{0.99}{1 - 0.99} = \frac{0.99}{0.01} = 99 \quad \text{(nearly 100)}$$

3-1g LEAKAGE CURRENT AND STABILITY

A reverse-biased *PN* junction always exhibits some leakage current as a result of minority-carrier combinations. In a forward-biased transistor circuit, as shown in Figure 3-8a, the leakage current, called I_{CBO} or simply I_{CO}, flows from collector to base. This I_{CO} is negligible in itself, but it increases with an increase in temperature, which in turn increases I_C and can cause erratic operation of a transistor circuit. A resistor is usually added to the emitter circuit to reduce the effect of I_{CO} and stabilize the transistor.

A simple test, as shown in Figure 3-8b, can determine if I_{CO} is excessive, which will reduce beta and cause improper operation when the transistor is used in a circuit.

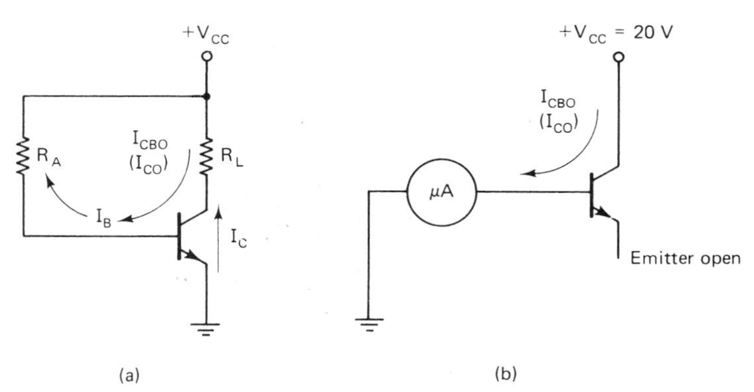

Figure 3–8 Leakage current (I_{CBO} or I_{CO}): (a) I_{CO} flow; (b) simple circuit for measuring I_{CBO}.

3-1h BIPOLAR TRANSISTOR DEFINITIONS

The following definitions are some of the more commonly used terms associated with bipolar transistors.

V_{CC}	supply voltage connected to the collector
V_{EE}	supply voltage connected to the emitter
V_{BB}	supply voltage connected to the base
V_C	collector voltage to ground
V_B	base voltage to ground
V_E	emitter voltage to ground

V_{CE}	voltage from collector to emitter
V_{EB} or V_{BE}	voltage from emitter to base
BV_{CEO}	dc breakdown voltage, collector to emitter, with base open
BV_{CBO}	dc breakdown voltage, collector to base, with emitter open
BV_{EBO}	dc breakdown voltage, emitter to base, with collector open
I_E	emitter current
I_B	base current
I_C	collector current
$I_{C\,max}$	maximum allowable collector current
I_{CBO} or I_{CO}	leakage current from collector to base
Maximum collector dissipation	maximum power a transistor can dissipate; the product of I_C and V_{CE}
f_{ab}	alpha cutoff frequency, the frequency at which the output of a transistor falls to 70% when in the common-base circuit configuration
f_{ae}	cutoff frequency for a common-emitter circuit

SECTION 3-2
BIPOLAR TRANSISTOR DEFINITION EXERCISE

Refer to the previous sections and write a brief definition of each term.

1. V_{CC} ______________________________

____

____

2. V_{EE} ______________________________

____

____

3. V_{BB} ______________________________

____

____

4. V_C ______________________________

____

____

5. V_B ______________________________

____

____

6. V_E

7. V_{CE}

8. V_{EB}

9. I_E

10. I_B

11. I_C

12. $I_{C\,max}$

13. I_{CO}

14. Maximum collector dissipation

15. Beta

16. Alpha

17. Load line

SECTION 3-3 EXERCISES AND PROBLEMS

Perform the following exercises before beginning Section 3-4.

1. Draw the structure of a *NPN* transistor. (Label all parts.)

2. Draw the schematic symbol of a *NPN* transistor. (Label the leads.)

3. Referring to Figure 3-3b, draw the *NPN* transistor with proper biasing. (Indicate polarities.)

4. Referring to Figure 3-3c, draw the circuit and use arrows to indicate the electron current paths through the transistor. (Lable the currents I_E, I_B, and I_C.)

5. Write the formulas that show the relationship of current flow through a bipolar transistor.

$I_C =$ ______________________

$I_B =$ ______________________

$I_E =$ ______________________

6. Draw the structure of a *PNP* transistor. (Label all parts.)

7. Draw the schematic symbol of a *PNP* transistor. (Label the leads.)

8. Referring to Figure 3-4b, draw a *PNP* transistor with proper biasing. (Indicate polarities.)

9. Referring to Figure 3-4c, draw a *PNP* transistor with proper biasing, but reverse the polarity of the voltages and compare it with the circuit in Exercise 8.

10. Referring to Figure 3-4d, draw the circuit and use arrows to indicate the electron flow in the circuit. (Label the currents I_E, I_B, and I_C.) Also show the direction of hole flow through the transistor.

11. Draw a sample of the output characteristic curves (I_C versus V_{CE}) for an *NPN* transistor, showing different values of I_B.

12. Draw a sample of the output characteristic curves (I_C versus V_{CE}) for a *PNP* transistor, showing different values of I_B.

13. A bipolar transistor that is turned on will have the base 0.6 V greater than the emitter. The V_{CC} is + or −6 V depending on the type of transistor. Indicate if the transistors are conducting (On) or not conducting (Off) by the voltages given at their leads.

a. $V_C = +3$ V
$V_B = +0.6$ V
$V_E = 0$ V

____

b. $V_C = +3$ V
$V_B = +1.9$ V
$V_E = +1.3$ V

____

c. $V_C = +6$ V
$V_B = 0$ V
$V_E = 0$ V

____

d. $V_C = +6$ V
$V_B = -0.6$ V
$V_E = 0$ V

____

e. $V_C = -6$ V
$V_B = 0$ V
$V_E = 0$ V

____

f. $V_C = -3$ V
$V_B = -0.6$ V
$V_E = 0$ V

____

g. $V_C = -6$ V
$V_B = 0$ V
$V_E = -0.6$ V

____

h. $V_C = -3$ V
$V_B = -1.5$ V
$V_E = -0.9$ V

____

14. Using Figure 3-3c as a guide, find the missing currents in the following problems by using the formulas $I_E = I_B + I_C$, $I_B = I_E - I_C$, and $I_C = I_E - I_B$.

a. $I_B = 20\ \mu$A
$I_E = 2$ mA

$I_C =$ _______

b. $I_C = 2.57$ mA
$I_E = 2.6$ mA

$I_B =$ _______

c. $I_C = 3.3$ mA
$I_B = 25\ \mu$A

$I_E =$ _______

d. $I_C = 2.4$ mA
$I_B = 10\ \mu$A

$I_E =$ _______

e. $I_C = 3.96$ mA
$I_E = 4$ mA

$I_B =$ _______

f. $I_B = 50\ \mu$A
$I_E = 5.15$ mA

$I_C =$ _______

15. Draw the circuit in Figure 3-14a and write in the voltage nomenclature (V_C, V_B, V_E, V_{EB}, and V_{CE}). Also indicate the current paths (I_E, I_B, and I_C) through the transistor using arrows.

16. From the common-emitter output characteristic curves shown below, find the I_C for each I_B that is given. (Use a V_{CE} of +12 V as a reference point.)

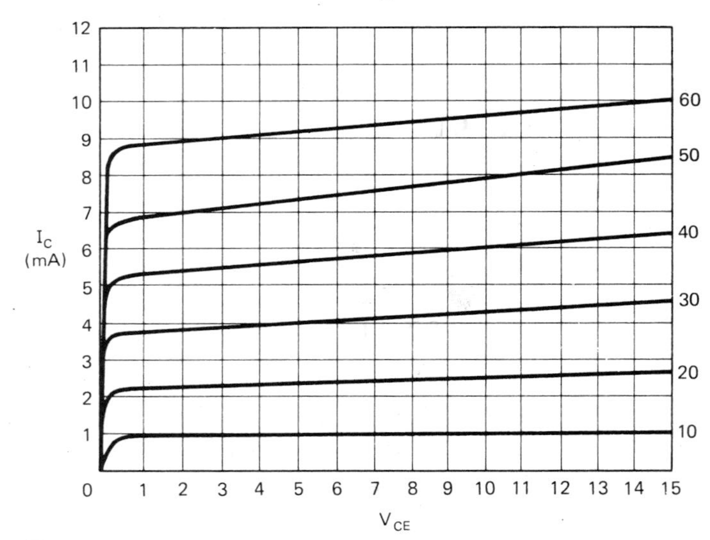

Figure 3–E.1

a. $I_B = 10\ \mu$A, $I_C =$ _____

b. $I_B = 20\ \mu$A, $I_C =$ _____

c. $I_B = 30\ \mu$A, $I_C =$ _____

d. $I_B = 40\ \mu$A, $I_C =$ _____

e. $I_B = 50\ \mu$A, $I_C =$ _____

f. $I_B = 60\ \mu$A, $I_C =$ _____

17. From the values given, find the dc beta (β_{dc} or H_{FE}) using the formula $\beta_{dc} = I_C / I_B$.

a. $I_C = 2$ mA
$I_B = 20$ μA

β_{dc} = ______

b. $I_C = 2.2$ mA
$I_B = 10$ μA

β_{dc} = ______

c. $I_C = 8$ mA
$I_B = 30$ μA

β_{dc} = ______

18. From the values given, find the ac beta (β_{ac} or h_{fe}) using the formula $\beta = \Delta I_C / \Delta I_B$.

a. ΔI_C = 2 to 4 mA
ΔI_B = 10 to 20 μA

β_{ac} = ______

b. ΔI_C = 3 to 5.5 ma
ΔI_B = 10.5 to 20.5 μA

β_{ac} = ______

c. ΔI_C = 4 to 6 mA
ΔI_B = 20 to 40 μA

β_{ac} = ______

19. From the values given, find the dc alpha (α_{dc} or H_{FB}) using the formula $\alpha_{dc} = I_C / I_E$.

a. $I_C = 2.2$ mA
$I_E = 2.24$ mA

α_{dc} = ______

b. $I_C = 3.4$ mA
$I_E = 3.45$ mA

α_{dc} = ______

c. $I_C = 5.2$ mA
$I_E = 5.5$ mA

α_{dc} = ______

20. If β is given, α can be found by the formula $\alpha = \beta/\beta + 1$. If α is given, β can be found by the formula $\beta = \alpha/1 - \alpha$. Find the respective factor when the other factor is given.

a. $\beta = 200$

α = ______

b. $\beta = 150$

α = ______

c. $\alpha = 0.995$

β = ______

d. $\alpha = 0.992$

β = ______

SECTION 3-4 EXPERIMENTS

EXPERIMENT 1. TESTING BIPOLAR TRANSISTORS

Objective:

To demonstrate a practical go/no go method of testing bipolar transistors with an ohmmeter.

Introduction:

For ohmmeter testing purposes, an *NPN* transistor is similar to two diodes back to back, as shown in Figure 3-9b. There exists two *PN* junctions, base–emitter and base–collector. When each of these junctions is forward biased by the ohmmeter, positive lead to *P* material and negative lead to *N* material (Figure 3-9c), there should be a low-resistance indication. There should be a high-resistance reading when these junctions are reverse biased, positive lead to *N* material and negative lead to *P* material (Figure 3-9d).

The PNP transistor can be tested with a similar method, except that the diodes are face to face as shown in Figure 3-9e. This simple test determines if the transistor is shorted or open on a *go* (no problems)/*no go* (it has problems) basis.

Materials Needed:

A standard or digital ohmmeter

One or several bipolar transistors, including both types, *NPN* and *PNP*

Procedure:

1. Set the ohmmeter to the midrange scale.
2. Refer to Figure 3-9c to connect the ohmmeter to an *NPN* transistor for each junction and record the readings in the indicated ohmmeter circles as high or low.
3. Refer to Figure 3-9d to connect the ohmmeter to the *NPN* transistor for

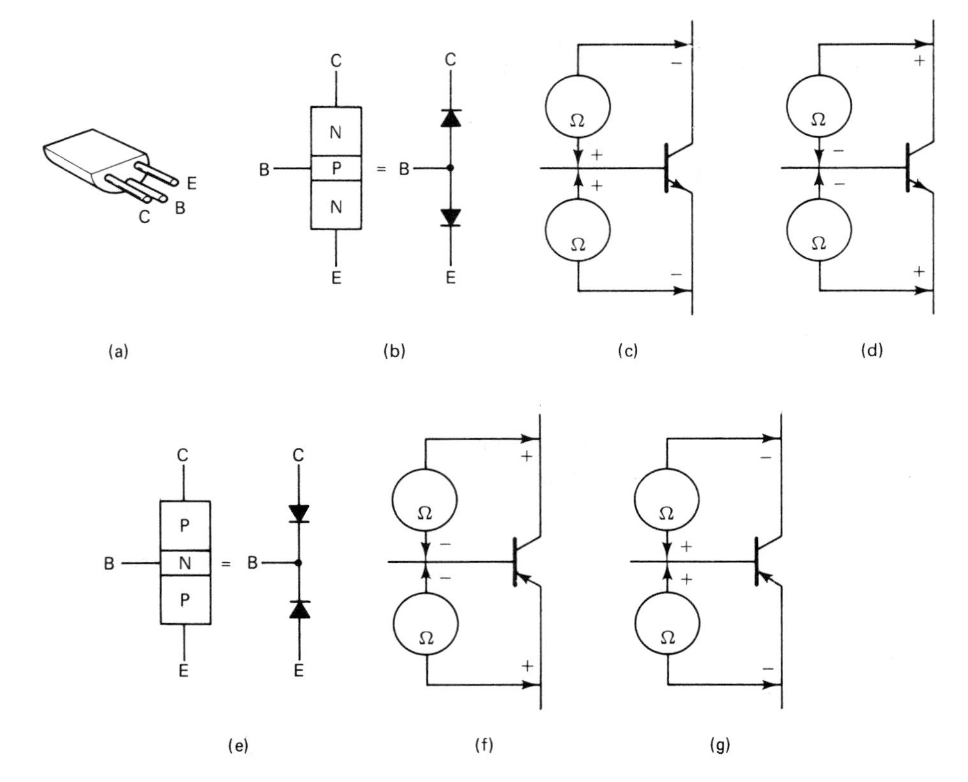

Figure 3–9 Testing bipolar transistors with an ohmmeter: (a) general package configuration; (b) *NPN* diode equivalent circuit; (c) and (d) *NPN* ohmmeter connections; (e) *PNP* diode equivalent circuit; (f) and (g) *PNP* ohmmeter connections.

each junction and record the readings in the indicated ohmmeter circles as high or low.

4. Using a *PNP* transistor, perform the same procedure as in steps 1 through 3, while referring to Figure 3-9f and g.

Fill-in Questions:

1. A forward-biased *PN* junction on a good bipolar transistor has Low resistance.
2. A reverse-biased *PN* junction on a good bipolar transistor has high resistance.
3. A forward-biased *PN* junction with a high ohmmeter reading indicates that the transistor is Open.
4. A reverse-biased *PN* junction with a low ohmmeter reading indicates that the transistor is Shorted.

EXPERIMENT 2. BIPOLAR TRANSISTOR AS A SWITCH

Objective:

To show how to recognize, by circuit voltage drops, whether a bipolar transistor is conducting or nonconducting when used as a switch.

Introduction:

Refer to Figure 3-10a and note that when switch S_1 is in position *A* the emitter–base junction is reverse biased ($V_B = 0$ V), the transistor is not conducting (off), and the total circuit voltage ($+V_{CC}$) will appear at the collector (V_C, or across collector and emitter). When S_1 is in position *B*, the emitter–base junction is forward biased ($V_B \approx 0.7$ V), the transistor is conducting (on), and the collector voltage will be near ground potential ($V_C \approx 0.2$ V). The transistor is saturated in this condition.

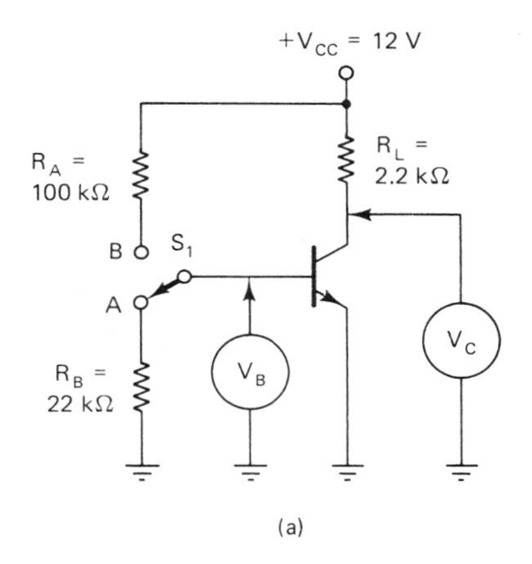

S_1 position	V_B	V_C	Condition of transistor (on or off)
A			
B			

(b)

Figure 3–10 Bipolar transistor used as a switch: (a) schematic diagram; (b) switch condition table.

Materials Needed:

1 Fixed +12-V power supply
1 Standard or digital voltmeter
1 2.2-kΩ resistor at 0.5 W (R_L)
1 22-kΩ resistor at 0.5 W (R_B)
1 100-kΩ resistor at 0.5 W (R_A)
1 2N2222 transistor or equivalent
1 Double-pole single-throw (DPST) switch (S_1)
1 Breadboard for constructing circuit

Procedure:

1. Construct the circuit shown in Figure 3-10a.
2. Make sure that S_1 is in position A.
3. Measure V_B and record the value in the data table (next to A).
4. Measure V_C and record the value in the data table (in the same row).
5. Indicate in the data table if the transistor is on or off (in the same row).
6. Move S_1 to position B.
7. Measure V_B and record the value in the data table (next to B).
8. Measure V_C and record the value in the data table (in the same row).
9. Indicate in the data table if the transistor is on or off (in the same row).
10. Calculate the approximate I_C from the formula

$$I_C = \frac{V_{R_L}}{R_L} = \frac{+V_{CC} - V_{C(\text{on})}}{R_L}$$

11. Record I_C here: ______ .

Fill-in Questions:

1. When the transistor is cut off (not conducting), the voltage at the collector (V_C) will equal Vcc .

2. When the voltage between base and emitter (V_B) is 0 V, the transistor is not CONDUCTING .

3. When V_C is near ground potential, the transistor is ON .

4. If an *NPN* silicon transistor is in saturation, the voltage drop from base to emitter will be about 0.7 V.

EXPERIMENT 3. BIPOLAR TRANSISTOR AS A VARIABLE RESISTOR

Objective:

To demonstrate how I_B directly controls I_C in a bipolar transistor and how the resistance of the transistor varies inversely.

Introduction:

Refer to Figure 3-11a and note that resistor R_A is changeable to produce different values of I_B. This, in turn, creates various values of

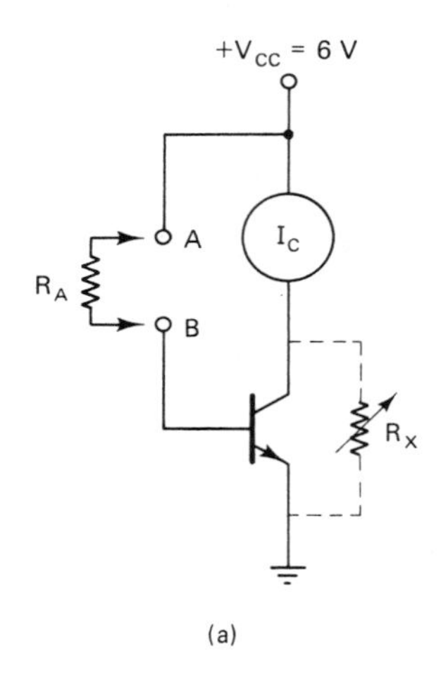

(a)

R_A (Ω)	I_C (mA)	$I_B \approx +V_{CC}/R_A$ (μA)	$R_X = +V_{CC}/I_C$ (Ω)	$\beta_{dc} = I_C/I_B$
1 M				
470 k				
220 k				
100 k				
47 k				
22 k				

(b)

Figure 3–11 Bipolar transistor used as a variable resistor: (a) schematic diagram; (b) data table.

I_C and in effect alters the resistance (R_X) of the transistor. The supply voltage $+V_{CC}$ should be kept constant for each change of R_A. Approximate values for I_B and R_X are found from the formulas given in the data table of Figure 3-11b. The dc beta is also approximated in the last column.

Materials Needed:

1 Fixed +6-V power supply
1 Standard or digital ammeter
1 Standard or digital voltmeter
1 2N2222 transistor or equivalent
1 1-MΩ resistor at 0.5 W
1 470-kΩ resistor at 0.5 W
1 220-kΩ resistor at 0.5 W
1 100-kΩ resistor at 0.5 W
1 47-kΩ resistor at 0.5 W
1 22-kΩ resistor at 0.5 W
1 Breadboard for constructing circuit

(The 470-kΩ through 22-kΩ resistors are bracketed together as R_A.)

Procedure:

1. Construct the circuit shown in Figure 3-11a, leaving points *A* and *B* open.
2. Place the 1-MΩ resistor at points *A* and *B*.
3. Make sure that $+V_{CC}$ is held at +6 V with each change of R_A.
4. Measure the value of I_C and record in the data table.
5. Calculate I_B from the formula $I_B \approx +V_{CC}/R_A$ and record the result in the data table.
6. Calculate R_X from the formula $R_X = +V_{CC}/I_C$ and record the result in the data table.
7. Calculate β_{dc} from the formula $\beta_{dc} = I_C/I_B$ and record the result in the data table.
8. Repeat steps 2 through 7 for the remaining R_A values given in the data table.

Fill-in Questions:

1. When I_B increases, ___Ic___ increases.

2. When I_C increases, ___Rx___ decreases.

3. The transistor is said to be in the active region when it operates as a ___variable___ resistance.

4. Amplification takes place in the ___active___ region.

5. The beta of a transistor is not ___constont___, but may vary considerably.

EXPERIMENT 4. CURRENT–VOLTAGE CHARACTERISTICS OF A BIPOLAR TRANSISTOR

Objective:

To show how I_B has more influence over I_C than the voltage across the collector–emitter (V_{CE}), and to construct a graph of output characteristics curves to understand better the operation of a bipolar transistor.

Introduction:

This experiment requires two ammeters and a voltmeter to obtain, as much as possible, accurate readings and facilitate the gathering of data. A separate 1.5-V battery for the base voltage supply (V_{BB}) also stabilizes the circuit considerably more than using a single power source. First, I_B is set to a given value; then V_{CE} is increased in 2-V steps as I_C is measured and recorded in the data table. This procedure is repeated for various values of I_B until the data table is complete. The data are then plotted on the graph to give a more visual indication of the relationships of I_B and V_{CE} to I_C.

Materials Needed:

1 Variable low-voltage power supply (up to 20 V)
2 Standard or digital ammeters
1 Standard or digital voltmeter
1 2N2222 transistor or equivalent

Figure 3–12 Current–voltage characteristics of a bipolar transistor: (a) schematic diagram; (b) data table; (c) graph for constructing output curves.

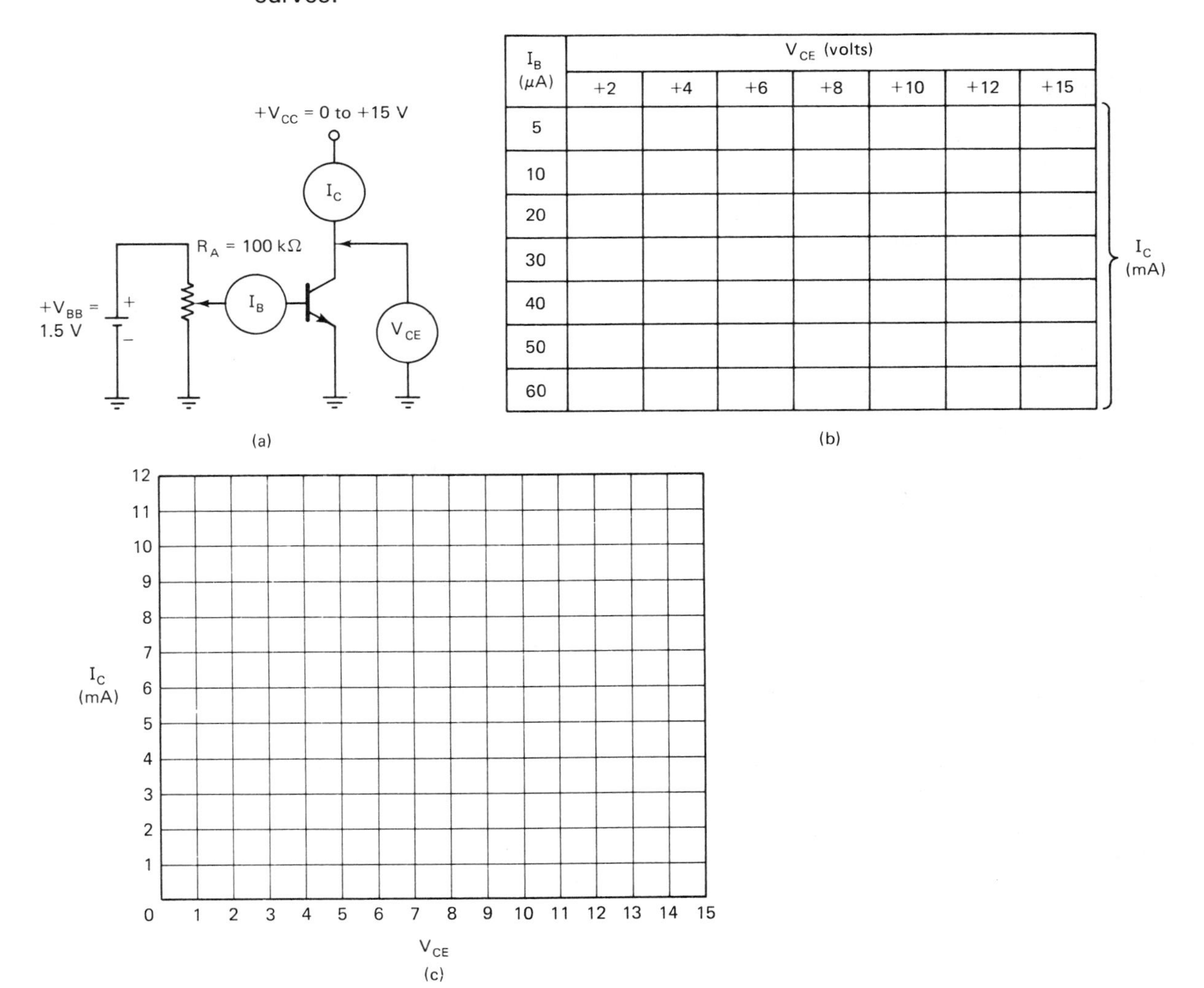

I_B (μA)	V_{CE} (volts)						
	+2	+4	+6	+8	+10	+12	+15
5							
10							
20							
30							
40							
50							
60							

1 100-k Ω potentiometer (R_A)
1 1.5-V battery
1 Breadboard for constructing circuit

Procedure:

1. Construct the circuit shown in Figure 3-12a.
2. Adjust R_A for an I_B reading of 5 μA.
3. Adjust the power supply for a V_{CE} of +2 V.
4. Measure I_C and record the value in the proper location of the data table.
5. Increase V_{CE} to +4 V.
6. Measure I_C and record the value in the proper location of the data table.
7. Repeat steps 5 and 6 for the V_{CE} values given in the data table up to +15 V.
8. Adjust the variable power supply to 0 V.
9. Adjust R_A for an I_B reading of 10 μA.
10. Repeat steps 3 through 8 to complete the readings for I_C.
11. Continue this procedure of increasing I_B and V_{CE} while measuring and recording I_C for the values given in the data table.
12. When the data table is complete, arrange the data on the graph to construct the *I–V* characteristic curves (refer to Figure 3-7).
    a. Using the first row of data, find the intersections of V_{CE} and the recorded I_C and mark dots horizontally along the graph. Next, connect the dots with a straight line. At the end of the line to the right, record the value of I_B for the line of data.
    b. Follow this method until all the rows of data are displayed on the graph.

Fill-in Questions:

1. When I_B increases, there is a substantial increase in ___Ic___.

2. When V_{CE} increases, there is a relatively small change in ___Ic___.

3. Current I_B has more control of ___Ic___ than does ___VCE___.

4. The graph shows the relationship of ___IB___, ___IC___, and ___VCE___.

5. From the graph, when I_B = 30 μA and V_{CE} = +6 V, then I_C = ___4___ mA.

EXPERIMENT 5. PLOTTING A LOAD LINE

Objective:

To demonstrate how to plot a load line for an amplifier, understand the quiescent point, and show how an input signal affects I_C and V_C.

Introduction:

The common-emitter amplifier shown in Figure 3-13a is the most used circuit with bipolar transistors. By constructing a load line on the *I–V* characteristic curves for a given transistor, an approximation of the operation of the amplifier with a given input signal can be determined. Refer to Figure 3-7b for construction of the load line.

Materials Needed:

1 Fixed +12-V power supply
1 Standard or digital voltmeter
1 Oscilloscope
1 Audio or function generator
1 2N2222 transistor or equivalent
1 220-Ω resistor at 0.5 W (R_E)
1 2.2-kΩ resistor at 0.5 W (R_L)
1 22-kΩ resistor at 0.5 W (R_B)
1 100-kΩ resistor at 0.5 W (R_A)
1 220-μF electrolytic capacitor at 25 WV dc (C_E)
2 1-μF capacitors (C_B and C_C)

Procedure:

1. Using the data from Experiment 4, construct the characteristic curves on the graph in Figure 3-13b. (They should appear the same as on the graph in Figure 3-12b.)
2. Make a dot on the graph at +12 V for $+V_{CC}$.

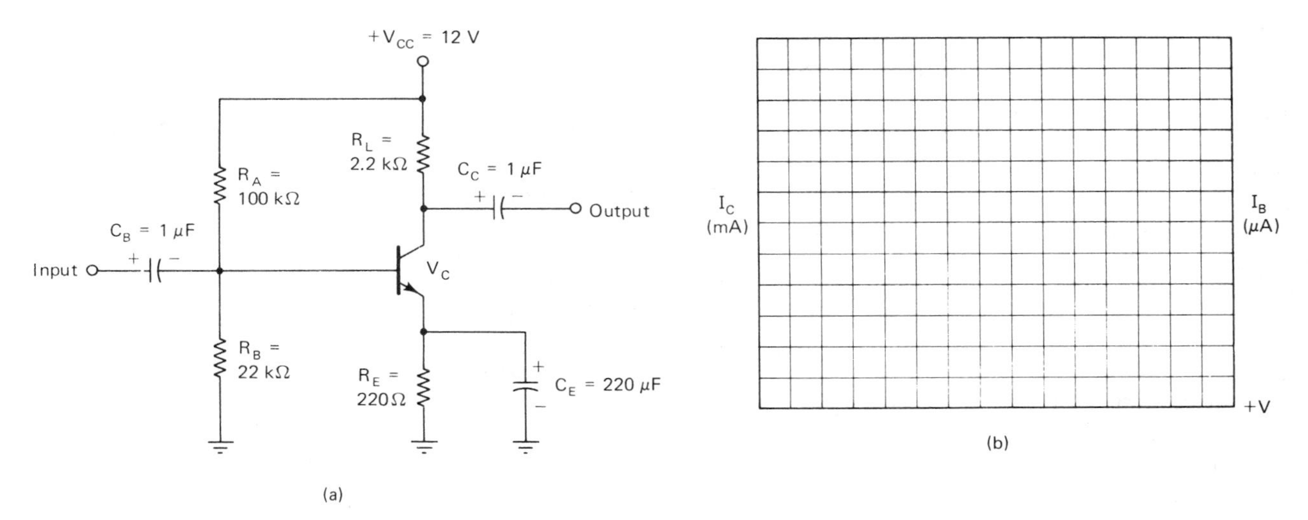

Figure 3–13 Plotting and understanding a load line: (a) schematic diagram of common-emitter amplifier; (b) graph for constructing curves and load line.

3. Calculate $I_{C\,max}$ from the formula $I_{C\,max} = +V_{CC}/(R_L + R_E)$.
4. Record $I_{C\,max}$ here: ______ .
5. Place a dot on the graph at $I_{C\,max}$.
6. Using a straightedge, draw a load line from the dot at $I_{C\,max}$ to the dot at $+V_{CC} = 12$ V.
7. Place a dot on the load line where it crosses $I_B = 20\ \mu$A.
8. From this point, draw a light line horizontally to the left to find I_{CQ}.
9. Also from this point, draw a light line vertically down to find V_{CQ}. These are the dc operating points when no input signal is present.
10. Draw a light line perpendicular to the load line where it crosses $I_B = 30\ \mu$A.
11. Draw a light line perpendicular to the load line where it crosses $I_B = 20\ \mu$A.
12. Draw a light line perpendicular to the load line where it crosses $I_B = 10\ \mu$A.
13. Draw a sine wave between these perpendicular lines with the positive alternation going toward the 30-μA indication. This represents the input signal voltage; label it 0.05 V p-p, v_{in}.
14. Draw a light line horizontally to the left toward the I_C axis from the 30-μA indication on the load line.
15. Draw a light line horizontally to the left toward the I_C axis from the 10-μA indication on the load line.
16. Draw a sine wave between these lines with the positive alternation going toward the higher value of I_C. This represents the change in I_C caused by the input signal voltage at the base.
17. Draw a light line vertically down toward the $+V$ axis from the 30-μA indication on the load line.
18. Draw a light line vertically down toward the $+V$ axis from the 10-μA indication on the load line.
19. Draw a sine wave between these lines with the negative alternation going to the lesser value of V_C. This represents the changes in V_C caused by the input signal voltage at the base. Label this v_{out}.
20. From the graph, find V_{CQ} and record it here: ______ .
21. From the graph, find I_{CQ} and record it here: ______ .
22. Find the maximum I_C swing and record it here: ______ mA p-p.
23. Find the maximum V_C swing and record it here: ______ V p-p.
24. Construct the circuit shown in Figure 3-13a.
25. With the voltmeter, measure V_C to ground and record it here: ______ . (It should be approximately half of $+V_{CC}$.)
26. Connect the signal generator at the input of C_B and ground.
27. Set the signal generator for 1 kHz with an amplitude of 0.05 V p-p. (Use the oscilloscope to make this measurement.)

28. Move the oscilloscope to the output of C_C and measure the output signal voltage and record it here: ______ V p-p.
29. Compare the output signal voltage of step 28 to the output signal voltage of step 23. There might be some difference.

Fill-in Questions:

1. A load line can help determine the operation of a transistor amplifier.
2. When v_{in} goes positive, I_C increases.
3. When v_{in} goes negative, I_C decreases.
4. When v_{in} goes positive, V_C goes negative.
5. When v_{in} goes negative, V_C goes positive.

EXPERIMENT 6. COMMON-EMITTER AMPLIFIER

Objective:

To demonstrate how a bipolar transistor is used in a common-emitter amplifier circuit configuration, and to understand some of its characteristics.

Introduction:

Experiments 4 and 5 are based on the common-emitter amplifier because it is the most popular circuit configuration. It is easily recognized since the input signal is between base and ground and the output signal is between collector and ground. The following is a summary of its features compared to the other two circuit configurations.

Input impedance: moderate (500 Ω to 1 kΩ)
Output impedance: moderate (≈ 50 kΩ)
Voltage gain: yes, $A_v = \Delta V_C/\Delta V_B$
Current gain: yes, $B = \Delta I_C/\Delta I_B$
Power gain: yes, highest
Signal phase inversion: yes (180°)
Used primarily as a voltage amplifier or switch

Materials Needed:

1 Fixed +12-V power supply
1 Standard or digital voltmeter
1 Oscilloscope (dual trace preferred)
1 Signal generator (100 Hz to 1 MHz)
1 2N2222 transistor or equivalent
1 220-Ω resistor at 0.5 W (R_E)
1 2.2-kΩ resistor at 0.5 W (R_L)
1 22-kΩ resistor at 0.5 W (R_B)
1 100-kΩ resistor at 0.5 W (R_A)
2 1-μF capacitors (C_B and C_C)
1 220-μF electrolytic capacitor at 15 WV dc (CE)
1 Breadboard for constructing circuit

Procedure:

1. Construct the circuit shown in Figure 3-14a.
2. Before connecting the signal generator, measure the dc operating voltages V_C, V_B, V_E, and V_{BE} and record in the appropriate places on the figure.
3. Connect the signal generator to the input and set it for 1 kHz with an amplitude of 0.05 V p-p.
4. Use the oscilloscope at the input (base to ground) to measure the input voltage.
5. Draw the input signal on graph A, indicating its voltage peak to peak, and record in its proper place the setting of the vertical attenuator of the oscilloscope marked V/div.
6. Using the oscilloscope, measure the output voltage (v_{out}) at C_C to ground.
7. Draw the output signal on graph B, indicating its voltage peak to peak, and record in its proper place the setting of the vertical attenuator of the oscilloscope marked V/div.
8. Calculate the voltage gain of the amplifier from the formula $A_v = v_{out}/v_{in}$ and record it in its proper place.

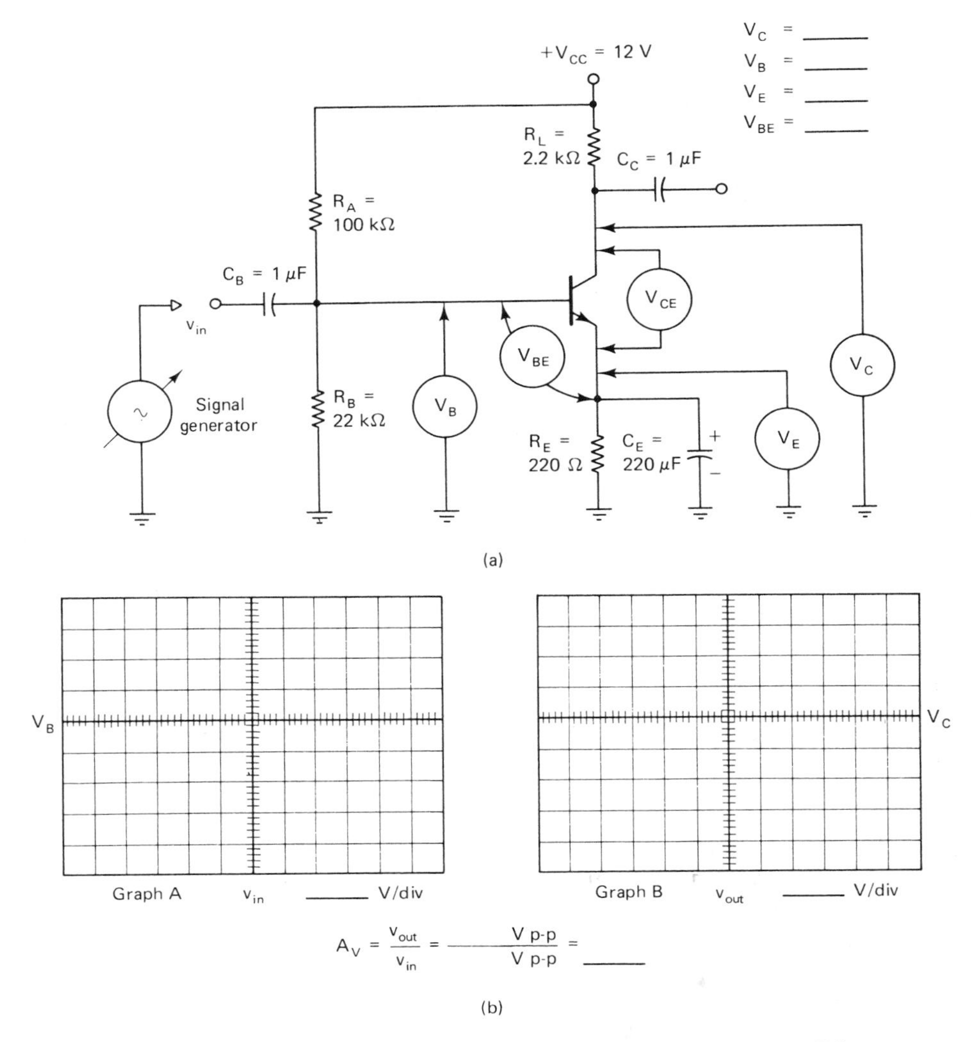

Figure 3–14 Common-emitter amplifier: (a) schematic diagram; (b) input/output voltage waveforms and gain.

Fill-in Questions:

1. The input signal of a common-emitter amplifier is between ___base___ and the ___ground___.
2. The output signal of a common-emitter amplifier is between the ___collector___ and the ___ground___.
3. The output signal is ___180°___ out of phase with the input signal.
4. The voltage gain is found by the formula ___$A_V = V_{out}/V_{in}$___.

EXPERIMENT 7. COMMON-COLLECTOR (EMITTER-FOLLOWER) AMPLIFIER

Objective:

To demonstrate how a bipolar transistor is used in a common-collector amplifier circuit configuration, and to understand some of its characteristics.

Introduction:

The common-collector amplifier is similar to the common-emitter amplifier except that R_L is missing and the collector is connected directly to $+V_{CC}$. The input signal is be-

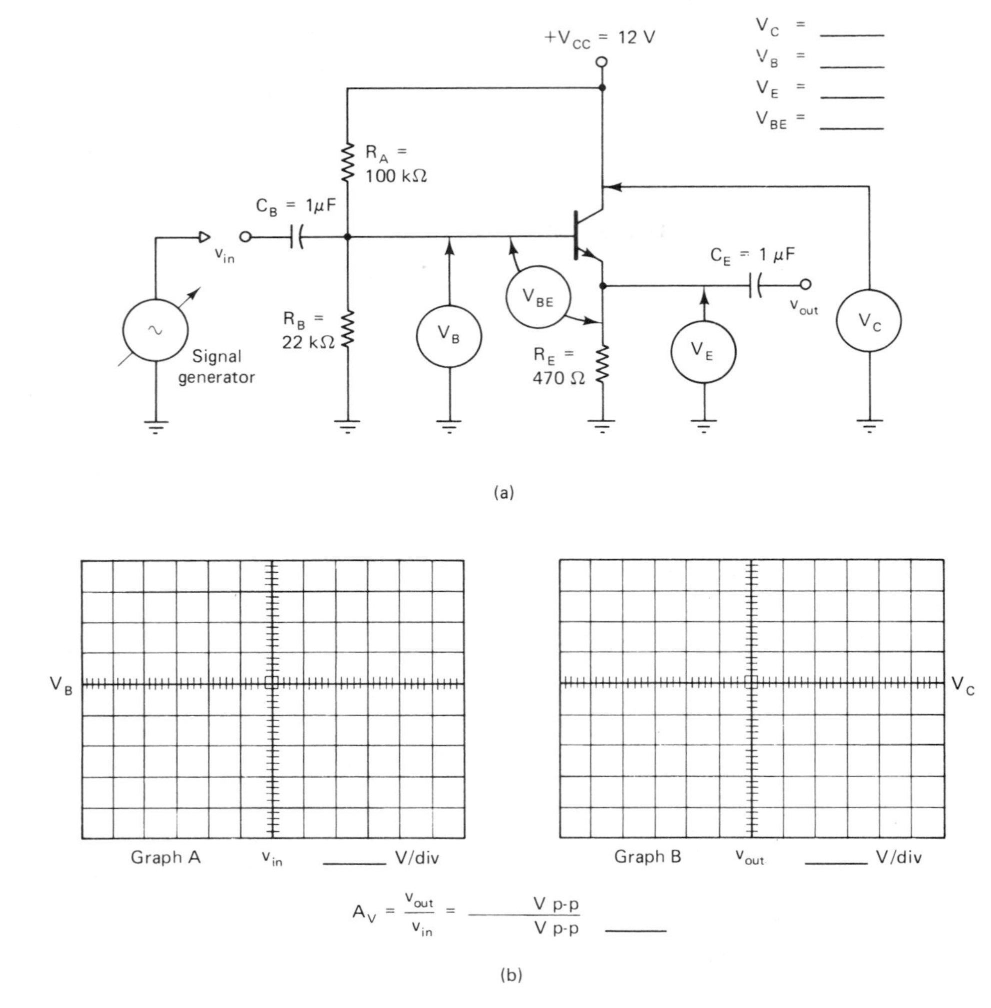

Figure 3–15 Common collector (emitter-follower) amplifier: (a) schematic diagram; (b) input/output voltage waveform and gain.

tween base and ground. The output signal is between the emitter and ground. Resistor R_E is not bypassed with a large electrolytic capacitor; therefore, when v_{in} goes positive, I_C increases and the voltage drop across V_E (v_{out}) goes positive. When v_{in} goes negative, I_C decreases and v_{out} goes negative. The output voltage follows the input voltage; hence the term *emitter follower*. The voltage gain of this amplifier is always less than 1.

Some of its features are:

Input impedance: highest (20 to 300 kΩ)
Output impedance: lowest (300 to 500 Ω)
Voltage gain: no, less than 1
Current gain: yes, gamma $= \Delta I_E / \Delta I_B$
Power gain: yes
Signal phase inversion: no (0°)
Used primarily as a high- to low-impedance matching circuit or buffer amplifier

Materials Needed:

1 Fixed +12-V power supply
1 Standard or digital voltmeter
1 Oscilloscope (dual trace preferred)
1 Signal generator (100 Hz to 1 MHz)
1 2N2222 transistor or equivalent
1 470-Ω resistor at 0.5 W (R_E)
1 22-kΩ resistor at 0.5 W (R_B)
1 100-kΩ resistor at 0.5 W (R_A)
2 1-μF capacitors (C_B and C_E)
1 Breadboard for constructing circuit

Procedure:

1. Construct the circuit shown in Figure 3-15a.
2. Before connecting the signal generator, measure the dc operating voltages V_C,

V_B, V_E, and V_{BE} and record in the appropriate places on the figure.

3. Connect the signal generator to the input and set it for 1 kHz with an amplitude of 1 V p-p.
4. Use the oscilloscope at the input (base to ground) to measure the input voltage.
5. Draw the input signal on graph A, indicating its voltage peak to peak, and record in its proper place the setting of the vertical attenuator of the oscilloscope marked V/div.
6. Using the oscilloscope, measure the output voltage (v_{out}) at C_E to ground.
7. Draw the output signal on graph B, indicating its voltage peak to peak, and record in its proper place the setting of the vertical attenuator of the oscilloscope marked V/div.
8. Calculate the voltage gain of the amplifier from the formula $A_v = v_{out}/v_{in}$ and record it in its proper place.

Fill-in Questions:

1. The input signal of a common-collector amplifier is between the Base and the Ground.
2. The output signal of a common-collector amplifier is between the Emitter and the Ground.
3. The output signal is in phase with the input signal.
4. The voltage gain is found by the formula Av = Vout/Vin.
5. The voltage gain of a common-collector amplifier is always less than 1.

EXPERIMENT 8. COMMON-BASE AMPLIFIER

Objective:

To demonstrate how a bipolar transistor is used in a common-base amplifier circuit configuration, and to understand some of its characteristics.

Introduction:

A common-base amplifier can be recognized by the input signal to the emitter and the output signal from the collector. The input signal varies V_{BE}, which in turn varies I_E. Since some of I_B flows out of the transistor, the amount of variation of I_C is less than I_E. Therefore, the current gain for this amplifier is less than 1; however, a voltage gain can still be accomplished because of the ratio of input resistance to output resistance.

Some of its features are:

Input impedance: lowest (50 to 500 Ω)
Output impedance: highest (300 kΩ to 1 MΩ)
Voltage gain: yes, $A_V = \Delta V_C/\Delta V_E$
Current gain: no, $\alpha = \Delta I_C/\Delta I_E$
Power gain: yes
Signal phase inversion: no (0°)
Used primarily as a low- to high-impedance matching circuit and for RF amplifiers

Materials Needed:

1 Fixed +12-V power supply
1 Standard or digital voltmeter
1 Oscilloscope (dual trace preferred)
1 Signal generator (100 Hz to 1 MHz)
1 2N2222 transistor or equivalent
1 220-Ω resistor at 0.5 W (R_E)
1 2.2-kΩ resistor at 0.5 W (R_L)
1 22-kΩ resistor at 0.5 W (R_B)
1 100-kΩ resistor at 0.5 W (R_A)
1 220-μF electrolytic capacitor at 15 WV dc (C_B)
2 1-μF capacitors (C_E and C_C)
1 Breadboard for constructing circuit

Procedure:

1. Construct the circuit shown in Figure 3-16a.
2. Before connecting the signal generator, measure the dc operating voltages V_C, V_E, V_B, and V_{BE} and record in their appropriate places on the figure.
3. Connect the signal generator to the input and set it for 1 kHz with an amplitude of 0.05 V p-p.
4. Use the oscilloscope at the input (emit-

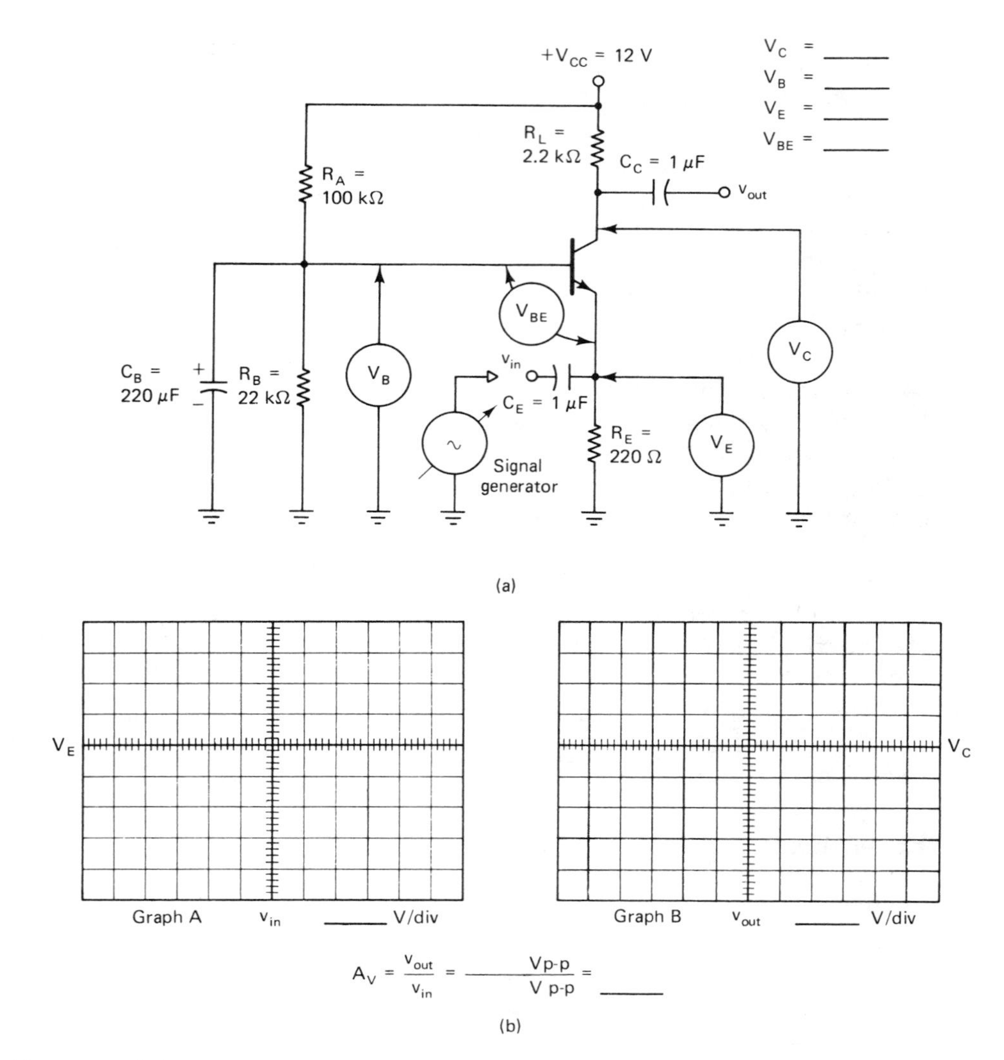

Figure 3–16 Common-base amplifier: (a) schematic diagram; (b) input/output voltage waveform and gain.

ter to ground) to measure the input voltage.

5. Draw the input signal on graph A, indicating its voltage peak to peak, and record the setting of the vertical attenuator of the oscilloscope marked V/div in its proper place.
6. Using the oscilloscope, measure the output voltage (v_{out}) at C_C to ground.
7. Draw the output on graph B, indicating its voltage peak to peak, and record in its proper place the setting of the vertical attenuator of the oscilloscope marked V/div.
8. Calculate the voltage gain of the amplifier from the formula $A_V = v_{out}/v_{in}$ and record it in its proper place.

Fill-in Questions:

1. The input signal of a common-base amplifier is between the Emitter and the ground.
2. The output signal of a common-base amplifier is between the Collector and the ground.
3. The output signal is In phase with the input signal.
4. The voltage gain is found by the formula $A_V = V_{out}/V_{in}$.
5. The current gain of a common-base amplifier is always less than 1.

EXPERIMENT 9. POWER AMPLIFIER

Objective:

To show that a power transistor is basically the same as a small-signal transistor but can handle greater amounts of power.

Introduction:

Power transistors operate on the same principle as small-signal transistors, except that they can handle higher values of voltage and current. They are usually used in the last stage or output of an electronic system and are the solid-state workhorses that operate speakers, relays, motors, and other devices requiring a specific amount of power. They are usually mounted to heat sinks or the chassis of equipment, which helps to dissipate the heat that is generated during operations. A mica insulator and silicone grease are usually required when they are mounted to a chassis, because the collector is connected internally to the transistor case. The circuit in this experiment is not designed to be the most efficient power amplifier, but shows how the power amplifier can be used to drive a speaker.

Materials Needed:

1 Fixed +12-V power supply
1 Standard or digital voltmeter
1 Oscilloscope
1 Signal generator (100 Hz to 1 MHz)
1 2N3055 power transistor or equivalent
1 10-kΩ resistor at 0.5 W (R_A)
1 1-μF capacitor
1 8-Ω loudspeaker
1 Breadboard for constructing circuit

Procedures:

1. Connect the signal generator to the speaker via capacitor C_B as shown in Figure 3-17a.
2. Set the signal generator at 1 kHz and adjust the amplitude for maximum output.
3. Notice the amount of loudness of the sound from the speaker.
4. Construct the circuit shown in Figure 3-17b.

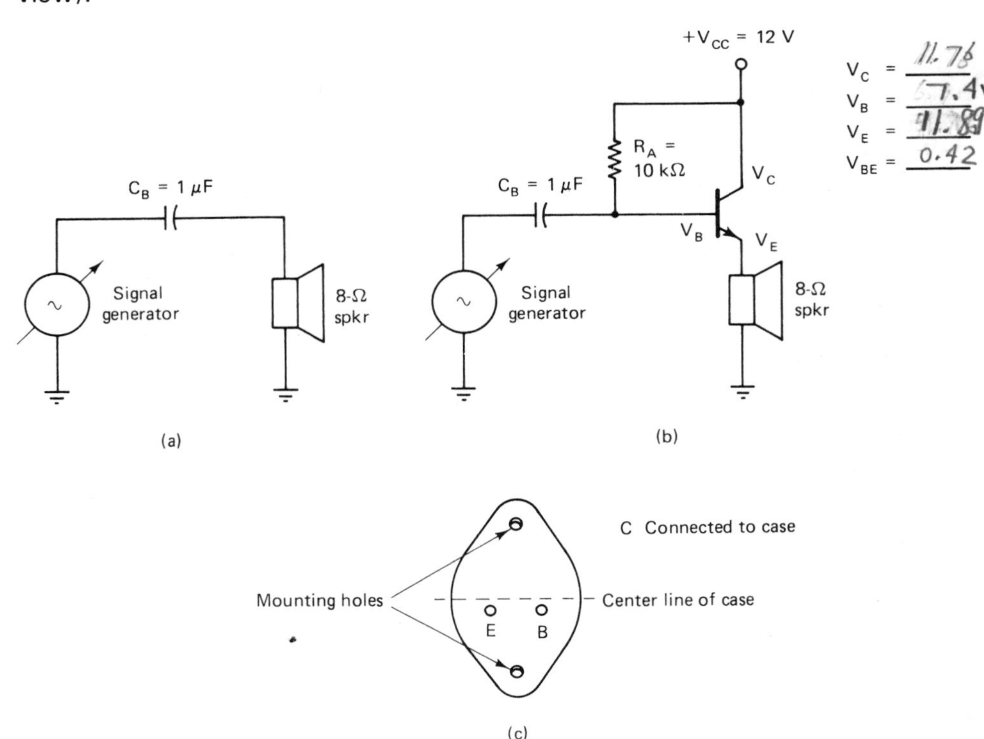

Figure 3–17 Simple power amplifier: (a) signal generator and speaker; (b) single transistor power amplifier; (c) 2N3055 lead identification (bottom view).

5. Before connecting the signal generator, measure the dc operating voltages V_C, V_B, V_E, and V_{BE} and record in their appropriate places on the figure.
6. Connect the signal generator to the input and set it for 1 kHz. A sound should be heard from the speaker.
7. Connect the oscilloscope across the speaker.
8. Adjust the amplitude of the signal generator until the signal is at maximum amplitude without any distortion.
9. Record the peak-to-peak voltage across the speaker here: 45mV V p-p.
10. Connect the oscilloscope to the base of the transistor and record the peak-to-peak voltage here: 4 V p-p. Is the sound louder with this circuit than with only the generator connected to the speaker? (Yes or No.) Although this circuit is an emitter follower with a voltage gain of less than 1, it does have considerable current gain, which is required to drive the speaker.

Fill-in Questions:

1. A power transistor is capable of handling MORE voltage and current than a small-signal transistor.

2. A power transistor usually requires a heat sink in a normal application.

3. A typical power transistor is not concerned so much with voltage amplification as with current amplification.

4. With most power transistors the Collectul is internally connected to the case.

EXPERIMENT 10. TWO-STAGE AMPLIFIER

Objective:

To show the relationship of amplifying a small signal with a voltage amplifier and of applying it to a power amplifier.

Introduction:

This experiment uses the common-emitter amplifier of Experiment 6, which is connected to the power amplifier of Experiment 9. A small signal is placed at the input of transistor Q_1, amplified, taken from the collector of Q_1, and applied to the base of Q_2 via coupling capacitor C_3. The circuit of Q_1 is a voltage amplifier and the circuit of Q_2 is a current amplifier. The normal voltage gain of Q_1 is reduced because of the loading effects of circuit Q_2. These loading effects, together with signal waveform distortion, are taken into account during the design of amplifier circuits, while attempting to achieve maximum gain.

Materials Needed:

1 Fixed +12-V power supply
1 Standard or digital voltmeter
1 Oscilloscope (dual trace preferred)
1 Signal generator (100 Hz to 1 MHz)
1 2N2222 transistor or equivalent
1 2N3055 transistor or equivalent
1 220-Ω resistor at 0.5 W (R_4)
1 2.2-kΩ resistor at 0.5 W (R_3)
1 10-kΩ resistor at 0.5 W (R_5)
1 22-kΩ resistor at 0.5 W (R_2)
1 100-kΩ resistor at 0.5 W (R_1)
2 1-μF capacitors (C_1 and C_3)
1 220-μF electrolytic capacitor at 15 WV dc (C_2)
1 8-Ω loudspeaker
1 Breadboard for constructing circuit

Procedure:

1. Construct the circuit shown in Figure 3-18.
2. Before connecting the signal generator, measure the dc operating voltages of Q_1 and Q_2 and record in the proper place on the figure.
3. Connect the signal generator to the input of circuit Q_1 and set it for 1 kHz with an amplitude of 0.5 V p-p. Use the oscilloscope for this measurement. A sound should be heard at the speaker.

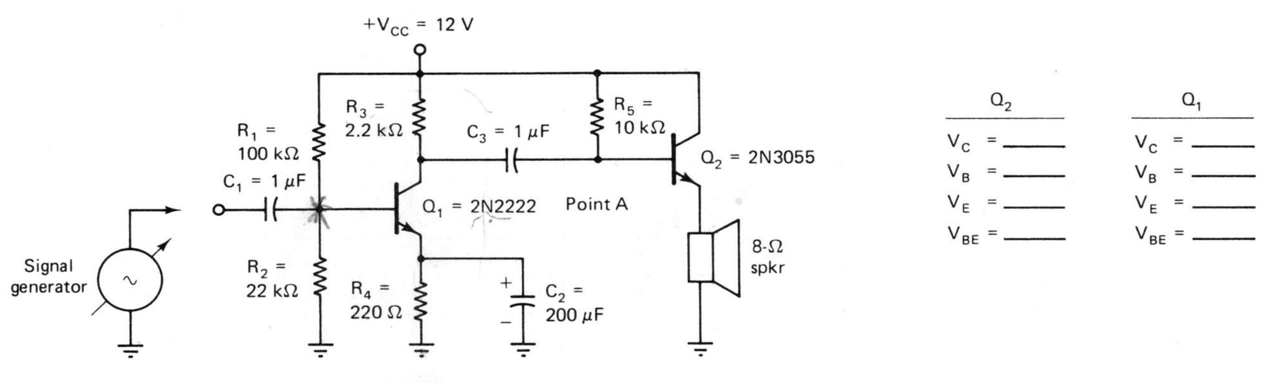

Figure 3–18 Two-stage amplifier.

4. Connect the oscilloscope to the collector of Q_1 and record the voltage amplitude here: ______ V p-p.
5. Disconnect C_3 at point A. No sound will be heard.
6. Now observe the signal at the collector of Q_1 and record the voltage amplitude here: ______ V p-p. Did this voltage increase? (Yes or No.) This demonstrates the loading effect that circuit Q_2 has on the circuit of Q_1.

Fill-in Questions:

1. The circuit of Q_1 is a voltage amplifier.
2. The circuit of Q_2 is a current amplifier.
3. The circuit of Q_2 has a loading effect on the circuit of Q_1.

SECTION 3-5 BIPOLAR TRANSISTOR BASIC TROUBLESHOOTING APPLICATION: COMMON-EMITTER AMPLIFIER

Construct this circuit using your own values or those of a circuit from a previous section. Open or short the components as listed and record the voltages in the proper place in the table. The abbreviations will help indicate the voltage conditions associated with each problem. All voltages are referenced to ground.

Figure 3–E.2

Table Abbreviations:

+V$_{CC}$	Power supply voltage
INC	Increase
DEC	Decrease
NOC	No change
Xistor	Transistor
GND	Ground

+V$_{CC}$ = 12 V
R$_L$
R$_A$
V$_{out}$
C$_B$
V$_B$
V$_C$
V$_{in}$
V$_E$
R$_B$
R$_E$
C$_E$

Condition	V_C	V_B	V_E	V_{out} (V p-p)	Comments
Normal					All voltages are proper
R_L open	DEC	DEC	DEC		Open R_L at $+V_{CC}$: no voltage applied to collector of Xistor
R_E open	INC	INC	INC		Open R_E at GND: no current through Xistor; C_E may charge higher
R_A open	INC	DEC	DEC		Open R_A at $+V_{CC}$: V_B pulled toward GND; Xistor cutoff
R_B open	DEC	INC	INC		Open R_B at GND: V_B pulled toward $+V_{CC}$; Xistor saturates
C_E open	NOC	NOC	NOC		DC voltage normal; reduced signal gain because of negative feedback

Fault					Remarks
C_E shorted	DEC	DEC	DEC		Forward bias increased; Xistor saturates
C open	$+V_{CC}$	DEC	DEC		Measure bottom of R_L for V_C: Xistor not conducting
B open	$+V_{CC}$	INC	DEC		Measure between R_A and R_B for V_B: Xistor not conducting
E open	$+V_{CC}$	INC	DEC		Measure top of R_E for V_E: Xistor not conducting
C-B shorted	DEC	INC	INC		V_B and V_C same voltage; Xistor conducts more
E-B shorted	$+V_{CC}$	DEC	DEC		V_B and V_E same voltage; Xistor cut off
C-E shorted	DEC	NOC	INC		Xistor not in circuit; voltage are normal for R_L and R_E in series

Figure 3–E.2 (cont'd.)

SECTION 3-6 MORE EXERCISES

Perform the following exercises before beginning Section 3-7.

1. Referring to Figure 3-9c and d, draw the proper ohmmeter connections for an *NPN* transistor and, assuming the transistor is good, indicate the proper high and low ohmic readings.

   a.

   b.

2. Referring to Figure 3-9f and g, draw the proper ohmmeter connections for a *PNP* transistor and, assuming the transistor is good, indicate the proper high and low ohmic readings.

   a.

   b.

3. If an ohmmeter's leads are connected to the transistors as indicated and a resulting reading occurs, determine if each transistor is open, shorted or normal.

   a. Type of transistor: *NPN*
   Positive lead to collector
   Negative lead to base
   Ohmic reading = *low*

   Conclusion: ______

   b. Type of transistor: *NPN*
   Positive lead to base
   Negative lead to emitter
   Ohmic reading = *high*

   Conclusion: ______

   c. Type of transistor: *PNP*
   Positive lead to collector
   Negative lead to base
   Ohmic reading = *high*

   Conclusion: ______

   d. Type of transistor: *PNP*
   Positive lead to emitter
   Negative lead to base
   Ohmic reading = *low*

   Conclusion: ______

4. In the active region a bipolar transistor behaves as a current-controlled variable resistor. A common-emitter amplifier is similar to a voltage-divider circuit, where R_L is fixed and the transistor (R_X) is variable. When I_C increases, V_{out} decreases, and when I_C decreases, V_{out} increases. Calculate I and V_{out} for the series circuits using the formulas $I = V_{CC}/(R_L + R_X)$ and $V_{out} = IR_X$.

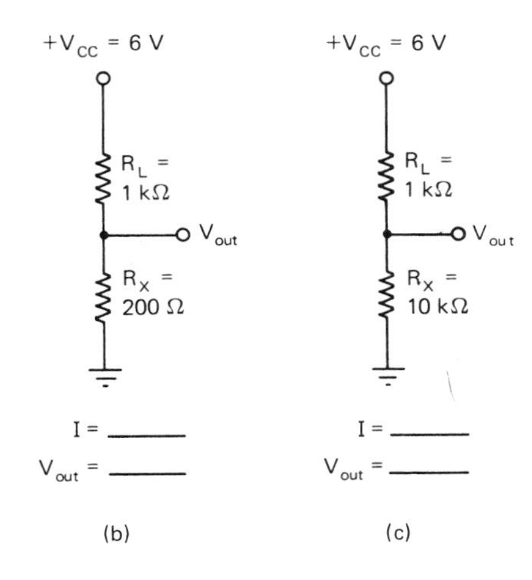

Figure 3–E.3

5. A bipolar transistor can be operated in three modes; cutoff (open switch), saturation (closed switch), or active region (variable resistance), where amplification takes place. The voltage from collector to ground (V_C) can give an indication of the circuit's operation. If $V_{CC} = +6$ V, from the voltage indications given, match the transistor circuits to their proper equivalent circuit.

	Choices
a. $V_C = +3$ V	**1.** Closed switch
b. $V_C = +6$ V	**2.** Open switch
c. $V_C = +0.2$ V	**3.** Variable resistance

6. Using Ohm's law, calculate the unknown values of V_C, V_B, V_E, and I_C. These formulas will aid in the calculations: $V_C = V_{CC} - V_{RL}$, $V_{R_L} = I_C R_L$, $V_B = V_E + 0.6$ V, $V_E = V_B - 0.6$ V, $V_E = I_C R_E$, $I_C = V_{R_L}/R_L$, $I_C \approx I_E$, and $I_E = V_E/R_E$.

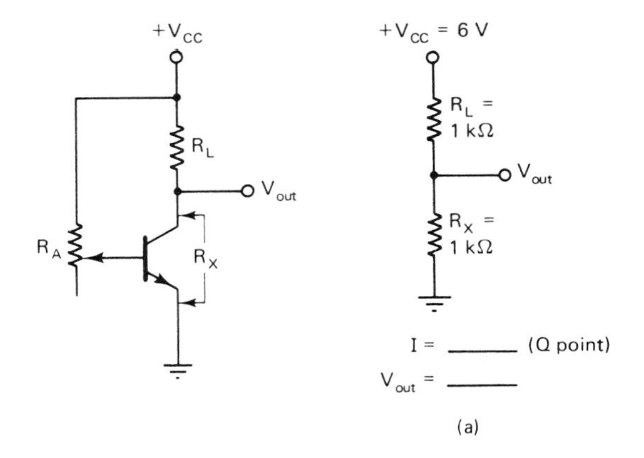

Figure 3–E.4

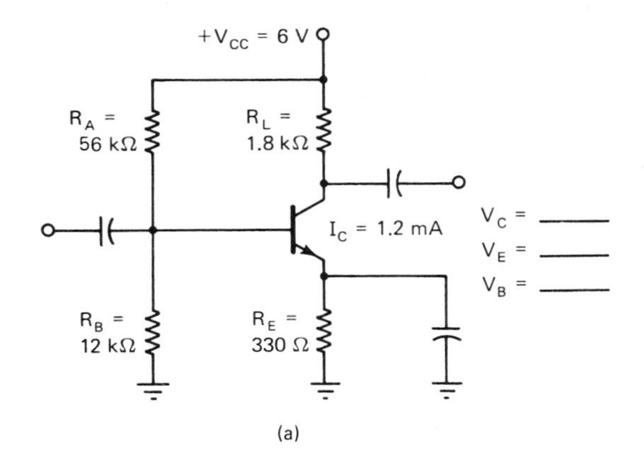

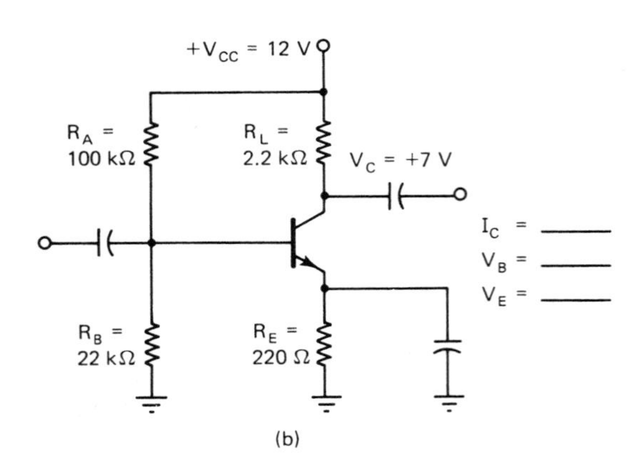

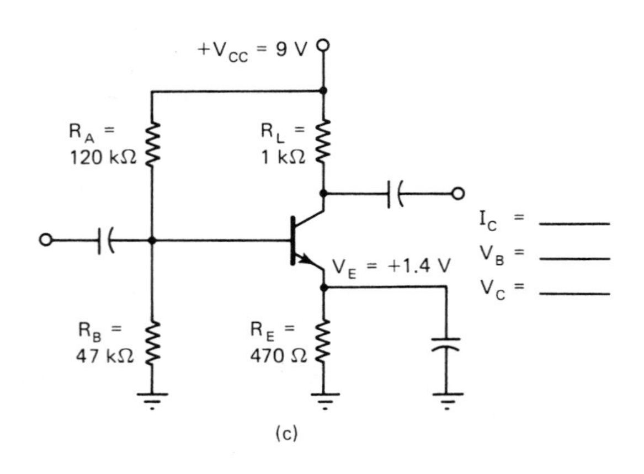

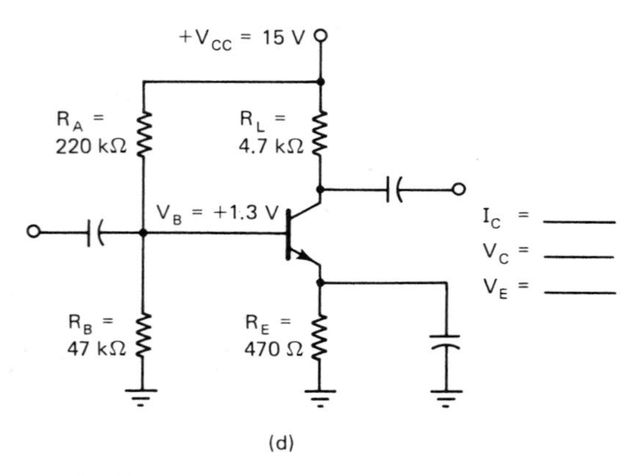

Figure 3–E.4 (cont'd.)

7. Calculate the voltage gains using the formula $A_V = v_{out}/v_{in}$.

a. $v_{out} = 2.8$ V p-p
$v_{in} = 0.025$ V p-p

$A_V =$ ______

b. $v_{out} = 4.2$ V p-p
$v_{in} = 0.05$ V p-p

$A_V =$ ______

c. $v_{out} = 0.56$ V p-p
$v_{in} = 0.01$ V p-p

$A_V =$ ______

d. $v_{out} = 1.08$ V p-p
$v_{in} = 1.2$ V p-p

$A_V =$ ______

8. Draw a simple circuit for each of the following amplifiers:

a. Common emitter

b. Common collector

c. Common base

9. From the following list, write the correct description in the table under each amplifier configuration.

1. Highest power gain
2. Moderate power gain
3. Lowest power gain
4. Voltage gain less than 1
5. Current gain less than 1
6. Moderate input impedance
7. Highest input impedance
8. Lowest input impedance
9. Highest output impedance
10. Lowest output impedance
11. Moderate output impedance
12. 180° signal phase inversion
13. No signal phase inversion
14. Beta
15. Alpha
16. Gamma
17. Input at emitter, output at collector
18. Input at base, output at emitter
19. Input at base, output at collector

Common-emitter Amplifier	*Common-collector Amplifier*	*Common-base Amplifier*

SECTION 3-7 BIPOLAR TRANSISTOR INSTANT REVIEW

The bipolar transistor is a current-operated device. The amount of current flowing in the base (I_B) directly controls the amount of current flowing from emitter to collector (I_C).

The emitter–base junction is normally forward biased, while the collector–base junction is reverse biased.

The bipolar transistor is a "normally off" device, when $V_{BE} = 0$ V. An *NPN* transistor conducts (turns on) when the base voltage is more positive than the emitter voltage. A *PNP* transistor conducts (turns on) when the base voltage is more negative than the emitter voltage.

SECTION 3-8 SELF-CHECKING QUIZZES

3-8a BIPOLAR TRANSISTOR: TRUE–FALSE QUIZ

Place a T for true or an F for false to the left of each statement.

____ **1.** Bipolar transistors are current-operated devices.

____ **2.** If an *NPN* silicon transistor is conducting, its base voltage will be at least +0.6 V greater than the emitter voltage.

____ **3.** The collector–base junction of a bipolar transistor is forward biased to turn it on.

____ **4.** An increase in I_B will cause a decrease in I_C.

____ **5.** The major current in a *PNP* transistor is hole flow.

____ **6.** Beta is expressed as I_C divided by I_B.

____ **7.** A forward-biased emitter–base junction has low resistance.

____ **8.** The common-base amplifier has a current gain of less than 1.

____ **9.** The common-collector amplifier has a phase inversion of 180° between the input signal and the output signal.

____ **10.** The emitter-follower amplifier has a voltage gain of less than 1.

3-8b BIPOLAR TRANSISTOR: MULTIPLE-CHOICE QUIZ

Circle the correct answer for each question.

1. If a transistor with a beta of 120 has an I_B of 30 μA, its I_C will be:

a. 3.6 mA **b.** 4 mA
c. 90 mA **d.** 150 mA

2. If a transistor with a beta of 150 has an I_C of 3 mA, its I_B will be:

a. 5 μA **b.** 20 μA
c. 50 μA **d.** 147 μA

3. If a transistor has a beta of 110, its alpha will be:

a. 111 **b.** 109
c. 1.0 **d.** 0.99

Questions 4 through 8 refer to Figure 3-A.

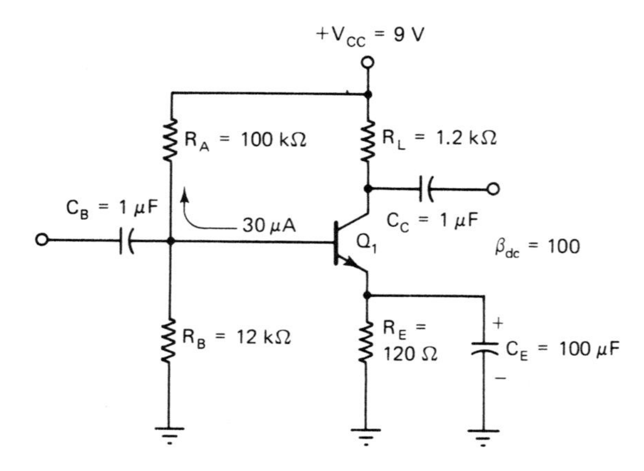

Figure 3-A

4. The collector current, I_C, of Q_1 should measure:

a. 0.3 mA **b.** 3 mA
c. 4 mA **d.** 6 mA

5. The collector voltage, V_C, will measure:

a. 3.6 V **b.** 4.5 V
c. 5.4 V **d.** 9 V

6. The emitter voltage, V_E, will measure:

a. 0.036 V **b.** 0.36 V
c. 0.48 V **d.** 0.72 V

7. The base voltage, V_B, will measure:

a. 0.636 V **b.** 0.96 V
c. 1.08 V **d.** 1.32 V

8. A sine-wave input signal causes I_B to increase and to decrease by ± 10 μA. The output signal voltage will be:

a. 1.2 V p-p and in phase
b. 1.2 V p-p and out of phase
c. 2.4 V p-p and in phase
d. 2.4 V p-p and out of phase

9. The term I_{CO} refers to leakage current between:

a. emitter–base
b. emitter–collector

c. collector–base

d. all of the above

10. For an *NPN* transistor common-emitter amplifier with V_{CC} = +6V, if V_C = +6 V, V_B = +1 V, and V_E = 0 V, the transistor can be considered:

a. operating normally

b. saturated

c. cutoff

d. none of the above

ANSWERS TO EXPERIMENTS AND QUIZZES

Experiment 1. (1) low (2) high (3) open (4) shorted

Experiment 2. (1) V_{CC} (2) conducting (3) on (or conducting) (4) +0.7

Experiment 3. (1) I_C (2) R_X (3) variable (4) active (5) constant

Experiment 4. (1) I_C (2) I_C (3) I_C, V_{CE} (4) I_B, I_C, V_{CE} (5) 4

Experiment 5. (1) operation (2) increases (3) decreases (4) negative (5) positive

Experiment 6. (1) base, ground (2) collector, ground (3) 180° (4) $A_v = v_{out}/v_{in}$

Experiment 7. (1) base, ground (2) emitter, ground (3) in phase (4) $A_V = v_{out}/v_{in}$ (5) one

Experiment 8. (1) emitter, ground (2) collector, ground (3) in phase (4) $A_V = v_{out}/v_{in}$ (5) one

Experiment 9. (1) more (2) heat sink (3) current (4) collector

Experiment 10. (1) voltage (2) current (3) loading effect

True–False. (1) T (2) T (3) F (4) F (5) T (6) T (7) T (8) T (9) F (10) T

Multiple Choice. (1) a (2) b (3) d (4) b (5) c (6) b (7) b (8) d (9) c (10) c

Unit 4

Junction Field-Effect Transistor

INTRODUCTION Junction field-effect transistor (JFET) characteristics resemble those of the vacuum tube, which was the original amplifying device in electronics.

UNIT OBJECTIVES Upon completion of this unit, you will be able to:

1. Identify JFET schematic symbols.
2. Explain the physical structure of a JFET.
3. Describe the operation of a JFET.
4. Draw the current–voltage characteristic curve for a JFET.
5. Define ohmic region and pinch-off region of a JFET.
6. Show the different biasing techniques for a JFET.
7. Measure leakage current.
8. Test a JFET with an ohmmeter.
9. Explain the three amplifier configurations for a JFET.
10. Define the terms source, gate, drain, channel, and electrostatic field.
11. Troubleshoot a JFET common-source amplifier.

SECTION 4-1 THEORY OF STRUCTURE AND OPERATION

4-1a STRUCTURE AND SCHEMATIC SYMBOLS

A junction field-effect transistor (JFET) consists of one type of semiconductor material with a channel of the opposite semiconductor material running through it, as shown in Figure 4-1. One end of the channel is designated as the source, which is similar to the emitter for a bipolar transistor and the other end, the drain, which is similar to the collector. The gate is the controlling lead and has a similar function as the base.

An *N*-channel JFET has *P* material for its gate and substrate. A *P*-channel JFET has *N* material for its gate and substrate. The gate and the substrate are electrically the same.

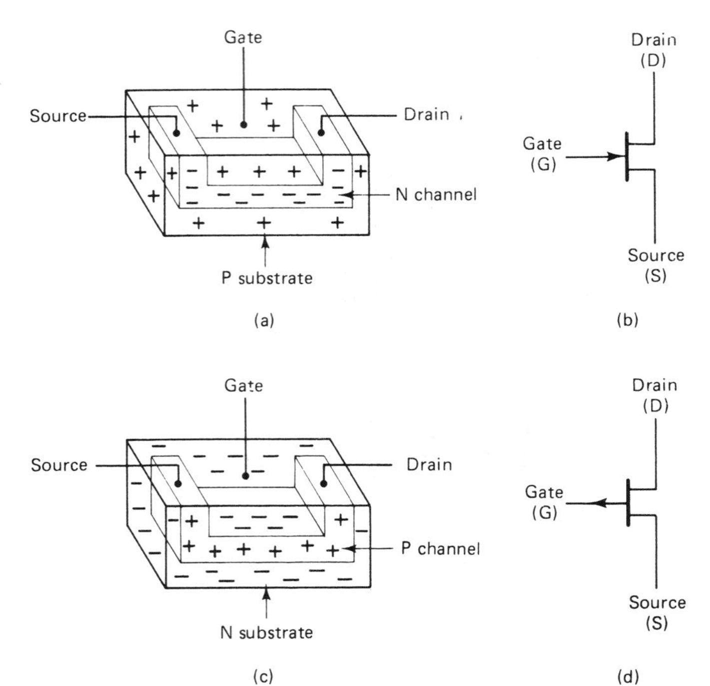

Figure 4–1 JFETs: (a) *N*-channel structure; (b) *N*-channel schematic symbol; (c) *P*-channel structure: (d) *P*-channel schematic symbol.

The *N*-channel JFET schematic symbol has the arrow pointing toward the device, and the *P*-channel JFET schematic symbol has the arrow pointing away from the device. The packages containing JFETs look the same as bipolar transistors.

4-1b THEORY OF OPERATION

An *N*-channel JFET is properly biased with a positive voltage applied to the drain and a negative voltage applied to the source, as shown in Figure 4-2. When the gate-to-source voltage (V_{GS}) is zero (Figure 4-2a), there is no control voltage and maximum electron current flows from the source, through the channel, to the drain, and is referred to as drain current (I_D). With this condition, the JFET is called a "normally on" device. When V_{GS} is reverse biased with a negative voltage (Figure 4-2b), the electrostatic field on the gate causes a depletion region to occur in the channel. The width of the channel decreases, which causes I_D to decrease. If V_{GS} is made more negative (Figure 4-2c), the channel width decreases further and less I_D flows. A point is reached where V_{GS} is sufficiently negative to cut off the channel and no I_D flows (Figure 4-2d). The reverse-biased junction of the gate and source has a high resistance, which accounts for the main advantage of the JFET, where high-input-impedance circuits are needed. Forward bias is seldom used, since the high impedance would be destroyed and the JFET would not operate properly.

A *P* channel operates the same way, except that the voltages are reversed and the majority carriers of the channel are holes.

Since the current through a JFET flows in one direction (no electron–hole combination as with a bipolar transistor), it is referred to as a unipolar device.

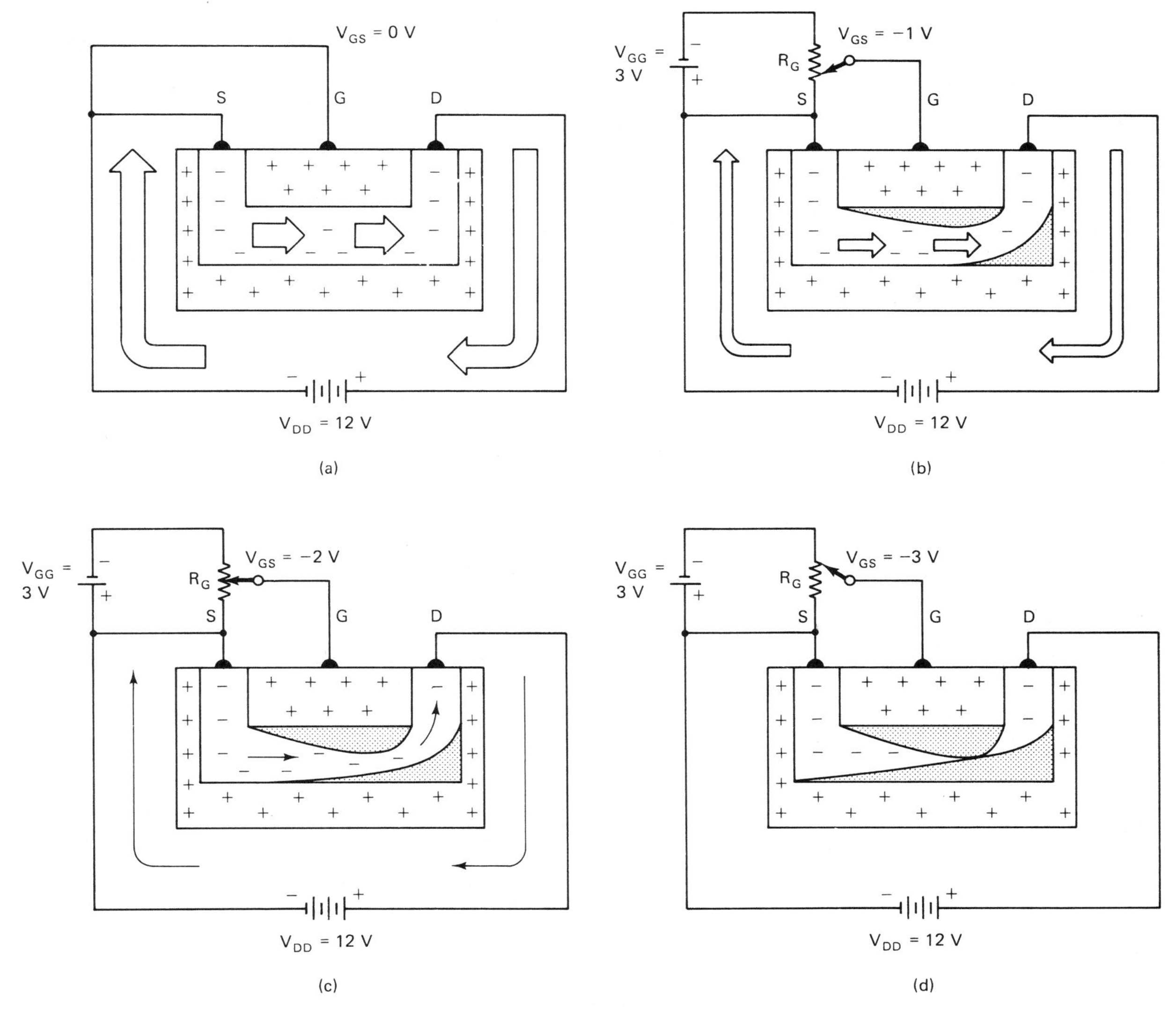

Figure 4–2 JFET operation: (a) maximum current; (b) $V_{GS} = -1$ V, less current; (c) $V_{GS} = -2$ V, lesser current; (d) $V_{GS} = -3$ V, JFET cut off, no current.

4-1c CURRENT–VOLTAGE CHARACTERISTICS AND AMPLIFICATION FACTOR

A test circuit can be constructed for an *N*-channel JFET to gather data for determining current–voltage characteristics, as shown in Figure 4-3a. Potentiometer R_G is adjusted to set a desired V_{GS}. A variable power supply is then set at various voltages and the I_D readings are recorded. This procedure is repeated several times, and the data are then plotted on a graph to illustrate the drain characteristics, as shown in Figure 4-3c.

Notice that, similar to a bipolar transistor, the voltage from drain to source has little effect on I_D, whereas it is controlled primarily by V_{GS}. With a specific value of V_{GS}, I_D will increase when V_{DS} is increased from 0 V (similar to a pure resistive circuit). However, a point is reached where an increase in V_{DS} causes an increase of the depletion region of the channel and I_D tends to level off. This point is called the pinch-off voltage (V_p). The useful operation of a JFET as an amplifier is above or beyond this V_p point. Remember that a JFET is operating

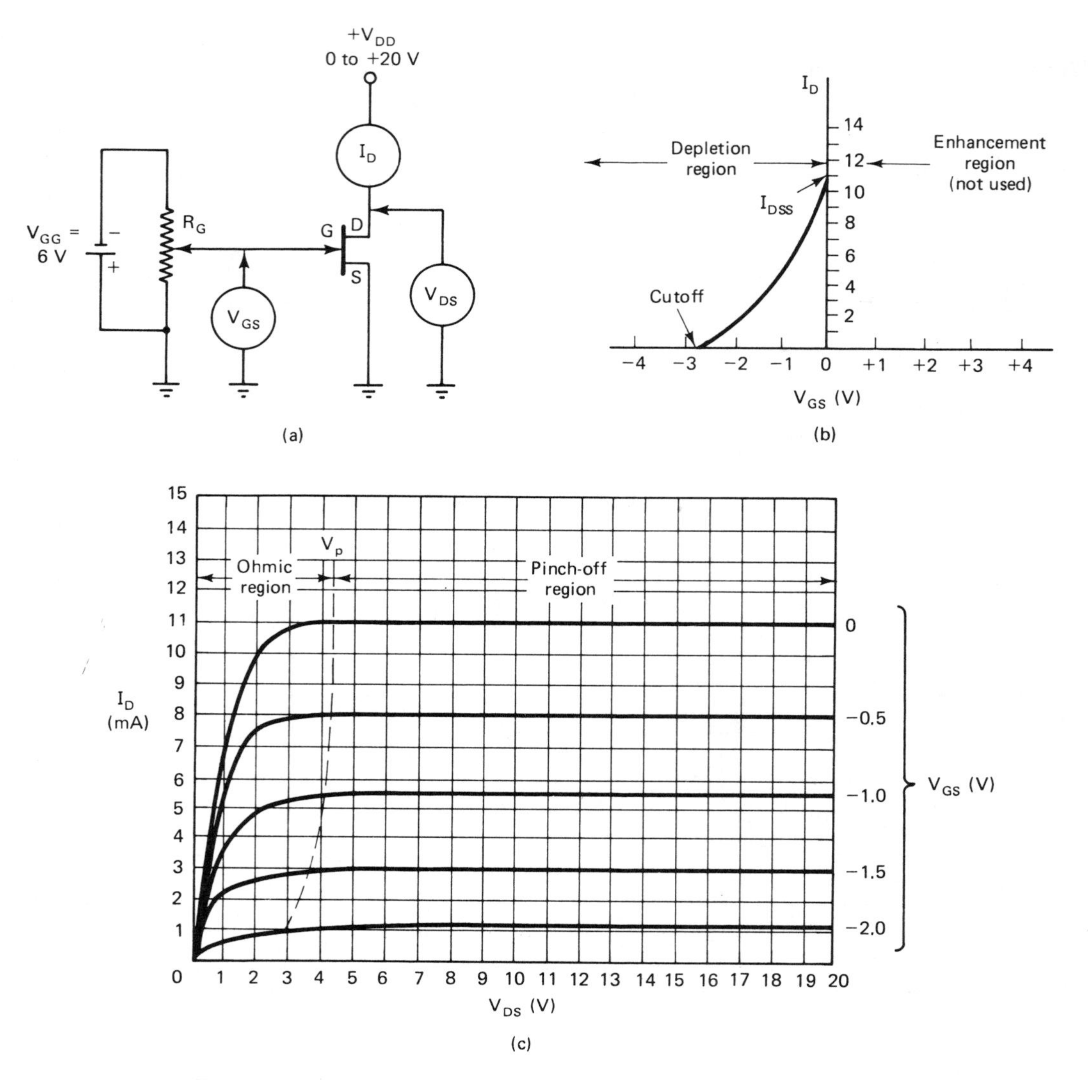

Figure 4–3 JFET current–voltage characteristics: (a) test circuit; (b) transfer characteristic curve (V_{GS} versus I_D); (c) output characteristic curve (I_D versus V_{DS}).

in the pinch-off region, but may not necessarily be cut off. If V_{DS} is made sufficiently high, avalanche breakdown could occur, where I_D increases sharply, and could possibly damage the JFET.

Another aid in understanding JFET operation is the transfer characteristic curve shown in Figure 4-3b. This curve shows the effects of V_{GS} on I_D. When $V_{GS} = 0$ V, maximum drain current flows and is indicated by I_{DSS}. Voltage V_{GS} is increased until it cuts off the drain current. This reverse-biased operation is called the *depletion region*. If forward bias is used, I_D increases, which enhances the JFET (makes it conduct more); however, a forward-biased junction has low resistance and is very seldom used with a JFET.

The amplification factor for a JFET is called *transconductance* and is given by the formula

$$g_m = \frac{\Delta I_D}{\Delta V_{GS}}$$

where a small change in V_{GS} causes a small change in I_D, and V_{DS} is held constant. Transconductance was expressed in units called mhos (the reciprocal of ohms), but is now referred to as siemens.

For example, referring to the graph of Figure 4-3c, note that if V_{DS} = 10 V and V_{GS} changes from −0.5 V to −1.0 V then I_D changes from 8.0 mA to 5.4 mA. With these values inserted in the formula, the G_m is

$$g_m = \frac{\Delta I_D}{\Delta V_{GS}} = \frac{8.0\ \text{mA} - 5.4\ \text{mA}}{1.0\ \text{V} - 0.5\ \text{V}} = \frac{2.6\ \text{mA}}{0.5\ \text{V}} = 5.2\ \text{mmhos or } 5200\ \mu\text{mhos}$$

4-1d JFET BIASING TECHNIQUES

A JFET may have its gate returned to a fixed bias voltage supply, as shown in Figure 4-4a. The gate is negative 1 V with respect to the source. The same condition can be accomplished with self-bias using a source resistor, as shown in Figure 4-4b. The current flowing through the JFET develops a positive 1 V at the source (V_S). Since the gate is returned to ground via R_g, the gate voltage (V_G) is 0 V. Voltage V_S is positive 1 V with respect to V_G; in other words, V_G is negative 1 V with respect to V_S.

Resistor R_S also provides some stability to the JFET with temperature increases. When the temperature rises, more current flows through the JFET. This causes the voltage drop across R_S to increase, which in turn reverse biases the JFET more and limits the current flow.

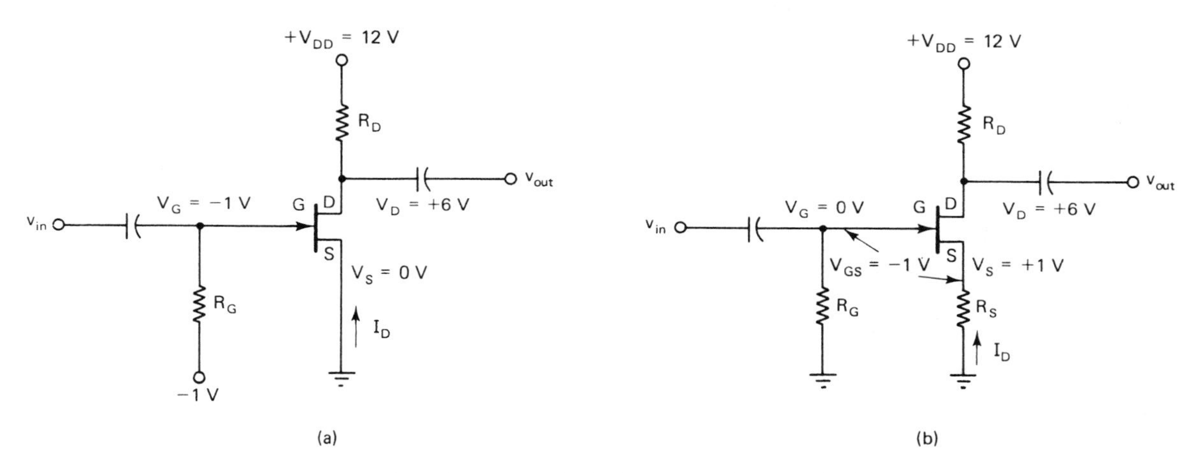

Figure 4–4 Types of JFET biasing: (a) fixed; (b) self-bias with source resistor.

4-1e LEAKAGE CURRENT

Because a *PN* junction is formed between the gate and the channel, there is some leakage current (I_{GSS}) when the JFET is in the reverse-biased condition. Excessive leakage current can cause erratic circuit operation. A simple method of measuring I_{GSS} is shown in Figure 4-5. This current should be extremely small.

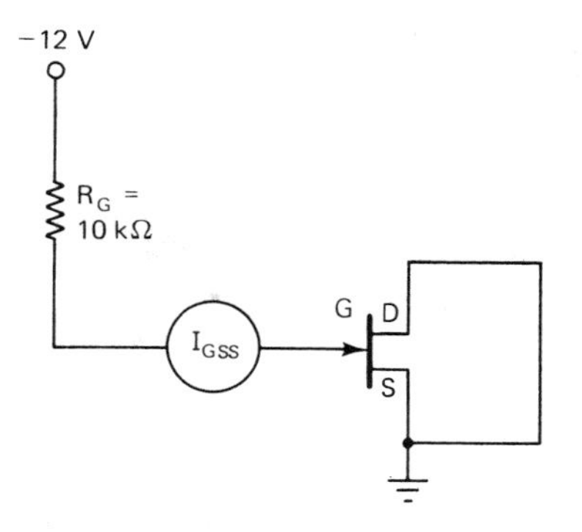

Figure 4–5 Measuring leakage current (I_{GSS}) of an *N*-channel JFET.

4-1f JFET DEFINITIONS

The following definitions are some of the more commonly used terms associated with JFETs.

V_{DD}	supply voltage connected to the drain
V_{SS}	supply voltage connected to the source
V_{GG}	supply voltage connected to the gate
V_D	voltage from drain to ground
V_G	voltage from gate to ground
V_S	voltage from source to ground
V_{DS}	voltage across drain and source
V_{GS}	voltage across gate and source
$V_{GS\,(off)}$	gate cutoff voltage
V_P	pinch-off voltage
I_D	drain current
I_{DSS}	saturation drain current when $V_{GS} = 0$ V
I_{GSS}	gate leakage current when $V_{DS} = 0$ V
g_m	transconductance or amplification factor
Y_{FS}	same as g_m
BV_{GSS}	gate-to-drain (reverse) breakdown voltage (an *N*-channel JFET with a BV_{GSS} rating of 25 V could not have more than -5 V on its gate with a V_D of $+20$ V)
BV_{DGS}	drain-to-source breakdown with the source shorted to the gate

SECTION 4-2 JFET DEFINITION EXERCISE

Refer to the previous sections and write a brief description for each term.

1. V_{DD} ______________________________

____

____

2. V_{SS} ______________________________

____

____

3. V_{GG} ______________________________

____

____

4. V_D ______________________________

____

____

5. V_G

6. V_S

7. V_{DS}

8. V_{GS}

9. $V_{GS\,(\text{off})}$

10. V_P

11. I_D

12. I_{DSS}

13. I_{GSS}

14. g_m

15. Y_{FS}

16. BV_{GSS}

17. BV_{DGS}

SECTION 4-3 EXERCISES AND PROBLEMS

Perform the following exercises before beginning Section 4-4.

1. Draw the structure of an *N*-channel JFET. (Label all parts.)

2. Draw the schematic symbol of an *N*-channel JFET. (Label the leads.)

3. Referring to Figure 4-3a, draw an *N*-channel JFET with proper biasing. (Indicate polarities.)

4. Referring to Figure 4-4b, draw an *N*-channel JFET amplifier using an arrow to indicate the electron current flow through the JFET.

5. Draw the structure of a *P*-channel JFET. (Label all parts.)

6. Draw the schematic symbol of a *P*-channel JFET. (Label all leads.)

7. Referring to Exercise 3, draw a *P*-channel JFET, but reverse the polarity of the voltages. (Indicate polarities.)

8. Referring to Exercise 4, draw a *P*-channel JFET amplifier, but reverse the polarity of the voltage and indicate hole flow through the JFET.

9. Draw a sample of the output characteristic curves (I_D versus V_{DS}) for an *N*-channel JFET, showing different values of V_{GS}.

10. Draw a sample of the output characteristic curves (I_D versus V_{DS}) for a *P*-channel JFET, showing different values of V_{GS}.

11. On the following circuits, indicate if the current I_D will be maximum, medium, or minimum.

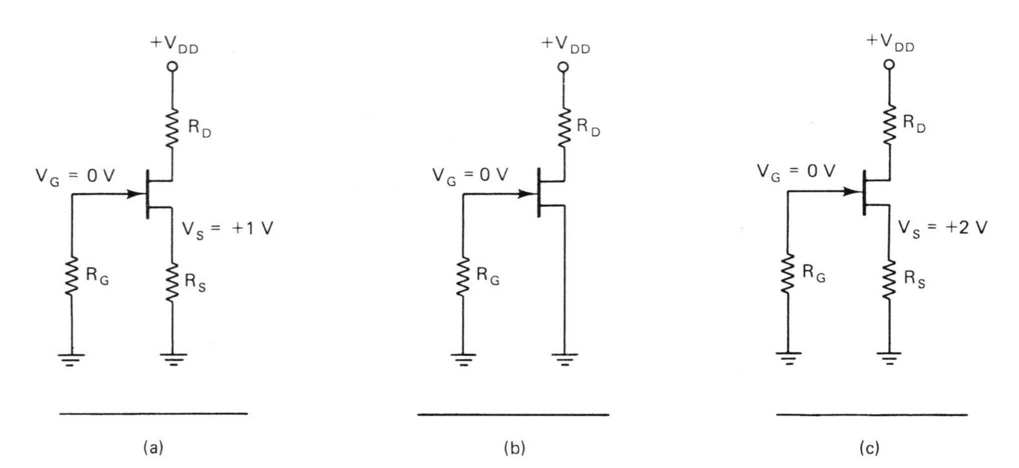

Figure 4–E.1

12. Find the g_m using the formula $g_m = \Delta I_D/\Delta V_{GS}$ from the values given.

a. $\Delta I_D = 1.8$ mA
$\Delta V_{GS} = 0.5$ V

$g_m =$ ______

b. $I_D = 1.6$ mA
$\Delta V_{GS} = 0.2$ V

$g_m =$ ______

c. $\Delta I_D = 2.1$mA
$\Delta V_{GS} = 0.3$ V

$g_m =$ ______

13. Draw the circuit in Figure 4-9a and write in the voltage nomenclature (V_D, V_G, V_S, V_{GS}, and V_{DS}). Also show with arrows the current path (I_S, I_D) through the JFET.

SECTION 4-4 EXPERIMENTS

EXPERIMENT 1. TESTING JFETS

Objective:

To demonstrate a practical go/no go method of testing JFETs with an ohmmeter. This is called a go/no go test.

Introduction:

For ohmmeter testing purposes, the *N*-channel JFET is similar to a diode with its cathode connected to the middle of a resistor, as

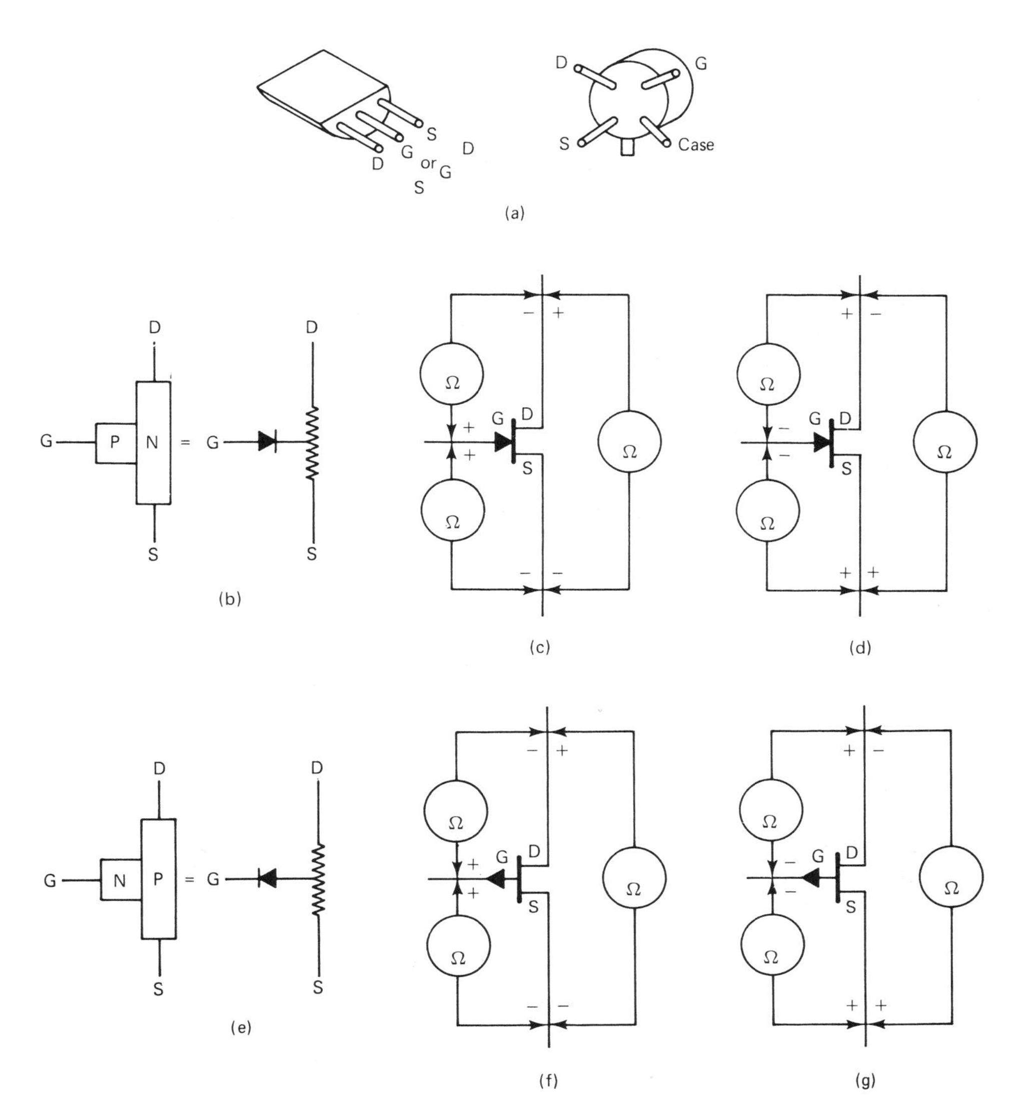

Figure 4–6 Testing JFETs with an ohmmeter: (a) general lead identification; (b) *N*-channel equivalent circuit; (c) and (d) *N*-channel ohmmeter connections; (e) *P*-channel equivalent circuit; (f) and (g) *P*-channel ohmmeter connections.

shown in Figure 4-6b. The ohmic resistance of the channel should be about the same regardless of the polarity of the ohmmeter lead connections from source to drain. With the positive lead on the gate, there should be a low-resistance reading when the negative lead is placed on the source or drain. The reading should be infinite when the negative lead is on the gate and the positive lead is placed on the source or drain. The same procedures are used for a *P*-channel JFET, except that the diode's anode is connected to the resistor and the ohmmeter polarities are reversed.

Materials Needed:

A standard or digital ohmmeter

One or several JFETs, including both *N*- and *P*-channel types

Procedure:

1. Set the ohmmeter to the midrange scale.
2. Refer to Figure 4-6c to connect the ohmmeter to an *N*-channel JFET and record the readings in the ohmmeter circles indicated.
3. Refer to Figure 4-6d to connect the ohmmeter to an *N*-channel JFET and record the readings in the ohmmeter circles indicated.
4. Using a *P*-channel JFET, perform the same procedures as in steps 1 through 3, while referring to Figure 4-6f and g.

Fill-in questions:

1. For an *N*-channel JFET, with the positive lead on the gate and the negative lead on the source, the ohmmeter should read ____________, compared to ____________ or ____________ when the leads are reversed.

2. For a *P*-channel JFET, with the positive lead on the gate and the negative lead on the drain, the ohmmeter should read ____________ or ____________, compared to ____________ when the leads are reversed.

3. If the positive lead is placed on the drain and the negative lead is placed on the source and the ohmmeter reads infinity, the JFET is ____________.

EXPERIMENT 2. OPERATION OF A JFET

Objective:

To show how to turn on and turn off a JFET and how to recognize these conditions by the voltage present at the drain.

Introduction:

Refer to Figure 4-7a and note that, when switch S_1 is in position A, $V_{GS} = 0$ V and the JFET is on or conducting. The voltage at the drain should be very low. When the switch is placed in position B, $V_{GS} = -3$ V and the JFET is off or not conducting. The voltage at the drain should be the same as $+V_{DD}$.

Materials Needed:

1 Adjustable dual ±12-V power supply
1 Standard or digital voltmeter
1 10-kΩ resistor at 0.5 W (R_D)
1 1-MΩ resistor at 0.5 W (R_G)
1 2N3823 JFET or equivalent
1 Single-pole double-throw (SPDT) switch (S_1)
1 Breadboard for constructing circuit

Procedure:

1. Construct the circuit shown in Figure 4-7a.
2. Make sure that S_1 is in position A.
3. Measure V_G and record the value in the data table next to A in Figure 4-7b.
4. Measure V_D and record the value in the data table (in the same row).
5. Indicate in the data table if the JFET is on or off (in the same row).
6. Move S_1 to position B.
7. Measure V_G and record the value in the data table next to B (in the second row).
8. Measure V_D and record the value in the data table (in the same row).
9. Indicate in the data table if the JFET is on or off (in the same row).
10. Calculate the approximate I_D from the formula

$$I_D = \frac{V_{R_D}}{R_D} = \frac{+V_{DD} - V_D(\text{on})}{R_D}$$

11. Record I_D here: ________.

Figure 4–7 Operation of an *N*-channel JFET: (a) test circuit; (b) data table.

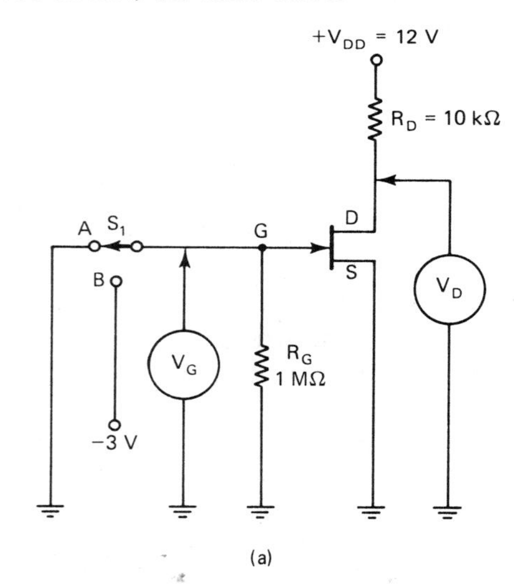

(a)

S_1 position	V_G	V_D	Condition of JFET (on or off)
A			
B			

(b)

Fill-in Questions:

1. When the JFET is conducting heavily, the voltage at the drain will be ______ .
2. When V_{GS} = 0 V, the JFET is ______ .
3. When the JFET is not conducting, the voltage at the drain will equal ______ .

EXPERIMENT 3. CURRENT–VOLTAGE CHARACTERISTICS OF A JFET

Objective:

To demonstrate how V_{GS} has more influence over I_D than the voltage across the drain–source (V_{DS}), and to construct a graph of output characteristic curves to understand better the operation of a JFET.

Introduction:

This experiment requires a variable positive voltage for V_{DS} and a negative voltage for V_{GS}. First, V_{GS} is set to a given value; then V_{DS} is increased to various voltages given in the data table of Figure 4-8b, as I_D

Figure 4–8 Current–voltage characteristics of a JFET: (a) test circuit; (b) data table; (c) graph for constructing output curves.

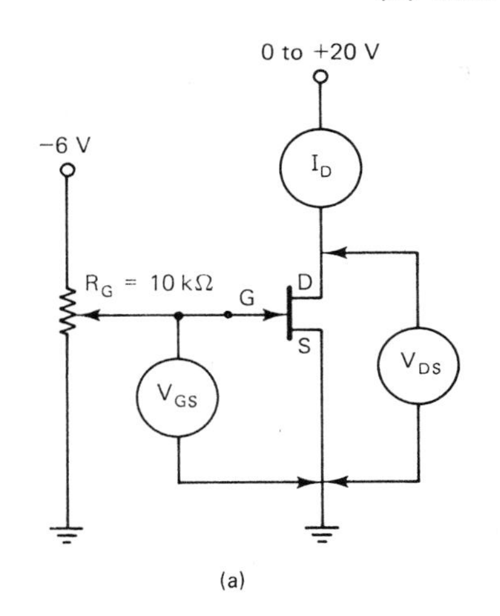

(a)

V_{GS}	V_{DS}								
	0	1	2	3	4	5	10	15	20
0									
−0.5									
−1.0									
−1.5									
−2.0									
−2.5									
−3.0									

I_D (mA) recorded in blocks

(b)

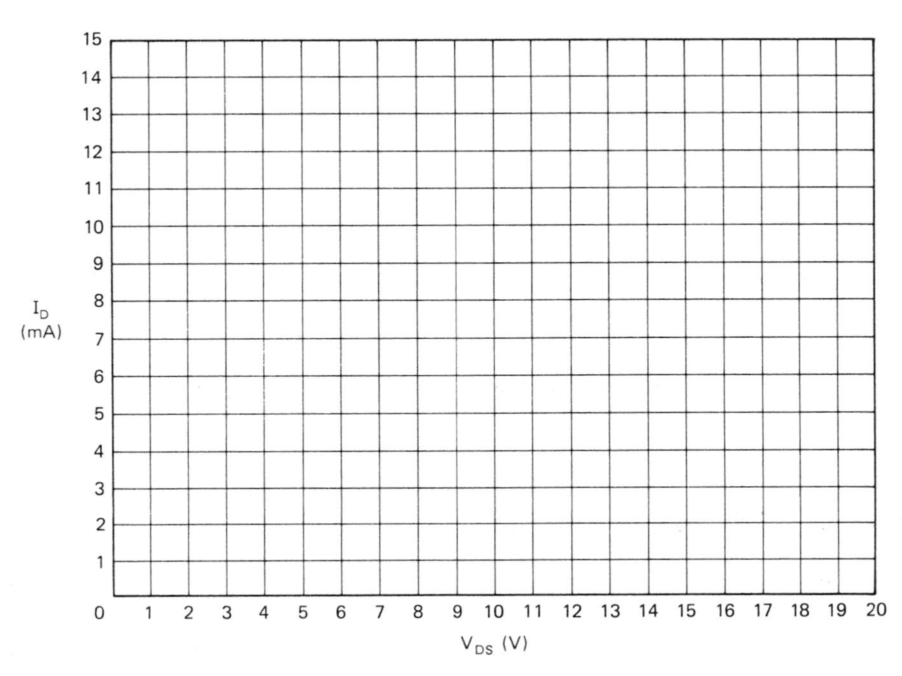

(c)

is measured and recorded. This procedure is repeated for various values of V_{GS} until the data table is complete. The data are then plotted on the graph to give a more visual indication of the relationship of V_{GS} and V_{DS} to that of I_D.

Materials Needed:

1 Adjustable dual ±20-V power supply
1 Standard or digital ammeter
1 Standard or digital voltmeter
1 2N3823 *N*-channel JFET or equivalent
1 10-kΩ potentiometer
1 Breadboard for constructing circuit

Procedure:

1. Construct the circuit shown in Figure 4-8a.
2. Adjust R_G for a V_{GS} reading of 0 V.
3. Adjust the positive power supply for a V_{DS} of 1 V.
4. Measure I_D and record the value in the proper location of the data table.
5. Increase V_{DS} to the next value given in the data table.
6. Measure I_D and record the value in the proper location of the data table.
7. Repeat steps 5 and 6 for the V_{DS} values given in the data table up to +20 V.
8. Adjust the variable positive power supply to 0 V.
9. Adjust R_G for the next value of V_{GS} given in the data table.
10. Repeat steps 3 through 8 to complete the readings for I_D.
11. Continue this procedure of increasing V_{GS} and V_{DS} while measuring and recording I_D for the values given in the data table.
12. When the data table is complete, arrange the data on the graph to construct the *I–V* characteristic curves (refer to Figure 4-3c).
    a. Using the first row of data, find the intersections of V_{DS} and the recorded I_D and mark dots horizontally along the graph. Next, connect the dots with a straight line. At the end of the line to the right, record the value of V_{GS} for the line of data.
    b. Follow this same method until all the rows of data are displayed on the graph.

Fill-in questions:

1. When bias voltage V_{GS} increases, there is a ____________ in I_D.

2. When V_{DS} increases, there is little or no change in ____________ .

3. Gate voltage has more control of ____________ than does ____________ .

4. The graph shows the relationship of ____________ , ____________ , and ____________ .

5. From the graph, when V_{GS} = −1.5 V and V_{DS} = +10 V, then I_D = ______ mA.

EXPERIMENT 4. COMMON-SOURCE AMPLIFIER

Objective:

To show how a JFET is used in a common-source amplifier configuration, and to understand some of its characteristics.

Introduction:

This circuit is recognized by the input to the gate and output taken at the drain. In terms of signals, the source is common to both input and output. It is similar to the bipolar transistor common-emitter amplifier. The following is a list of this circuit's features compared to the other two configurations.

Input impedance: high
Output impedance: moderate
Voltage gain: yes
Current gain: yes
Power gain: yes
Signal phase inversion: yes (180°)
Used primarily as a voltage amplifier or switch

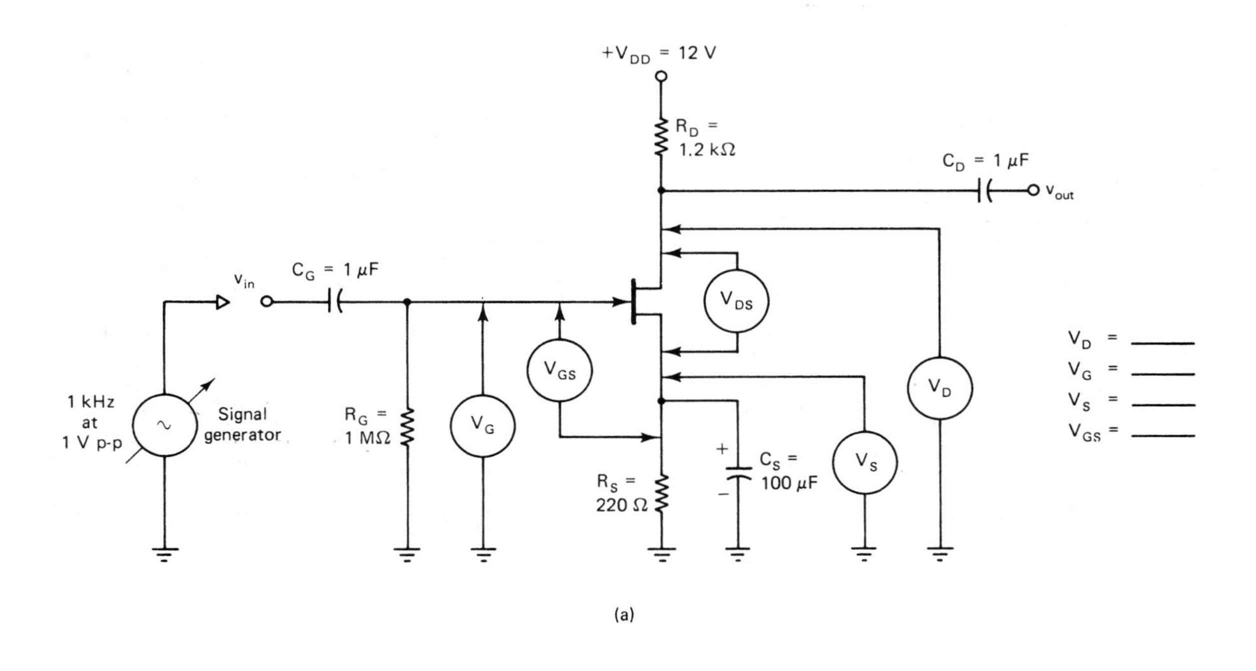

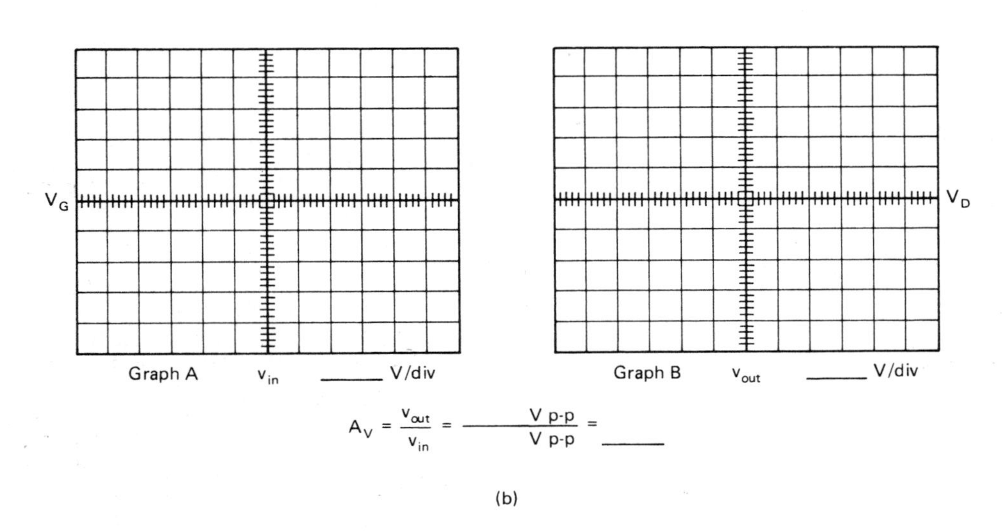

Figure 4–9 Common-source amplifier: (a) schematic diagram; (b) input/output voltage waveforms and gain.

Materials Needed:

1 Fixed +12-V power supply
1 Standard or digital voltmeter
1 Oscilloscope (dual trace preferred)
1 Signal generator (100 Hz to 1 MHz)
1 2N3823 *N*-channel JFET or equivalent
1 220-Ω resistor at 0.5 W (R_S)
1 1.2-kΩ resistor at 0.5 W (R_D)
1 1-MΩ resistor at 0.5 W (R_G)
2 1-μF capacitors at 12 WV dc (C_G and C_D)
1 100-μF electrolytic capacitor at 12 WV dc (C_S)
1 Breadboard for constructing circuit

Procedure:

1. Construct the circuit shown in Figure 4-9a.
2. Before connecting the signal generator, measure the dc operating voltages V_D, V_G, V_S, and V_{GS} and record in the appropriate places on the figure.
3. Connect the signal generator to the input and set it for 1 kHz with an amplitude of 0.5 V p-p.
4. Use the oscilloscope at the input (gate to ground) to measure the input voltage.
5. Draw the input signal on graph A, indi-

cating its voltage peak to peak, and record in its proper place the setting of the vertical attenuator of the oscilloscope marked V/div.

6. Using the oscilloscope, measure the output voltage (v_{out}) at C_D to ground.
7. Draw the output signal on graph B, indicating its voltage peak to peak, and record in its proper place the setting of the vertical attenuator of the oscilloscope marked V/div.
8. Calculate the voltage gain of the amplifier from the formula $A_v = v_{out}/v_{in}$ and record it in its proper place.

Fill-in Questions:

1. The input signal of a common-source amplifier is between the ____________ and the ____________ .

2. The output signal of a common-source amplifier is between the ____________ and the ____________ .

3. The output signal is ____________ out of phase with the input signal.

4. The voltage gain is found by the formula ____________ .

EXPERIMENT 5. COMMON-DRAIN (SOURCE-FOLLOWER) AMPLIFIER

Objective:

To demonstrate how a JFET is used in a common-drain amplifier configuration, and to understand some of its characteristics.

Introduction:

With the common-drain amplifier, the drain is connected directly to V_{DD}. The input signal is to the gate and the output signal is taken from the source. Its operation is similar to a bipolar transistor common-collector amplifier and its voltage gain is less than 1.

Some of its features are:

Input impedance: very high
Output impedance: low
Voltage gain: no, less than 1
Current gain: yes
Power gain: yes
Signal phase inversion: no (0 °)
Used primarily as a high- to low-impedance matching circuit or buffer amplifier

Materials Needed:

1 Fixed +12-V power supply
1 Standard or digital voltmeter
1 Oscilloscope (dual trace preferred)
1 Signal generator (100 Hz to 1 MHz)
1 2N3823 *N*-channel JFET or equivalent
1 10-kΩ resistor at 0.5 W (R_S)
1 1-MΩ resistor at 0.5 W (R_G)
2 1-μF capacitors at 12 WV dc (C_G and C_S)
1 Breadboard for constructing circuit

Procedure:

1. Construct the circuit shown in Figure 4-10a.
2. Before connecting the signal generator, measure the dc operating voltages V_D, V_G, V_S, and V_{GS} and record in their appropriate places on the figure.
3. Connect the signal generator to the input and set it for 1 kHz with an amplitude of 1 V p-p.
4. Use the oscilloscope at the input (gate to ground) to measure the input voltage.
5. Draw the input signal on graph A, indicating its voltage peak to peak, and record in its proper place the setting of the vertical attenuator of the oscilloscope marked V/div.
6. Using the oscilloscope, measure the output voltage (v_{out}) at C_S to ground.
7. Draw the output signal on graph B, indicating its voltage peak to peak, and record in its proper place the setting of the vertical attenuator of the oscilloscope marked V/div.
8. Calculate the voltage gain of the amplifier from the formula $A_v = v_{out}/v_{in}$ and record it in its proper place.

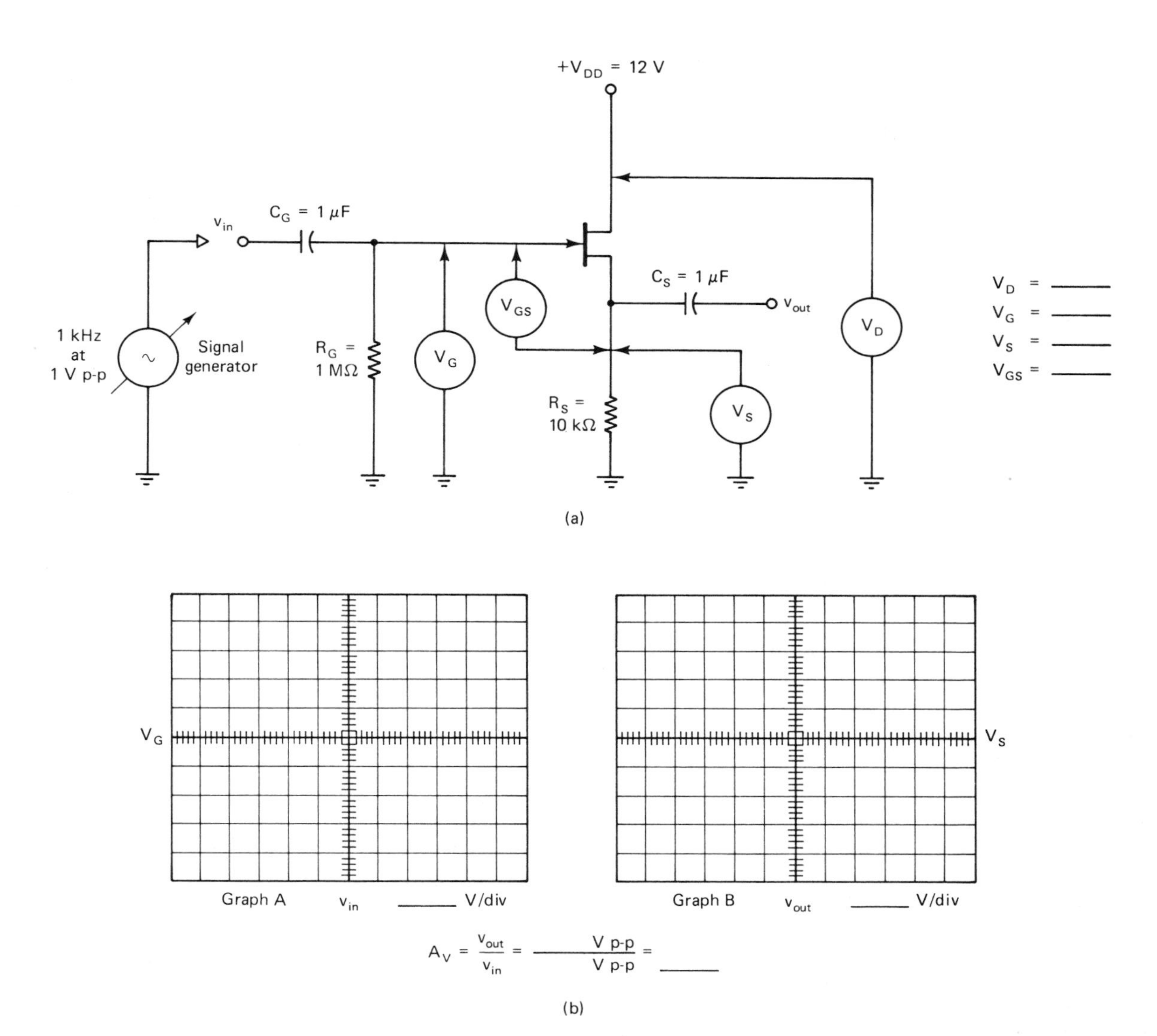

Figure 4–10 Common-drain (source follower) amplifier: (a) schematic diagram; (b) input/output voltage waveforms and gain.

Fill-in Questions:

1. The input signal of a common-drain amplifier is at the ____________ .
2. The output signal of a common-drain amplifier is at the ____________ .
3. The output signal is in phase with the ____________ signal.
4. The formula $A_V = v_{out}/v_{in}$ is used to find ____________ ____________ of the amplifier.
5. The voltage gain of a common ____________ amplifier is always less than 1.

EXPERIMENT 6. COMMON-GATE AMPLIFIER

Objective:

To show how a JFET is used in a common-gate amplifier, and to understand some of its characteristics.

Introduction:

A common-gate amplifier can be recognized by the input signal to the source and the output signal from the drain. It can be compared to the bipolar transistor common-base amplifier.

Some of its features are:

Input impedance: low
Output impedance: moderate
Voltage gain: moderate
Current gain: yes
Power gain: yes
Signal phase inversion: no (0°)
Used primarily as a low- to high-impedance matching circuit

Materials Needed:

1 Fixed +12-V power supply
1 Standard or digital voltmeter
1 Oscilloscope (dual trace preferred)
1 Signal generator (100 Hz to 1 MHz)
1 2N3823 *N*-channel JFET or equivalent
1 220-Ω resistor at 0.5 W (R_S)
1 1-kΩ resistor at 0.5 W (R_D)
1 1-MΩ resistor at 0.5 W (R_G)
1 100-μF electrolytic capacitor at 12 WV dc (C_G)
2 1-μF capacitors at 12 WV dc (C_S and C_D)
1 Breadboard for constructing circuit

Procedure:

1. Construct the circuit shown in Figure 4-11a.
2. Before connecting the signal generator, measure the dc operating voltages V_D,

Figure 4-11 Common-gate amplifier: (a) schematic diagram; (b) input/output voltage waveforms and gain.

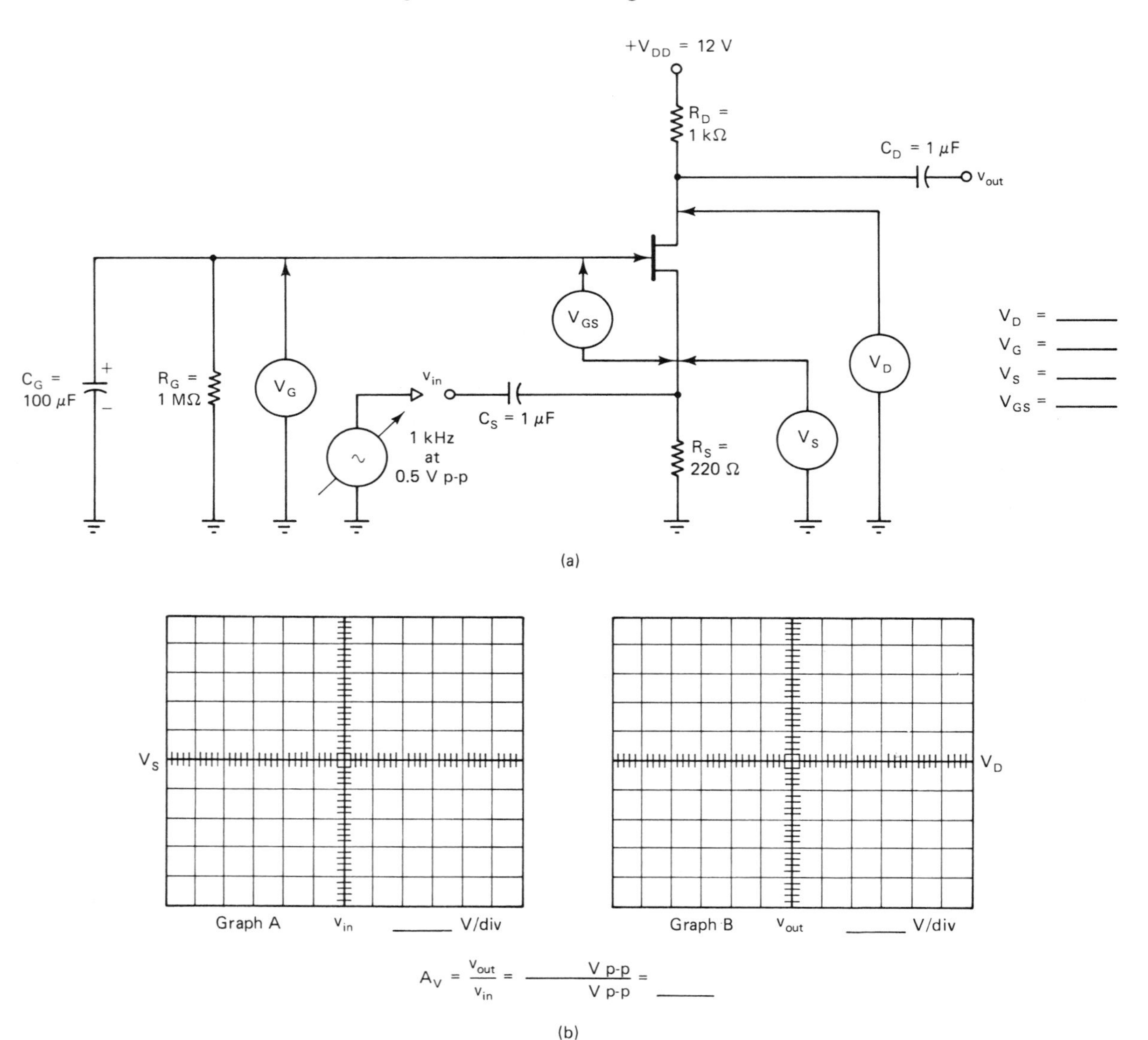

V_G, V_S, and V_{GS} and record in their appropriate places on the figure.

3. Connect the signal generator to the input and set it for 1 kHz with an amplitude of 0.5 V p-p.
4. Use the oscilloscope at the input (source to ground) to measure the input voltage.
5. Draw the input signal on graph A, indicating its voltage peak to peak, and record in its proper place the setting of the vertical attenuator of the oscilloscope marked V/div.
6. Using the oscilloscope, measure the output voltage (v_{out}) at C_D to ground.
7. Draw the output on graph B, indicating its voltage peak to peak, and record in its proper place the setting of the vertical attenuator of the oscilloscope marked V/div.
8. Calculate the voltage gain of the amplifier from the formula $A_V = v_{out}/v_{in}$ and record it in its proper place.

Fill-in Questions:

1. The input signal of a common-gate amplifier is at the ____________ .
2. The output signal of a common-gate amplifier is at the ____________ .
3. The ____________ signal is in phase with the input signal.
4. The voltage gain is found by v_{out} ____________ by v_{in}.

SECTION 4-5 JFET BASIC TROUBLESHOOTING APPLICATION: COMMON-SOURCE AMPLIFIER

Construct this circuit using your own values or those of a circuit from a previous section. Open or short the components as listed and record the voltages in the proper place in the table. The abbreviations will help indicate the voltage conditions associated with each problem. All voltages are referenced to ground.

Figure 4–E.2

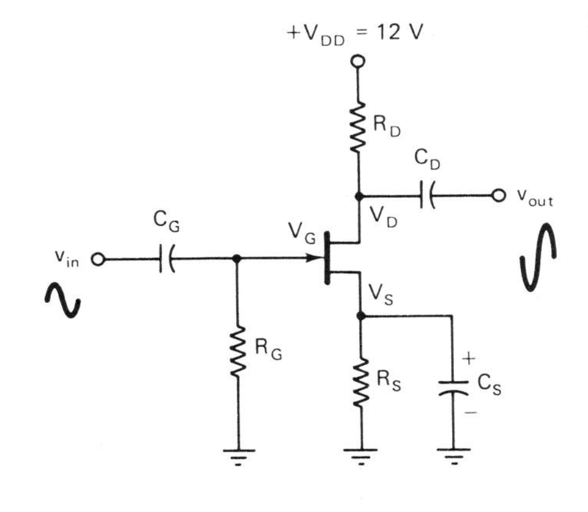

Table Abbreviations

V_{DD}	Power supply
INC	Increase
DEC	Decrease
NOC	No change
GND	Ground
NEG	Negative

Condition	V_D	V_G	V_S	V_{out} (vp-p)	Comments
Normal					All voltages proper.
R_G open					No V_{GND} reference. The gate floats high and I_D increases.
	DEC	INC	INC		
R_D open					No $+V_{DD}$ applied to circuit.
	DEC	NOC	DEC		
R_S open					No current flow through circuit.
	V_{DD}	INC	INC		
C_S open					d.c. voltages normal. Reduced signal gain because of NEG feedback.
	NOC	NOC	NOC		
C_S shorted					Reverse bias decreased, JFET saturates.
	DEC	NOC	DEC		
D open					JFET not conducting.
	V_{DD}	NOC	DEC		
G open					No control voltage. Reverse bias decreased and I_D increases.
	DEC	NOC	INC		
S open					JFET not conducting.
	V_{DD}	INC	DEC		

D-G shorted	DEC	INC	INC		V_D and V_G same voltage. JFET conducts more.
S-G shorted	DEC	INC	INC		V_G and V_S same voltage. JFET conducts more.
D-S shorted	DEC	NOC	INC		JFET not in circuit. Voltages are normal for R_D and R_S in series.

Figure 4–E.2 continued

SECTION 4-6 MORE EXERCISES

Perform the following exercises before beginning Section 4-7.

1. Referring to Figure 4-6c and d, draw the proper ohmmeter connections for an *N*-channel JFET and, assuming the JFET is good, indicate the proper high and low ohmic readings.

2. Referring to Figure 4-6f and g, draw the proper ohmmeter connections for a *P*-channel JFET and, assuming the JFET is good, indicate the proper high and low ohmic readings.

3. If an ohmmeter's leads are connected to the JFETs as indicated and a resulting reading occurs, determine if each JFET is open, shorted, or normal.

a. Type of JFET: *N-channel*
Positive lead to drain
Negative lead to gate
Reading is infinite

Conclusion: ______

b. Type of JFET: *N-channel*
Positive lead to source
Negative lead to gate
Reading is low

Conclusion: ______

c. Type of JFET: *P-channel*
Positive lead to gate
Negative lead to drain
Reading is low

Conclusion: ______

d. Type of JFET: *P-channel*
Positive lead to source
Negative lead to gate
Reading is infinite

Conclusion: ______

4. With a common-source amplifier, when $V_D = V_{DD}$, the JFET is cutoff, and when V_D is less than V_{DD}, the JFET is on or conducting. From the voltage readings given, indicate if each JFET is on or off.

a. $V_{DD} = +12$ V
$V_D = +6$ V
$V_S = +2$ V

Condition of JFET: ______

b. $V_{DD} = +12$ V
$V_D = +12$ V
$V_S = 0$ V

Condition of JFET: ______

c. $V_{DD} = -12$ V
$V_D = -12$ V
$V_S = 0$ V

Condition of JFET: ______

d. $V_{DD} = -12$ V
$V_D = -6$ V
$V_S = -1$ V

Condition of JFET: ______

5. Find the value of I_D and V_S for the following circuits using the formulas $I_D = V_{RD}/R_D = V_{DD} - V_D/R_D$, and $V_S = I_D R_S$.

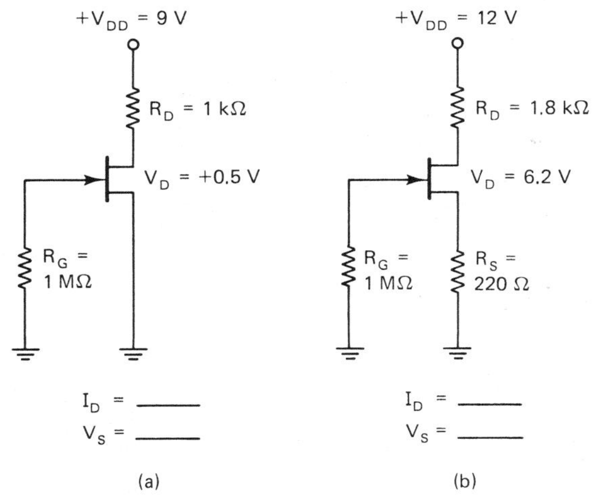

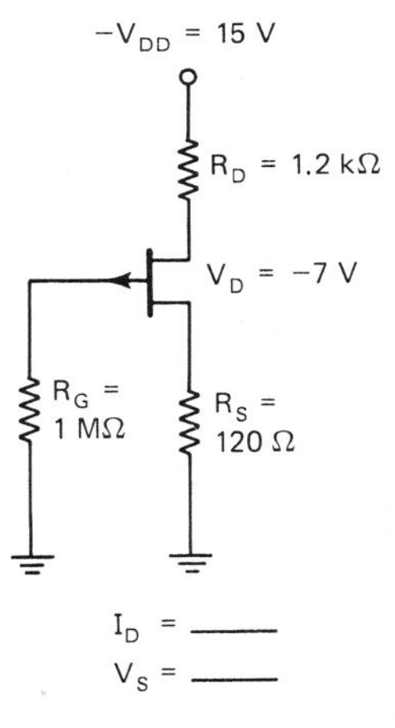

Figure 4-E.3

6. Draw a simple JFET circuit for each of following amplifiers:

a. Common source

b. Common drain

c. Common gate

SECTION 4-7 JFET INSTANT REVIEW

The JFET is a "normally on" unipolar device. Maximum current flows from source to drain through the channel when the gate-to-source bias voltage (V_{GS}) is zero. Current flow through the JFET is reduced when reverse bias is applied to the gate, and it is said to be operating in the depletion mode. The cross-sectional channel area or depletion region is controlled by an electrostatic field as a result of V_{GS}; therefore, it is a voltage-controlled device.

SECTION 4-8 SELF-CHECKING QUIZZES

4-8a JFET: TRUE–FALSE QUIZ

Place a T for true or an F for false to the left of each statement.

____ **1.** A JFET is a unipolar device.

____ **2.** An *N*-channel JFET must have the gate more positive thar the source to turn on.

____ **3.** The JFET is a voltage-controlled device.

____ **4.** A *P*-channel JFET has electrons for majority charge carriers.

____ **5.** The main advantage of a JFET is its high input resistance.

____ **6.** The JFET is a "normally on" device.

____ **7.** The transconductance of a JFET is found by the formula $g_m = \Delta V_{GS}/\Delta I_D$.

____ **8.** I_{DSS} refers to the maximum I_D when $V_{GS} = 0$ V.

____ **9.** A JFET normally operates in the depletion region.

____ **10.** When V_{DS} increases, but there is no increase in I_D, the JFET is operating in the pinch-off region.

4-8b JFET: MULTIPLE-CHOICE QUIZ

Circle the correct answer for each question.

1. When testing an *N*-channel JFET with an ohmmeter and placing the negative lead on the gate and the positive lead on the source results in a low ohmic reading, the JFET is:

a. normal
b. open
c. shorted
d. none of the above

2. The current flowing through a JFET is controlled by:

a. the gate current
b. V_{DS}
c. V_{GS}
d. none of the above

3. If a particular JFET has a 0.5-V change in V_{GS} and a corresponding 2.2-mA change in I_D, its g_m is:

a. 227 μmhos **b.** 4.4 μmhos
c. 4400 μmhos **d.** 1100 μmhos

4. I_{DSS}:

a. is called saturation current
b. occurs when the gate and source are shorted
c. occurs when $V_{GS} = 0$ V
d. all of the above
e. none of the above

5. I_{GSS}:

a. is called gate leakage current
b. occurs when $V_{DS} = 0$ V
c. both of the above
d. none of the above

Questions 6 through 10 refer to Figure 4-A.

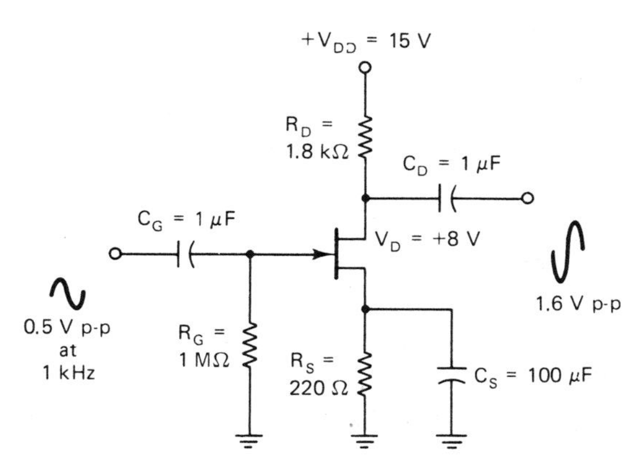

Figure 4–A

6. This circuit is a:

a. common-gate amplifier
b. common-drain amplifier
c. common-source amplifier
d. none of the above

7. The drain current, I_D, is approximately equal to:

a. 3.4 mA **b.** 3.9 mA
c. 7.4 mA **d.** 8.3 mA

8. The source voltage, V_S, is about:

a. 0.54 V **b.** 0.86 V
c. 1.4 V **d.** 2.2 V

9. The gate voltage, V_G, is about:

a. 0 V **b.** 0.5 V
c. 0.86 V **d.** 1.36 V

10. The voltage gain of the circuit is:

a. 1.8 **b.** 2.1
c. 3.2 **d.** 4.5

ANSWERS TO EXPERIMENTS AND QUIZZES

Experiment 1.	**(1)** low, high, infinite **(2)** high, infinite, low **(3)** open
Experiment 2.	**(1)** low **(2)** on **(3)** V_{DD}
Experiment 3.	**(1)** decrease **(2)** I_D **(3)** I_D, V_{DS} **(4)** V_{GS}, V_{DS}, I_D **(5)** 3
Experiment 4.	**(1)** gate, ground **(2)** drain, ground **(3)** 180° **(4)** $A_V = v_{out}/v_{in}$
Experiment 5.	**(1)** gate **(2)** source **(3)** input **(4)** voltage gain **(5)** drain
Experiment 6.	**(1)** source **(2)** drain **(3)** output **(4)** divided
True–False.	**(1)** T **(2)** F **(3)** T **(4)** F **(5)** T **(6)** T **(7)** F **(8)** T **(9)** T **(10)** T
Multiple Choice.	**(1)** c **(2)** c **(3)** c **(4)** d **(5)** c **(6)** c **(7)** b **(8)** b **(9)** a **(10)** c

Unit 5

Metal-Oxide Semiconductor Field-Effect Transistor

INTRODUCTION The metal-oxide semiconductor field-effect transistor (MOSFET) has wide applications in digital and computer circuits.

UNIT OBJECTIVES Upon completion of this unit, you will be able to:

1. Draw the schematic symbol for a depletion-type and enhancement-type MOSFET.
2. Explain the operation and differences of a depletion-type and enhancement-type MOSFET.
3. Describe the three modes of biasing for the MOSFET.
4. Explain the current–voltage characteristic curve for each type of MOSFET.
5. Describe the uses and advantages of a VMOSFET.
6. Show how a complementary MOSFET can be turned on and off.
7. List the safety precautions required when working with MOSFETs.
8. Construct a MOSFET amplifier.
9. Demonstrate how a MOSFET driver operates.
10. Define the terms depletion, enhancement, S_iO_2, gate-protection diodes, and CMOS.
11. Troubleshoot a MOSFET circuit.

SECTION 5-1 THEORY OF STRUCTURE AND OPERATION

5-1a TYPES OF MOSFETs

There are two types of metal-oxide semiconductor field-effect transistors (MOSFETs). The depletion type operates in the depletion and enhancement regions and is a "normally on" device. The enhancement type operates only in the enhancement region and is a "normally off" device. The gate of a MOSFET is insulated from the channel by a thin layer of silicon dioxide (SiO_2), but the bias voltage placed on the gate creates an electrostatic field effect that controls the current flow from source to drain. Because of the insulated gate, the input impedance to a MOSFET is much higher than that of a JFET. Typically, this impedance is on the order of 10^{14} Ω. Other names given to the MOSFET are surface field-effect transistor, insulated-gate field-effect transistor (IGFET), and metal-oxide semiconductor transistor (MOST). The lead identification and current and voltage nomenclature for MOSFETs are the same as those for JFETs.

5-1b DEPLETION-TYPE MOSFET STRUCTURE AND SCHEMATIC SYMBOL

A depletion-type MOSFET has a foundation or substrate of one type of semiconductor material with a channel of the opposite material running through it, as shown in Figure 5-1. The metal gate is insulated from the channel by a thin layer of silicon dioxide (SiO_2).

The *N*-channel depletion-type MOSFET schematic symbol has the arrow pointing toward the gate, and the *P*-channel depletion-type MOSFET schematic symbol has the arrow pointing away from the gate.

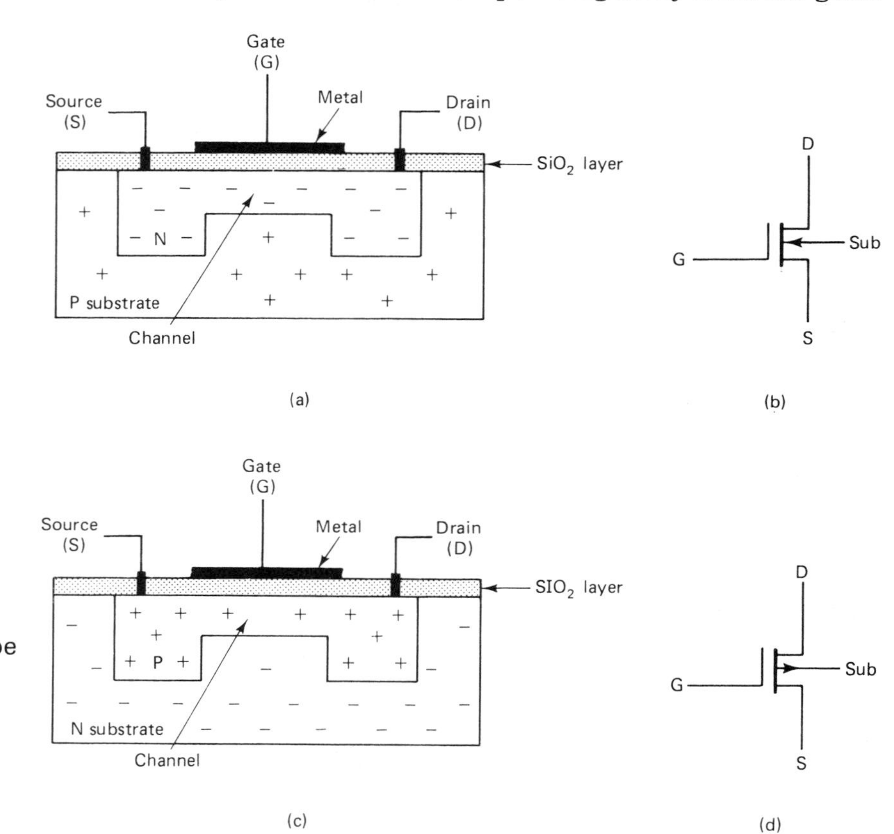

Figure 5–1 Depletion-type MOSFETs: (a) *N*-channel structure; (b) *N*-channel schematic symbol; (c) *P*-channel structure; (d) *P*-channel schematic symbol.

5-1c THEORY OF OPERATION FOR A DEPLETION-TYPE MOSFET

An *N*-channel depletion-type MOSFET is properly biased with a positive voltage applied to the drain and a negative voltage applied to the source, as shown in Figure 5-2. When the gate control voltage $V_{GS} = 0$ V (Figure 5-2a), there is little or no restriction to current flow I_D from source to drain. As V_{GS} is made negative (Figure 5-2b), electrons are repelled from the SiO_2 layer and a depletion region develops, which decreases the channel width and less I_D flows. If V_{GS} is made sufficiently negative (Figure 5-2c), the depletion region increases until the channel

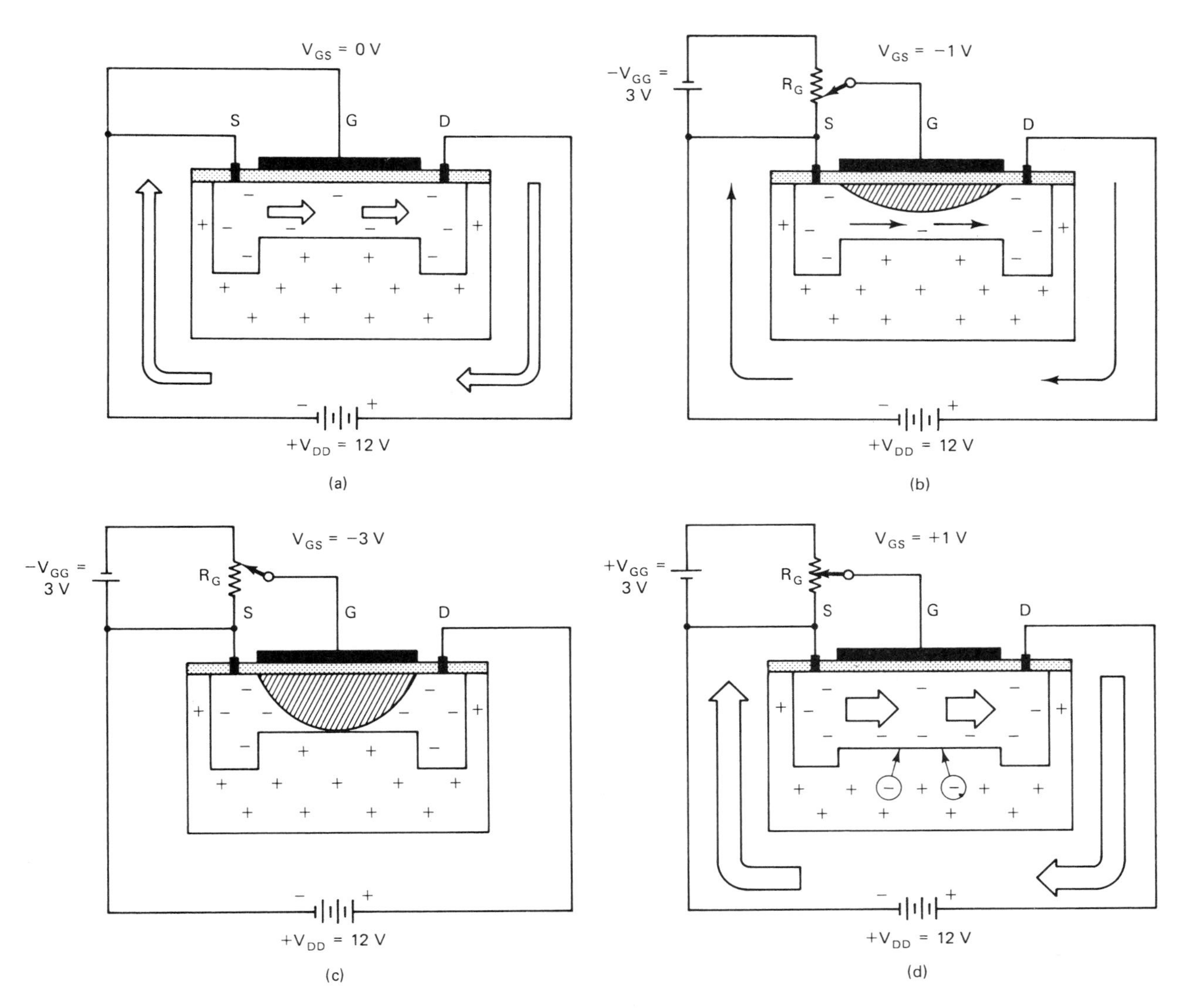

Figure 5–2 *N*-channel depletion-type MOSFET operation: (a) $V_{GS} = 0$ V, large current; (b) $V_{GS} = -1$ V, less current; (c) $V_{GS} = -3$ V, current cutoff; (d) $V_{GS} = +1$ V, maximum current.

is cut off and no current flows. The MOSFET is operating in the depletion region when reverse bias decreases the I_D. However, this MOSFET can have forward bias applied to its gate (Figure 5-2d). In this case the positive voltage attracts electrons from the substrate into the channel region. The width of the channel increases and more I_D will flow. Because the channel width has increased, the MOSFET is said to be operating in the enhancement region. Since the SiO_2 layer is an insulator, there should never be any appreciable gate current.

The *P*-channel depletion-type MOSFET operates in the same manner, except that the voltages are reversed and the current carriers of the channel are holes.

5-1d CURRENT–VOLTAGE CHARACTERISTICS OF A DEPLETION-TYPE MOSFET

The current–voltage characteristics for a depletion-type MOSFET are similar to those for a JFET, except that the bias voltage may be positive or negative, as shown in Figure 5-3. Again, it is seen how V_{GS} controls I_D. Voltage V_{DS} has little control over I_D, and this current remains fairly constant when the FET is operating in the pinch-off region.

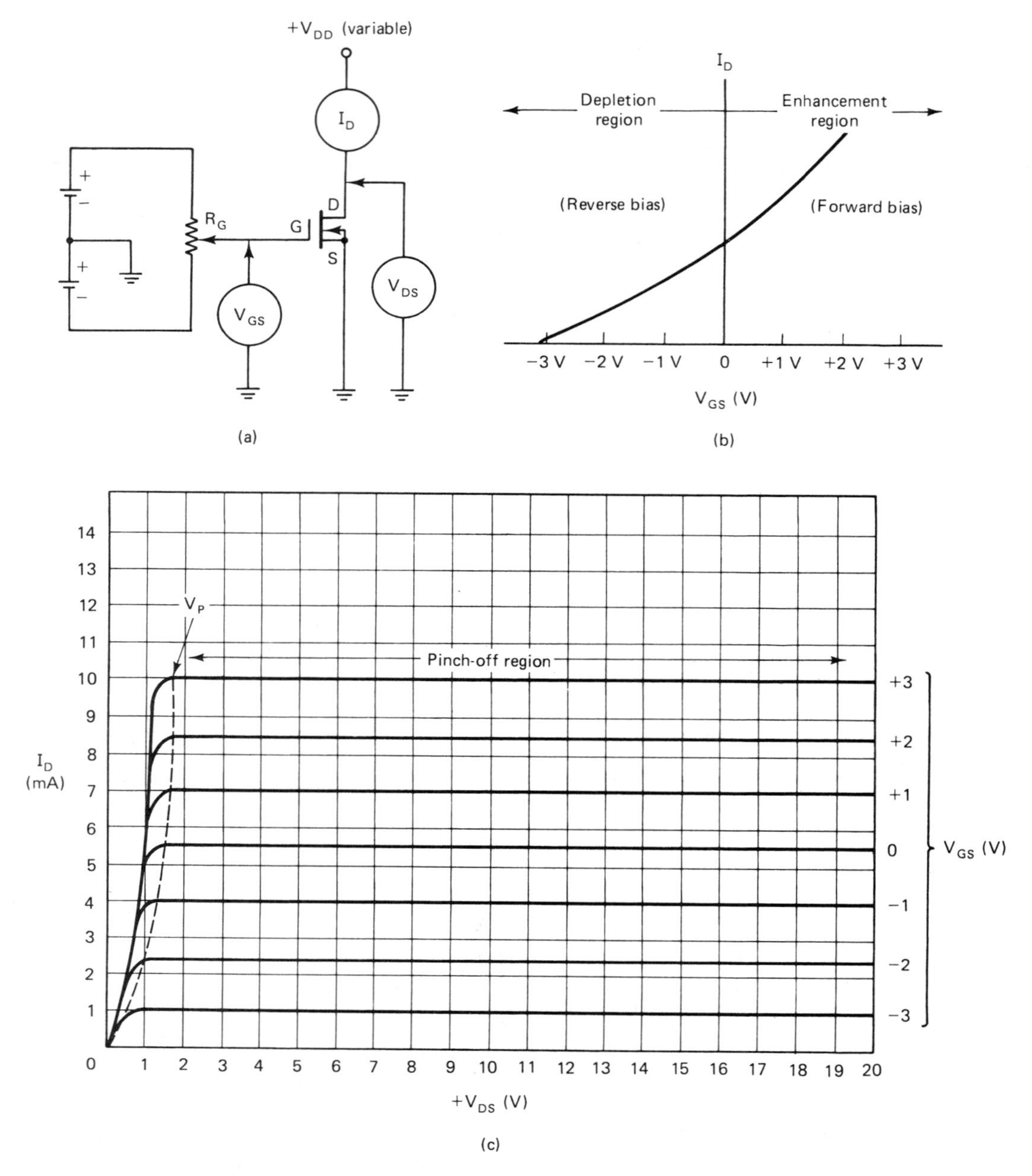

Figure 5–3 *N*-channel depletion-type MOSFET current–voltage characteristics: (a) test circuit; (b) transfer characteristic curve (V_{GS} versus I_D); (c) output characteristic curve (I_D versus V_{DS}).

5-1e DEPLETION-TYPE MOSFET BIASING TECHNIQUES

A depletion-type MOSFET can be biased three ways: with zero, negative, or positive voltage, as shown in Figure 5-4. When zero bias is used (Figure 5-4a), the source is connected to ground and the gate is electrically connected to ground via R_G. Negative bias (Figure 5-4b) uses a self-biasing source resistor R_S. When current I_D flows through this resistor, the top of it connected to the source is more positive than ground. However, the gate connected to ground via R_G is electrically at zero volts; therefore, the gate is negative with respect to the source. Positive bias (Figure 5-4c) may use a voltage-divider arrangement, such as R_A and R_G. The source is connected directly to ground, but the top of R_G, at the gate, is more positive when current flows through R_G and R_A toward $+V_{DD}$.

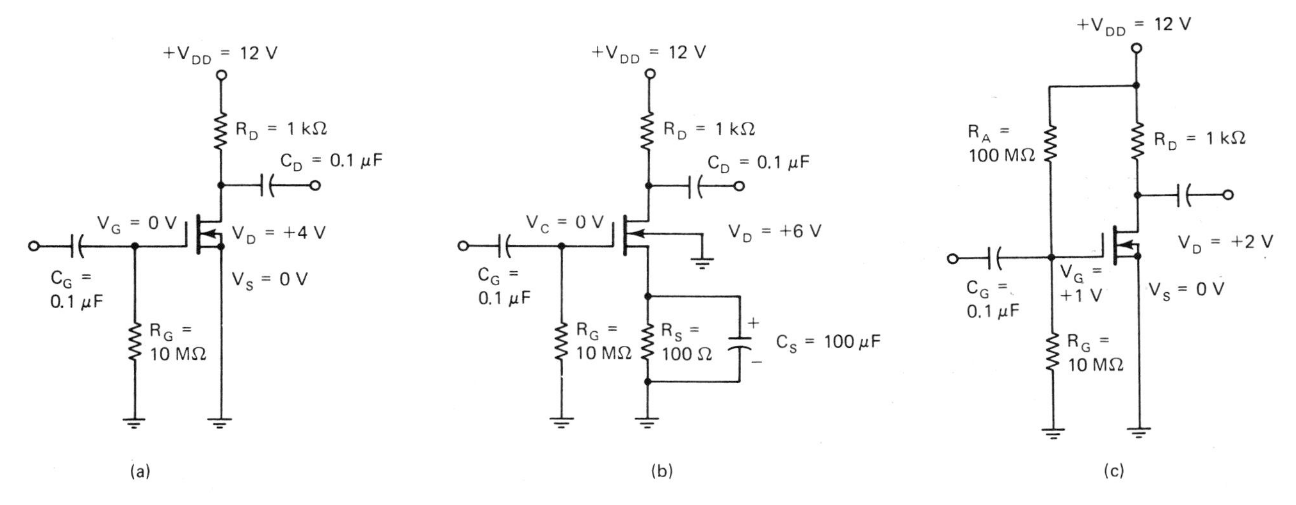

Figure 5–4 Self-bias arrangements for an *N*-channel depletion-type MOSFET: (a) $V_{GS} = 0$ V; (b) $V_{GS} = -1$ V; (c) $V_{GS} = +1$ V.

5-1f ENHANCEMENT-TYPE MOSFET STRUCTURE AND SCHEMATIC SYMBOL

An enhancement-type MOSFET has a substrate of one type of semiconductor material with a source and drain region of the opposite material. There is no channel between them, as shown in Figure 5-5. The metal gate is insulated from the entire structure by a layer of SiO_2.

The schematic symbol for an enhancement-type MOSFET has a broken vertical line joining the source, substrate, and drain, with the arrow pointing toward the gate for an *N* channel and the arrow pointing away from the gate for a *P* channel.

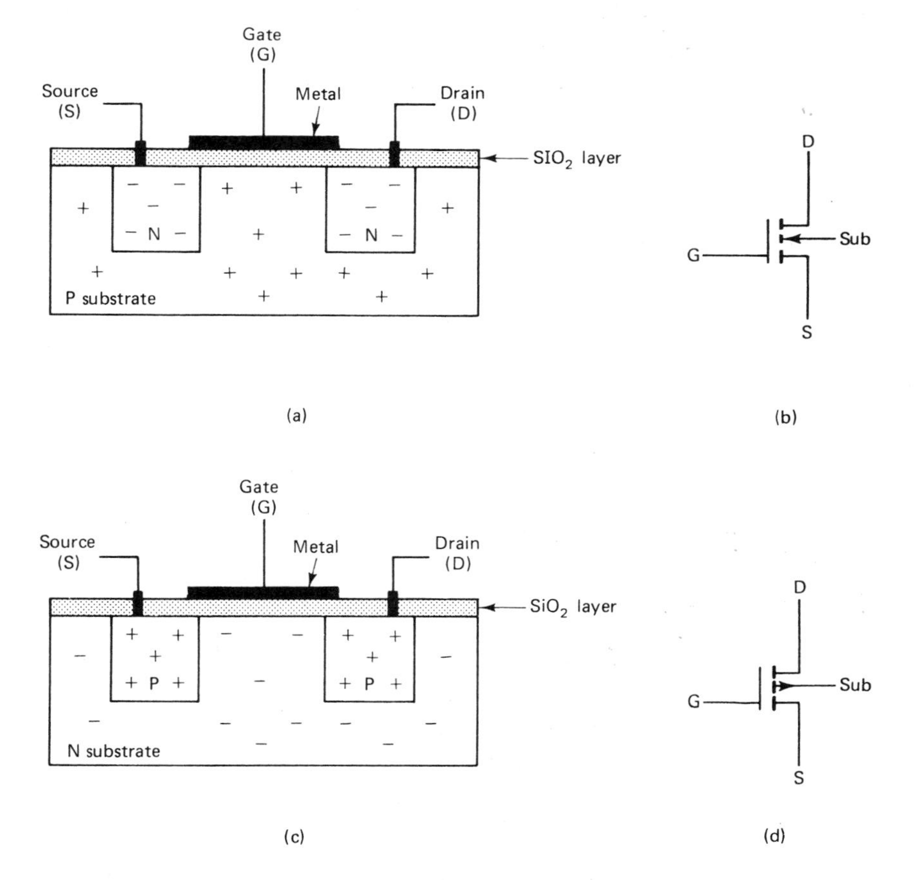

Figure 5–5 Enhancement-type MOSFETs: (a) *N*-channel structure; (b) *N*-channel schematic symbol; (c) *P*-channel structure; (d) *P*-channel schematic symbol.

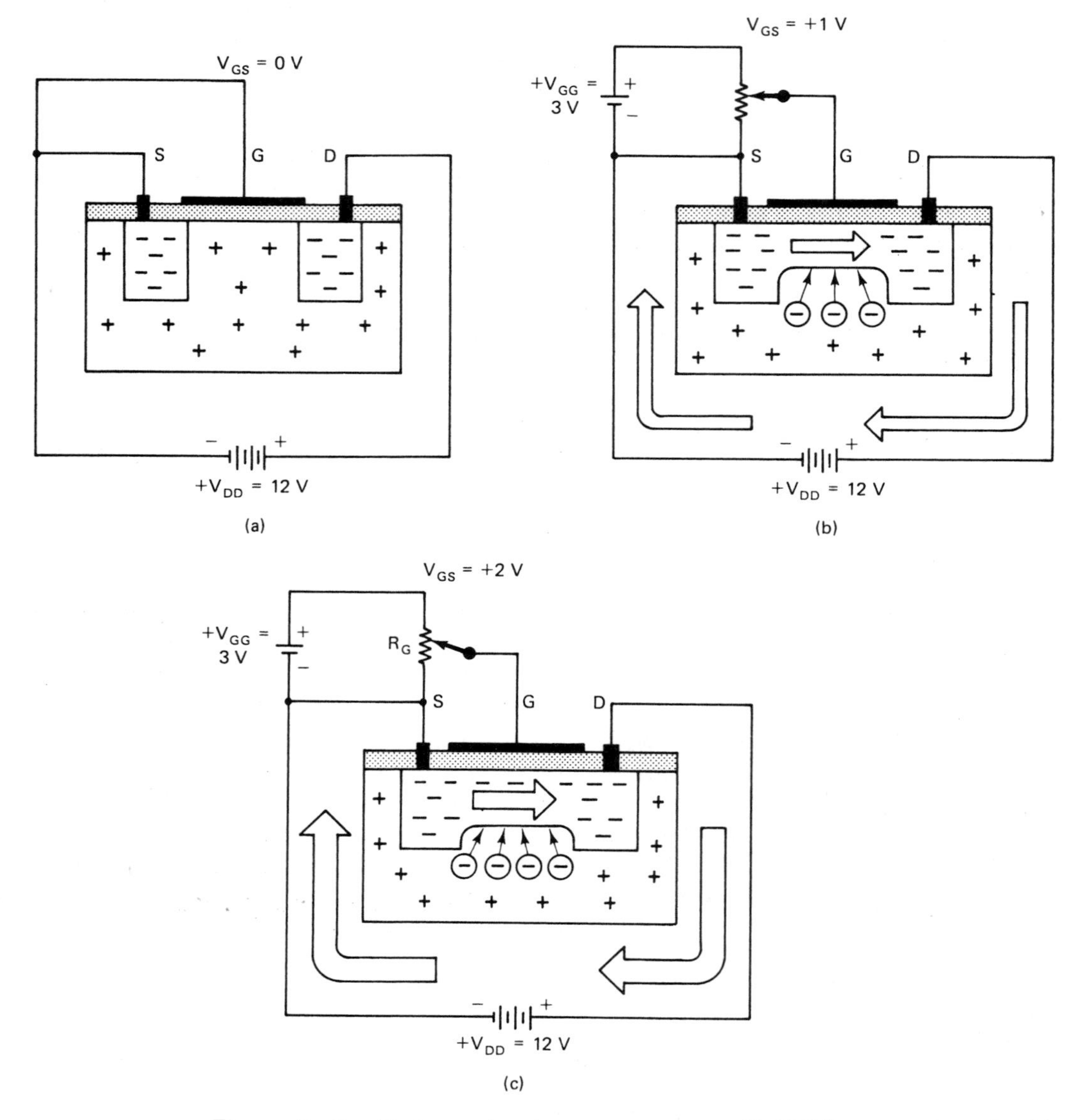

Figure 5–6 *N*-channel enhancement-type MOSFET operation: (a) V_{GS} = 0 V, no current; (b) V_{GS} = +1 V, some current; (c) V_{GS} = +2 V, more current.

5-1g THEORY OF OPERATION FOR AN ENHANCEMENT-TYPE MOSFET

An *N*-channel enhancement-type MOSFET is properly biased with positive voltage applied to the drain and negative voltage applied to the source, as shown in Figure 5-6. When the gate control voltage $V_{GS} = 0$ V (Figure 5-6a), nothing occurs and there is no current flow I_D from source to drain. If V_{GS} is made positive (Figure 5-6b), electrons flow from the substrate to the underside of the SiO_2 layer and create a channel from source to drain, which allows current I_D to flow. When V_{GS} is made more positive (Figure 5-6c), more electrons flow from the substrate, creating a larger channel width, which allows I_D to increase. This forward bias creates a wider channel area, which "enhances" the conductivity of the MOSFET. Reverse bias keeps the MOSFET cut off.

The *P*-channel enhancement-type MOSFET operates in the same manner, except that the voltages are reversed and the current carriers of the channel are holes.

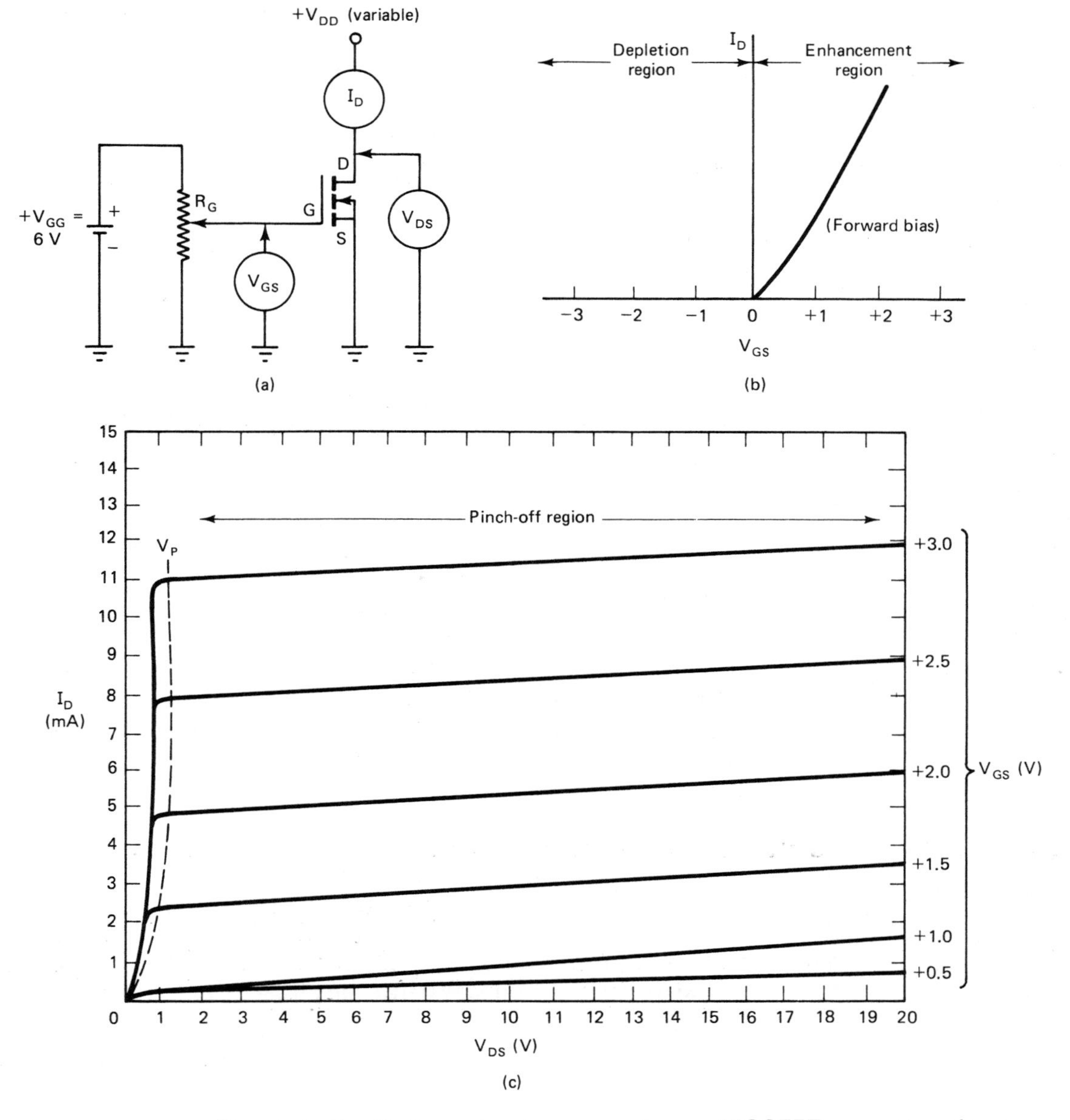

Figure 5–7 *N*-channel enhancement-type MOSFET current–voltage characteristics: (a) test circuit; (b) transfer characteristic curve (V_{GS} versus I_D); (c) output characteristic curve (I_D versus V_{DS}).

5-1h CURRENT–VOLTAGE CHARACTERISTICS OF AN ENHANCEMENT-TYPE MOSFET

The current–voltage characteristics for an enhancement-type MOSFET are similar to those of the other types of FETs, except that forward bias is used to turn it on, as shown in Figure 5-7. An increase in forward-bias voltage V_{GS} causes an increase in I_D; however, V_{DS} has little effect on I_D and it remains fairly constant within the pinch-off region.

5-1i ENHANCEMENT-TYPE MOSFET BIASING TECHNIQUES

Figure 5-8 shows three possible biasing arrangements for an *N*-channel enhancement-type MOSFET. With zero bias (Figure 5-8a), the source is connected to ground and the gate is electrically at the same potential via R_G. The MOSFET is off and a positive voltage is needed at the gate

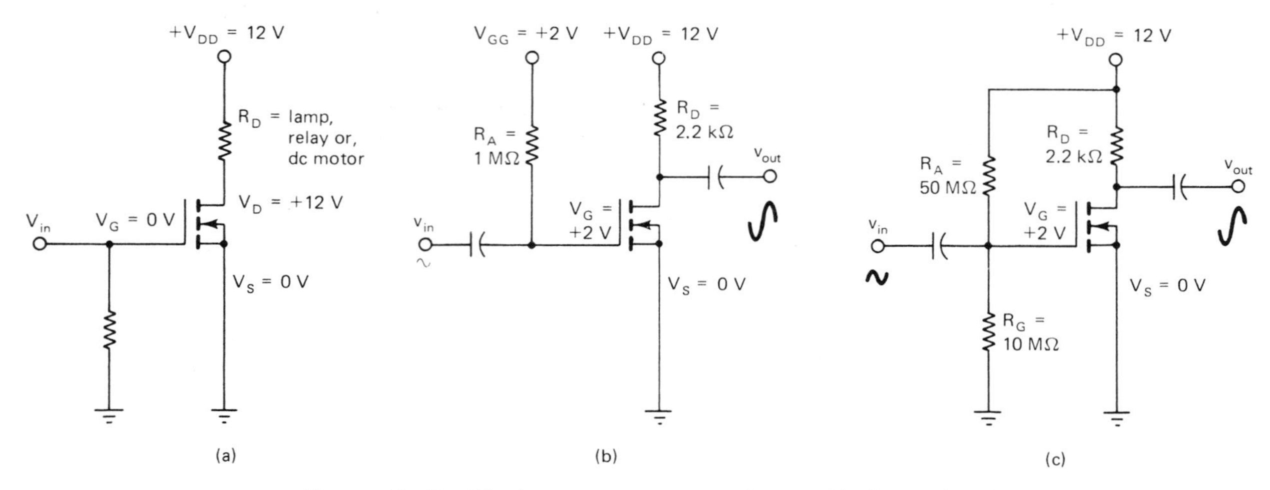

Figure 5–8 Biasing arrangements for an *N*-channel enhancement-type MOSFET: (a) V_{GS} = 0 V; (b) fixed bias, V_{GS} = +2 V; (c) voltage-divider bias, V_{GS} = +2 V.

to turn it on. This circuit could be used as a driver stage for a lamp, dc motor, or relay. The gate could be connected to a fixed bias voltage (Figure 5-8b) and the MOSFET would be on. An input signal voltage would vary I_D, thereby producing a larger output voltage. Forward bias voltage can be developed with the use of a voltage divider R_A and R_G (Figure 5-8c). The top of R_G would be positive. Again, the MOSFET would be on and an input signal voltage would vary I_D, causing a larger output voltage.

5-1j POWER MOSFET

A power MOSFET has a higher current capability than a regular MOSFET because of its different structure, as shown in Figure 5-9a. The substrate consists of *N*-type semiconductor material, which also serves as the drain electrode. Within the substrate there exists a *P*-type material body region and a heavily doped *N*-type region ($N+$). The gate in the form of a V is insulated from the semiconductor material by a thin layer of SiO_2. The source lead contacts the *P*-type body and $N+$ region.

Figure 5–9 VMOSFET: (a) structure; (b) schematic symbol.

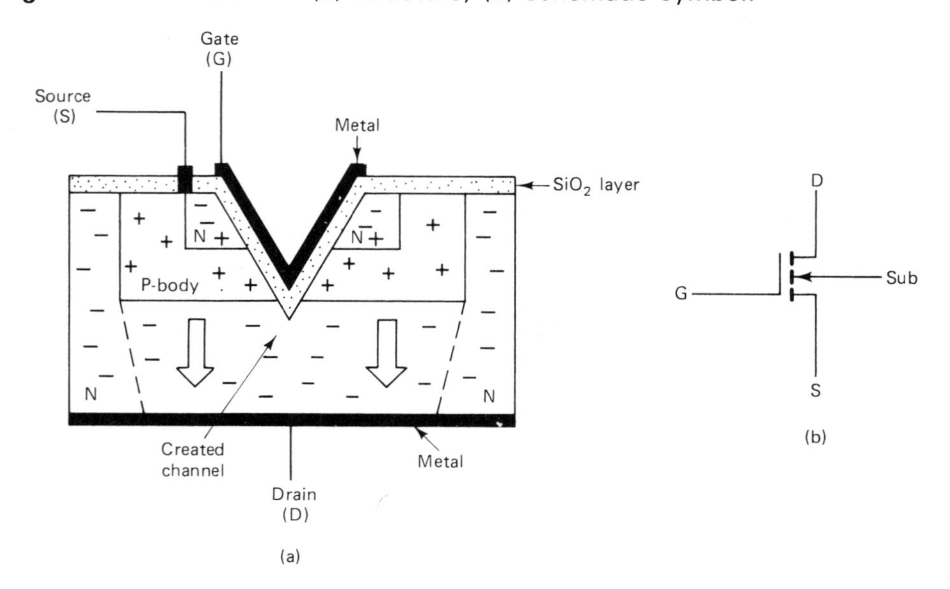

The operation of the power MOSFET is similar to that of a normal *N*-channel enhancement-type MOSFET, except that the control channel is created on both surfaces of the body region facing the V-shaped gate. Current flows from the source through the control channel to the substrate drain. The area of the control channel is larger than that of a normal MOSFET and more current is able to flow. Since the current flows vertically through the device rather than horizontally across its surface, it is called a "vertical" MOSFET" or simply a VMOSFET.

VMOSFETs can be used to drive speakers, lamps, LEDs, relays, dc motors, and so on.

5-1k COMPLEMENTARY METAL-OXIDE SEMICONDUCTOR

A complementary metal-oxide semiconductor (CMOS) consists of a *P*- and *N*-channel enhancement-type MOSFET fabricated on a single substrate, as shown in Figure 5-10. The gate of each MOSFET is usually connected internally, and their drains can be connected to form a single output. When the CMOS is biased as shown in Figure 5-10b and $V_{in} = 0$ V, the *P*-channel MOSFET is on (*N*-channel MOSFET is off) and V_{out} is about equal to $+V_{DD}$. When $V_{in} = +3$ V, the *N*-channel MOSFET is on (*P*-channel MOSFET is off) and V_{out} is about equal to 0 V. This circuit arrangement produces an inverter. If the input goes negative, the output goes positive, and vice versa. CMOS devices are usually used in switching circuits and are commonly found in digital integrated circuits and semiconductor memory devices.

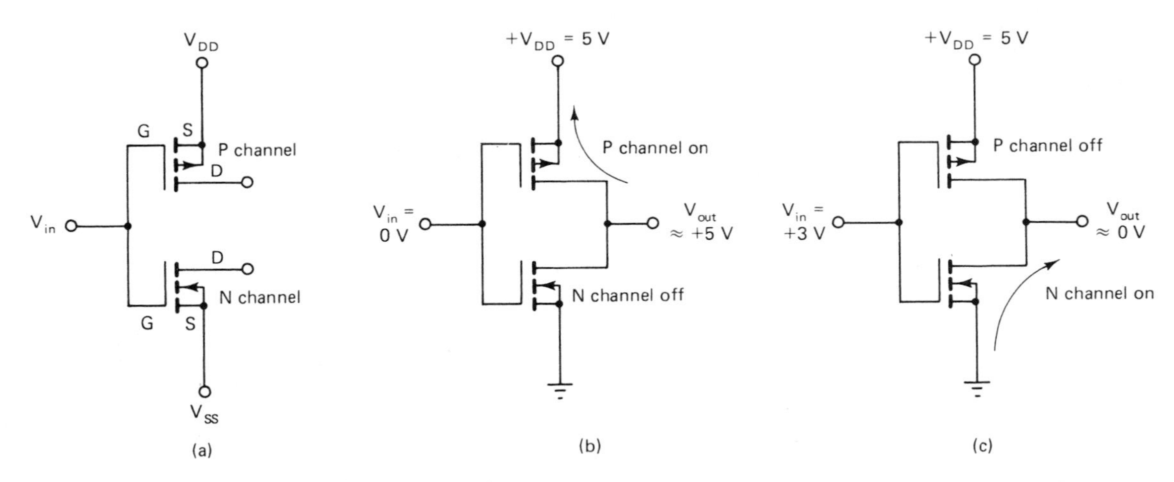

Figure 5–10 Complementary metal-oxide semiconductor (CMOS): (a) schematic diagram; (b) *P* channel on; (c) *N* channel on.

5-1l MOSFET SAFETY PRECAUTIONS

Special care must be used in handling MOSFETs as opposed to other semiconductor devices. The SiO_2 layer is sufficiently thin that, if a static charge of even less than 100 V is applied across the gate and channel, the oxide layer may rupture. This usually results in gate leakage current, which renders the MOSFET useless.

Static charges may occur on the leads of a MOSFET by rubbing on plastic bags or fabrics during shipping, from ungrounded soldering devices, and from the static charge built up in the body of the person handling the MOSFET. Manufacturers ship MOSFETs with their leads shorted together by means of a shorting ring, wire, or foil, and/or they

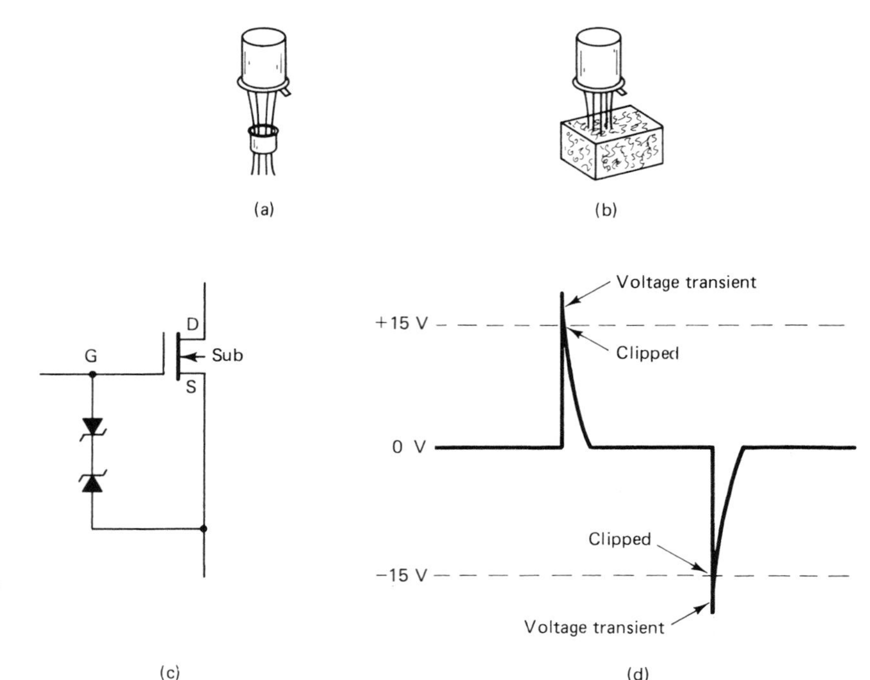

Figure 5–11 MOSFET safety precautions: (a) shorting ring; (b) antistatic foam; (c) zener diode gate protection; (d) example of gate protection.

are pressed into a conducting foam material, as shown in Figure 5-11a and b. These shorting devices should not be removed until the MOSFET is installed in its respective circuit. The following precautions should be taken when handling MOSFETs:

1. Turn off the power to the circuit being worked on. Voltage transients developed by separating components or opening circuits may damage the MOSFET.
2. Use a special "grounding wristband" or connect a clip lead from your metal watchband to the chassis or ground point of the circuit.
3. The soldering iron should be properly grounded or a clip lead connected from the tip of the tool to the chassis. Soldering guns are not recommended for use on MOSFETs.

Most newer types of MOSFETs have zener-diode protection from the gate to source, as shown in Figure 5-11c and d. Typically, these diodes will break down and conduct when the voltage between gate and source exceeds 15 V, thereby protecting the MOSFET from in-circuit transient voltages and out-of-circuit handling operations. However, it is still recommended that caution be used when handling all MOSFET devices.

SECTION 5-2 MOSFET DEFINITION EXERCISE

The same current, voltage, and lead identification nomenclature that applies to JFETs also applies to MOSFETs (see Section 4-2). A few additional terms are used with MOSFETs. Refer to the previous sections and write a brief definition for each term.

1. SiO_2 ____

____

____

2. Depletion ______________________________

____

____

3. Enhancement ______________________________

____

____

4. Gate-protection diodes ______________________________

____

____

5. Pinch-off voltage ______________________________

____

____

6. VMOSFET ______________________________

____

____

7. CMOS ______________________________

____

____

SECTION 5-3 EXERCISES AND PROBLEMS

Perform the following exercises before beginning Section 5-4.

1. Draw the structure of an *N*-channel depletion-type MOSFET. (Label all parts.)

2. Draw the schematic symbol of an *N*-channel depletion-type MOSFET. (Label the leads.)

3. Draw the structure of a *P*-channel depletion-type MOSFET. (Label all parts.)

4. Draw the schematic symbol of a *P*-channel depletion-type MOSFET. (Label the leads.)

5. Referring to Figure 5-3a, draw an *N*-channel depletion-type MOSFET with proper biasing. (Indicate polarities.)

6. Referring to Exercise 5, draw a *P*-channel depletion-type MOSFET, but reverse the polarity of voltages. (Indicate polarities.)

7. Draw a sample of the output characteristic curves (I_D versus V_{DS}) for an *N*-channel depletion-type MOSFET, showing different values of V_{GS}.

8. Draw a sample of the output versus characteristic curves (I_D versus V_{DS}) for a *P*-channel depletion-type MOSFET, showing different values of V_{GS}.

9. On the following circuits, indicate if the current I_D will be maximum, medium, or minimum.

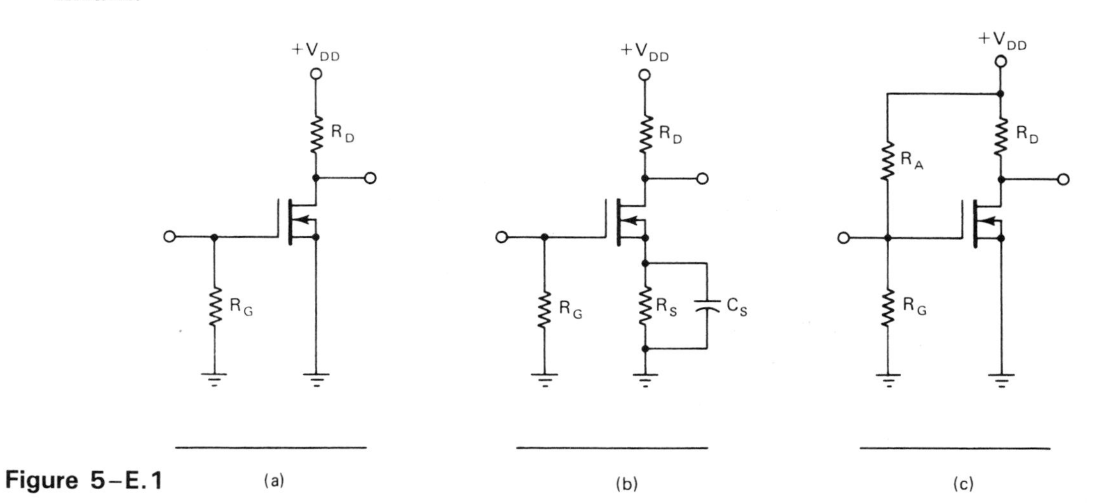

Figure 5–E.1 (a) (b) (c)

10. On the following circuits, calculate I_D and V_{GS} using the formulas $I_D = (+V_{DD} - V_D)/R_D$, $V_{GS} = V_G - V_S$, and $V_S = I_D R_S$.

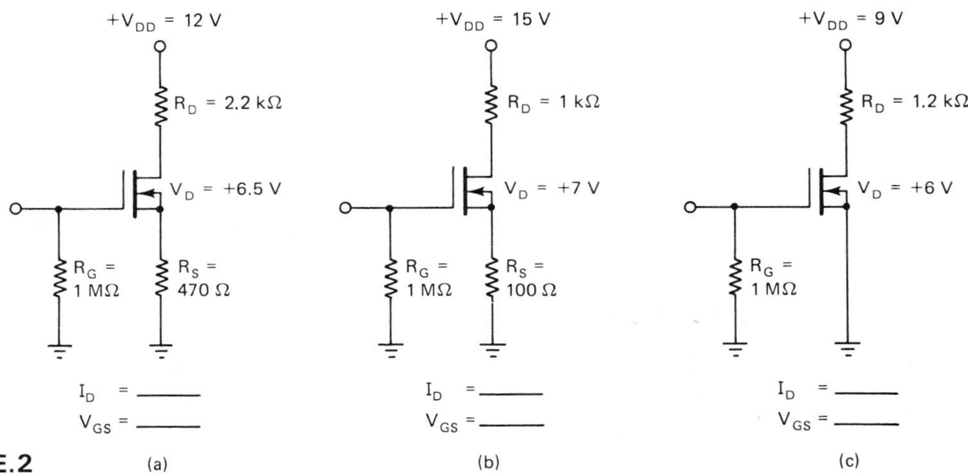

Figure 5–E.2 (a) (b) (c)

11. Draw the structure of an *N*-channel enhancement-type MOSFET. (Label all parts.)

12. Draw the schematic symbol of an *N*-channel enhancement-type MOSFET. (Label all leads.)

13. Referring to Figure 5-7a, draw an *N*-channel enhancement-type MOSFET with proper biasing. (Indicate polarities.)

14. Draw a sample of the output characteristic curves (I_D versus V_{DS}) for an *N*-channel enhancement-type MOSFET, showing different values of V_{GS}.

15. Draw the structure of a *P*-channel enhancement-type MOSFET. (Label all parts.)

16. Draw the schematic symbol of a *P*-channel enhancement-type MOSFET. (Label all leads.)

17. Referring to Exercise 13, draw a *P*-channel enhancement-type MOSFET, but reverse the polarity of voltages. (Indicate polarities.)

18. Draw a sample of the output characteristic curves (I_D versus V_{DS}) for a *P*-channel enhancement-type MOSFET, showing different values of V_{GS}.

19. On the following circuits, indicate if the current I_D will be maximum, medium, or minimum. (*Hint:* Find V_G.)

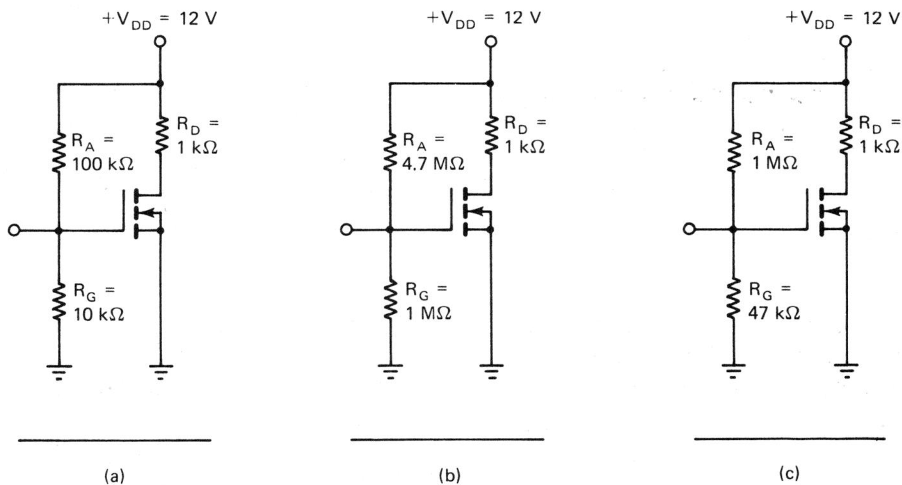

Figure 5–E.3 (a) (b) (c)

20. From the following list, write the correct description under each type of MOSFET.

1. Operates in the enhancement region only.
2. Has the highest input impedance.
3. "Normally on" device.
4. Operates in the depletion and enhancement regions.
5. Usually has less complicated biasing circuits.
6. "Normally off" device.
7. A channel must be created.
8. Has a channel from source to drain.

Depletion-type MOSFET	*Enhancement-type MOSFET*

SECTION 5-4 EXPERIMENTS

CAUTION: All MOSFET devices should be handled with care, even if they have diode gate protection. Follow the safety procedures given in Section 5-11.

EXPERIMENT 1. OPERATION OF A DEPLETION-TYPE MOSFET

Objective:

To show how a depletion-type MOSFET is a "normally on" device, and how it operates in the depletion and enhancement region.

Introduction:

This experiment uses a dual-gate, diode-protected, *N*-channel, depletion-type MOSFET. The gates are connected together to form a single gate. When $V_{GS} = 0$ V, the MOSFET is conducting (on) and the power supply voltage ($+V_{DD}$) divides proportionately across R_D and the MOSFET. When V_{GS} goes positive, more current flows through the MOSFET. Its effective resistance decreases; therefore, the voltage drop across it also decreases. It is operating in the enhancement region. If V_{GS} is made negative, less current flows through the MOSFET. Its effective resistance increases and so does the voltage drop across it. It is operating in the depletion region. When V_{GS} is made sufficiently negative to cut off the current flow through the MOSFET, the voltage at the drain equals $+V_{DD}$. The MOSFET is now considered off. Resistors R_A and R_B form a voltage divider to develop the +3 V used for V_{GS}.

Materials Needed:

1 Adjustable dual ±12-V power supply
1 Standard or digital voltmeter
1 1-kΩ resistor at 0.5 W (R_D)
1 3.3-kΩ resistor at 0.5 W (R_B)
1 10-kΩ resistor at 0.5 W (R_A)
1 1-MΩ resistor at 0.5 W (R_G)
1 40673 MOSFET or equivalent
1 Triple-pole, single-throw (TPST) switch (S_1)
1 Breadboard for constructing circuit

Procedure:

1. Construct the circuit shown in Figure 5-12a.
2. Place S_1 in position A.
3. Measure V_G and record the value in the data table next to A in Figure 5-12b.
4. Measure V_D and record value in the data table (in the same row).
5. Indicate in the data table if the MOSFET is on or off (in the same row).
6. Place S_1 in position B.
7. Measure and record V_G and V_D in their proper places in the data table. Indicate in its proper place in the data table if the MOSFET is on or off.
8. Place S_1 in position C.
9. Measure and record V_G and V_D in their proper places in the data table. Indicate in its proper place in the data table if the MOSFET is on or off.

Fill-in Questions

1. When $V_{GS} = 0$ V, the depletion-type MOSFET is ____________ .

2. When V_{GS} goes positive, the MOSFET conducts ____________ and is operating in the ____________ region.

3. When V_{GS} goes negative, the MOSFET conducts ____________ and is operating in the ____________ region.

4. When V_D is equal to $+V_{DD}$, the MOSFET is ____________ .

Figure 5–12 Operation of an *N*-channel depletion-type MOSFET: (a) test circuit; (b) data table.

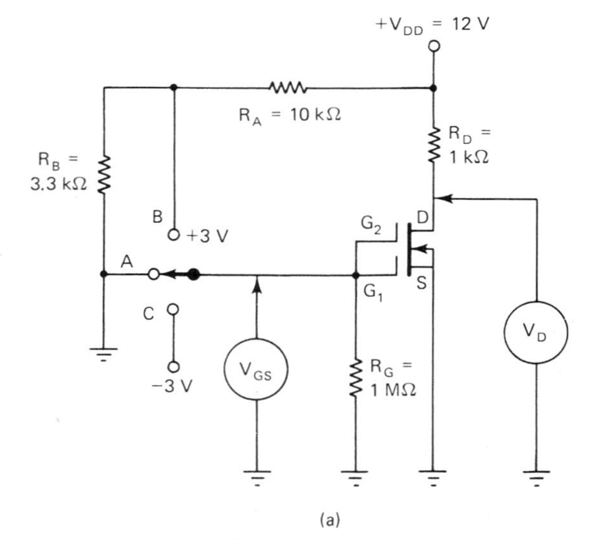

S_1 position	V_{GS}	V_D	Condition of MOSFET (on or off)
A			
B			
C			

(b)

EXPERIMENT 2. CURRENT–VOLTAGE CHARACTERISTICS OF A DEPLETION-TYPE MOSFET

Objective:

To demonstrate how a positive and negative V_{GS} influence I_D more than V_{DS}, and to construct a graph of output characteristic curves to understand better the operation of a depletion-type MOSFET.

Introduction:

This experiment requires a positive variable voltage for V_{DS} and a variable positive and negative voltage for V_{GS}. First, V_{GS} is set to a given value; then V_{DS} is increased to various voltages given in the data table of Figure 5-13b, as I_D is measured and recorded. This procedure is repeated for various values of V_{GS} until the data table is complete. The data are then plotted on the graph to give a more visual indication of the relationships of V_{GS} and V_{DS} to I_D.

Materials Needed:

1 Adjustable dual ±20-V power supply
1 Standard or digital ammeter
1 Standard or digital voltmeter

Figure 5–13 Current–voltage characteristics of a depletion-type MOSFET: (a) test circuit; (b) data table; (c) graph for constructing output characteristic curves.

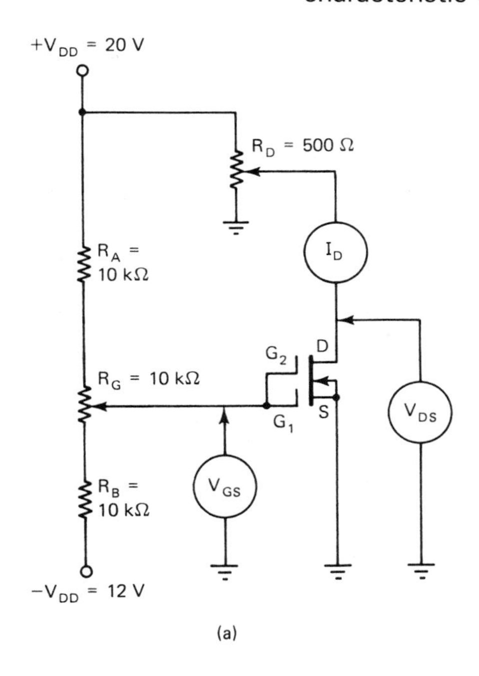

(a)

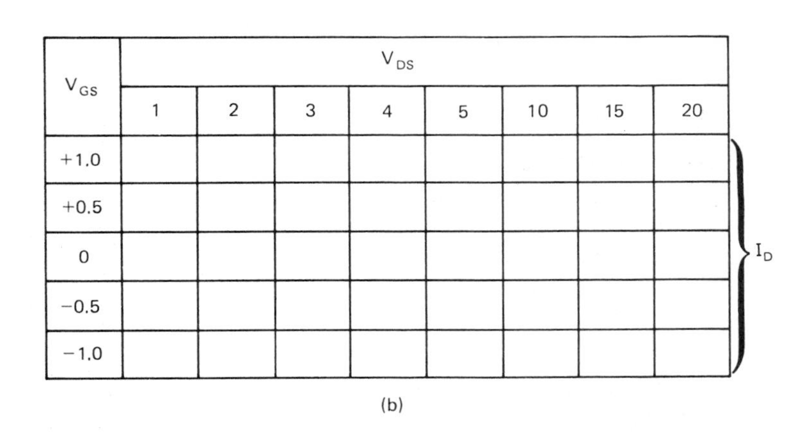

V_{GS}	V_{DS}							
	1	2	3	4	5	10	15	20
+1.0								
+0.5								
0								
−0.5								
−1.0								

(I_D)

(b)

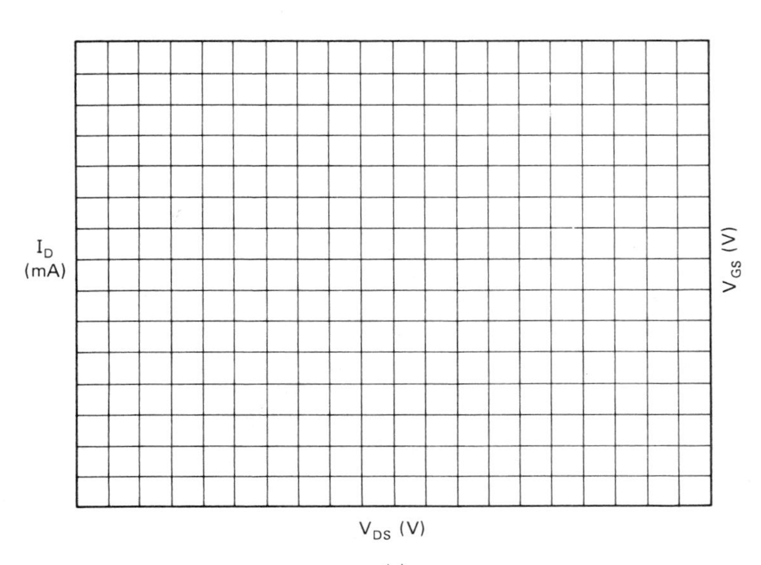

(c)

1 40673 MOSFET or equivalent
2 10-kΩ resistors at 0.5 W (R_A and R_B)
1 500-Ω potentiometer (R_D)
1 10-kΩ potentiometer (R_G)
1 Breadboard for constructing circuit

(*Note:* Zener diodes not shown are part of this MOSFET.)

Procedure:

1. Construct the circuit shown in Figure 5-13a.
2. Adjust R_G for a V_{GS} reading of +1 V.
3. Adjust R_D for a V_{DS} reading of +1 V.
4. Measure I_D and record the value in the proper location of the data table.
5. Increase V_{DS} to the next value given in the data table.
6. Measure I_D and record the value in the proper location of the data table.
7. Continue this procedure of adjusting V_{GS} and V_{DS} while measuring and recording I_D for all the values given in the data table.
8. When the data table is complete, arrange the data on the graph to construct the I–V characteristic curves (refer to Figure 5-3c).
   a. Using the first row of data, find the intersections of V_{DS} and the recorded I_D and mark dots horizontally along the graph. Next, connect the dots with a straight line. At the end of the line to the right, record the value of V_{GS} for the line of data.
   b. Follow this method until all the rows of data are displayed on the graph.

Fill-in Questions:

1. The graph shows the relationship of ____________, ____________, and ____________.

2. When V_{DS} increases and there is little or no change in ____________, the MOSFET is operating in the ________ region.

3. Gate voltage has more control of ____________ than does ____________.

4. From the graph, when V_{GS} = 0 V and V_{DS} = +10 V, then I_D = ______ mA.

5. Since no gate current flows and V_{GS} controls I_D, the MOSFET is a ____________ controlled device.

EXPERIMENT 3. OPERATION OF AN ENHANCEMENT-TYPE MOSFET

Objective:

To show how an enhancement-type MOSFET is a "normally off" device, and how it operates in the enhancement region.

Introduction:

When V_{GS} = 0 V, the MOSFET is cut off and no current is flowing through it. The voltage at the drain (V_D) will equal $+V_{DD}$. When V_{GS} goes positive, the MOSFET turns on and current flows from source to drain, through R_D to $+V_{DD}$. The power supply voltage ($+V_{DD}$) will divide proportionately across R_D and the MOSFET, depending on the amount of forward bias of V_{GS}. The MOSFET is operating in the enhancement region. (*Note:* this MOSFET has internal zener-diode gate protection.)

Materials Needed:

1 Fixed +12-V power supply
1 Standard or digital voltmeter
1 1-kΩ resistor at 0.5 W (R_D)
1 4.7-MΩ resistor at 0.5 W (R_G)
1 10-MΩ resistor at 0.5 W (R_A)
1 VN10KM or VN67AF MOSFET or equivalent
1 DPST switch (S_1)
1 Breadboard for constructing circuit

Procedure:

1. Construct the circuit shown in Figure 5-14a.
2. Place S_1 in position A.

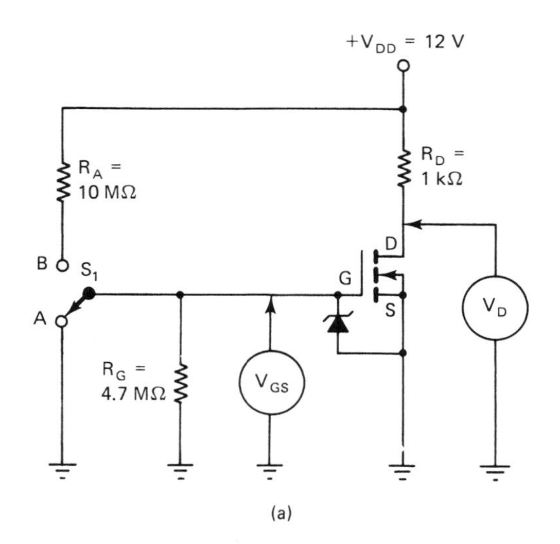

S_1 position	V_{GS}	V_D	Condition of MOSFET (on or off)
A			
B			

(b)

Figure 5–14 Operation of an *N*-channel enhancement-type MOSFET: (a) test circuit; (b) data table.

3. Measure V_{GS} and record the value in the data table next to A in Figure 5-14b.
4. Measure V_D and record the value in the data table (in the same row).
5. Indicate in the data table if the MOSFET is on or off (in the same row).
6. Place S_1 in position B.
7. Measure and record V_{GS} and V_D in their proper place in the data table. Indicate in its proper place in the data table if the MOSFET is on or off.

Fill-in questions:

1. When $V_{GS} = 0$ V, the enhancement-type MOSFET is ____________ .

2. When V_{GS} goes positive, this MOSFET ____________ and is operating in the ____________ region.

3. When V_D is equal to $+V_{DD}$, the MOSFET is ____________ .

4. When the MOSFET is on, voltage at the ____________ will be less than ____________ .

EXPERIMENT 4. CURRENT–VOLTAGE CHARACTERISTICS OF AN ENHANCEMENT-TYPE MOSFET

Objective:

To demonstrate how forward bias voltage V_{GS} controls I_D more than V_{DS}, and to construct a graph of output characteristic curves to understand better the operation of an enhancement-type MOSFET.

Introduction:

As in previous experiments, this one involves setting the bias voltage V_{GS} to a specific value and then increasing V_{DS} while recording the various values of I_D. The data are then plotted on a graph to give a more visual indication of how forward bias turns on the MOSFET and controls I_D, whereas V_{DS} has little or no control.

Materials Needed:

1 Fixed +20-V power supply
1 Standard or digital ammeter
1 Standard or digital voltmeter
1 VN10KM or VN67AF MOSFET or equivalent
1 4.7-MΩ resistor at 0.5 W (R_A)
1 100-Ω potentiometer at 5 W (R_D)
1 1-MΩ potentiometer (R_G)
1 Breadboard for constructing circuit

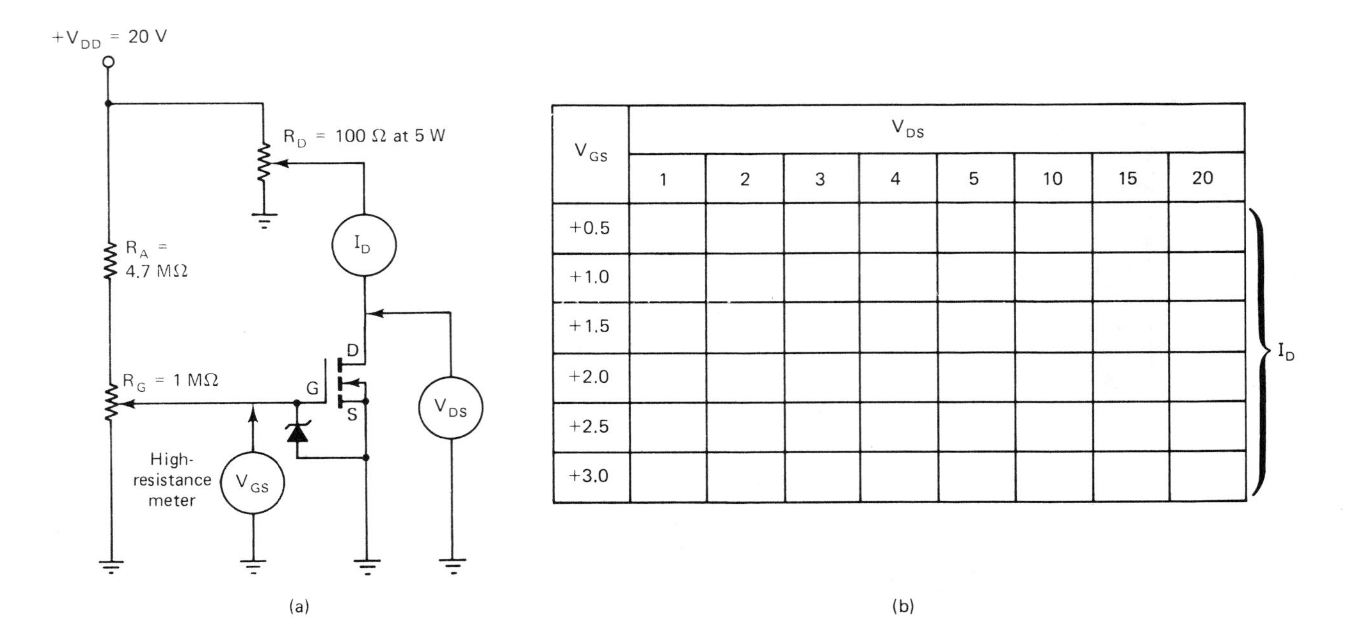

V_{GS}	V_{DS}							
	1	2	3	4	5	10	15	20
+0.5								
+1.0								
+1.5								
+2.0								
+2.5								
+3.0								

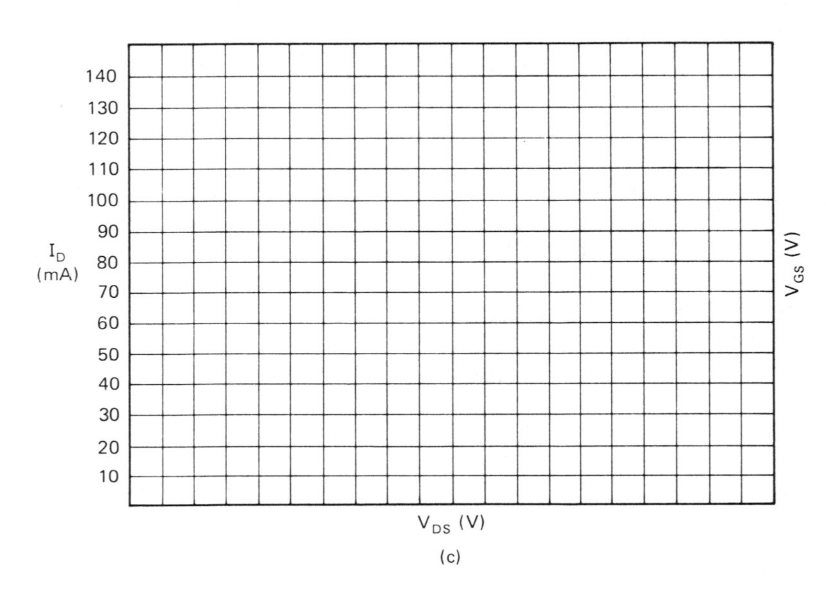

Figure 5–15 Current–voltage characteristics of an enhancement-type MOSFET: (a) test circuit; (b) data table; (c) graph for constructing output characteristic curves.

Procedure:

1. Construct the circuit shown in Figure 5-15a.
2. Adjust R_G for a V_{GS} reading of +0.5 V.
3. Adjust R_D for reading of +1 V for V_{DS}.
4. Measure I_D and record the value in the proper location of the data table.
5. Increase V_{DS} to the next value given in the data table.
6. Measure I_D and record the value in the proper location of the data table.
7. Continue this procedure of adjusting V_{GS} and V_{DS} while measuring and recording I_D for all the values given in the data table.
8. When the data table is complete, arrange the data on the graph to construct the I–V characteristic curves (refer to Figure 5-7c).
   a. Using the first row of data, find the intersections of V_{DS} and the recorded I_D and mark dots horizontally along the graph. Next, connect the dots with a straight line. At the end of the line to the right, record the value of V_{GS} for the line of data.

b. Follow this same method until all the rows of data are displayed on the graph.

Fill-in Questions:

1. The graph shows the relationship of ______________, ______________, and ______________.

2. From the graph, when $V_{GS} = +2$ V and $V_{DS} = +12$ V, then $I_D =$ ______ mA.

3. When $V_{GS} = 0$ V, there is no ______________ region between source and drain for an enhancement-type MOSFET.

4. Current I_D is controlled less by ______________ than by ______________.

5. Since no gate current flows and V_{GS} controls I_D, the MOSFET is a voltage-______________ ______________.

EXPERIMENT 5. DUAL-GATE DEPLETION-TYPE MOSFET AMPLIFIER

Objective:

To show how a depletion-type MOSFET is used in a common-source amplifier, and how a dc bias voltage on one gate can affect the gain of the amplifier with an input signal placed on the other gate.

Introduction:

A circuit such as this can be used in radio circuits for automatic gain control (AGC) where a controlled dc bias voltage on one gate can keep the gain of an amplifier at a relatively constant level. It can also be used as a frequency mixer, where one frequency is placed on one gate and a second frequency is placed on the other gate.

Materials Needed:

1 Adjustable dual ±12-V power supply
1 Standard or digital voltmeter
1 Oscilloscope (dual trace preferred)
1 Signal generator (100 Hz to 1 MHz)
1 40673 MOSFET or equivalent
1 1-kΩ resistor at 0.5 W (R_D)
1 1-MΩ resistor at 0.5 W (R_G)
1 10-kΩ potentiometer (R_A)
1 0.01-μF capacitor (C_{G2})
2 0.1-μF capacitors ($C_{G1}$and C_D)
1 Breadboard for constructing circuit

Procedure:

1. Construct the circuit shown in Figure 5-16a.
2. Set R_A to ground or $V_{G2} = 0$ V.
3. Before connecting the signal generator, measure and record V_D, V_{G1}, V_{G2}, and V_S in their proper places on the figure.
4. Connect the signal, generator, and, using the oscilloscope, set it for 1 MHz at 1 V p-p.
5. Place the oscilloscope at v_{out} and measure the peak-to-peak signal voltage. Record this in the proper place in Table 1.
6. Calculate the gain of the amplifier for these readings and place the result in the proper place in Table 1.
7. Adjust R_A so that $V_{G2} = -0.25$ V.
8. Measure v_{out} and record the peak-to-peak signal in its proper place in Table 1.
9. Calculate the gain of the amplifier for these readings and place the result in the proper place in Table 1.
10. Repeat steps 7 through 9 for the remaining values of V_{G2} given in Table 1.
11. Remove power from the circuit.
12. Switch the circuits on the two gates, where the signal generator C_{G1} and R_G are connected to G_2, and R_A and C_{G2} are connected to G_1.
13. Apply power to the circuit.
14. Using similar methods as with Table 1, complete the data and calculations for Table 2.

Fill-in questions:

1. A dc voltage placed on one gate of a dual-gate MOSFET will control the

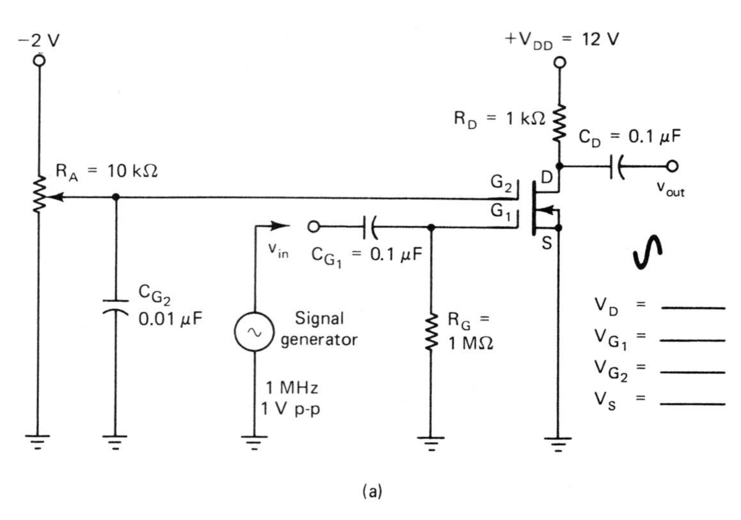

Table 1

V_{G_2}	V_{out} (V p-p)	$A_V = \frac{V_{out}}{V_{in}}$
0		
−0.25		
−0.5		
−0.75		
−1.0		

(b)

Table 2

V_{G_1}	V_{out} (V p-p)	$A_V = \frac{V_{out}}{V_{in}}$
0		
−0.25		
−0.5		
−0.75		
−1.0		

(c)

Figure 5–16 Dual-gate depletion-type MOSFET amplifier: (a) schematic diagram: (b) and (c) data tables.

signal ____________ when a signal is placed on the other gate.

2. When the dc bias voltage goes more negative for this circuit, the gain ____________ .

3. When the dc bias voltage goes less negative for this circuit, the gain ____________ .

4. When $V_{G1} = 0$ V, the gain of the amplifier is ______ .

5. When $V_{G2} = 0$ V, the gain of the amplifier is ______ .

EXPERIMENT 6. VMOSFET DRIVER AMPLIFIER

Objective:

To demonstrate how a power MOSFET can be used to operate a relay that controls another circuit, and how it provides isolation between the two circuits.

Introduction:

When $V_G = 0$ V, the VMOSFET is off and no current flows. When V_G goes positive, the MOSFET turns on and current flows through the VMOSFET and relay coil. The relay energizes and the relay contacts close, thereby enabling the controlled circuit. There is no electrical connection between the relay coil and its contacts, which provides isolation for the two circuits.

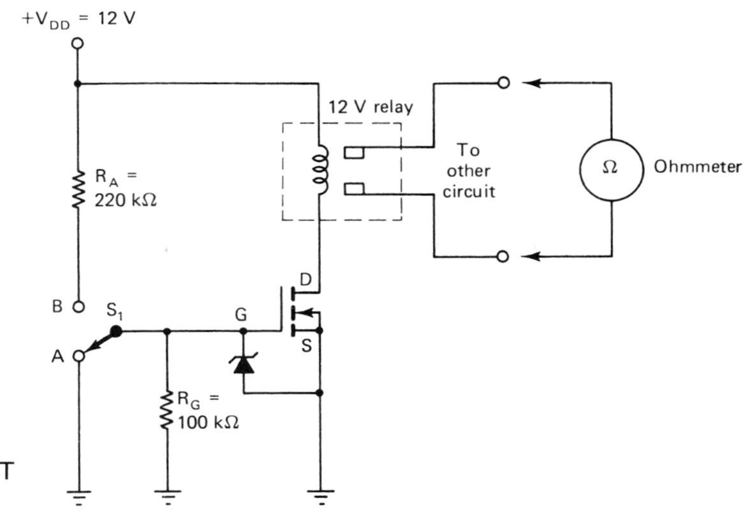

Figure 5-17 VMOSFET driver amplifier.

Materials Needed:

1 Fixed +12-V power supply
1 Standard or digital ohmmeter
1 VN10KM or VN67AF VMOSFET or equivalent
1 12-V relay requiring less than 1 A
1 100-kΩ resistor at 0.5 W (R_G)
1 220-kΩ resistor at 0.5 W (R_A)
1 DPST switch (S_1)

Procedure:

1. Construct the circuit shown in Figure 5-17.
2. Place switch S_1 in position A.
3. Connect the ohmmeter to the relay normally open (NO) contacts.
4. Record the reading of the ohmmeter here: ______ Ω.
5. Place S_1 in position B.
6. Record the reading of the ohmmeter here: ______ Ω.

Fill-in Questions:

1. When V_G = 0 V, the VMOSFET is ____________ and the relay contacts are ____________ .

2. When V_G goes positive, the VMOSFET is ____________ and the relay contacts are ____________ .

3. The relay provides electrical ____________ between the two circuits.

4. The ohmmeter indicates ______ Ω when the VMOSFET is off, and ______ Ω when the VMOSFET is on.

SECTION 5-5 MOSFET BASIC TROUBLESHOOTING APPLICATION: DRIVER AMPLIFIER

Construct the circuit as shown. Open the components as listed and record the voltages in the proper place in the table. Because of the delicate nature of the MOSFET, no opening or shorting of the case leads will be performed. The abbreviations will help indicate the voltage conditions associated with each problem. All voltages are referenced to ground.

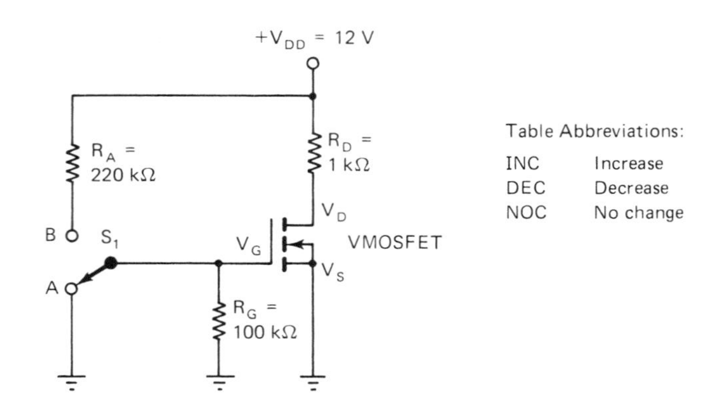

Position of S_1	Condition	V_D	V_G	V_S	Comments
A	Normal				All voltages proper (no input voltage)
A	R_G open	NOC	NOC	NOC	Without input voltage, unable to determine if problem exists
A	R_A open	NOC	NOC	NOC	Same as above
A	R_D open	DEC	NOC	NOC	No $+V_{DD}$ applied to MOSFET
B	Normal				All voltages proper (input voltage present)
B	R_G open	NOC	INC	NOC	V_G high; may cause instability
B	R_A open	INC	DEC	NOC	No input voltage applied to gate
B	R_D open	DEC	NOC	NOC	No $+V_{DD}$ applied to MOSFET

Figure 5–E.4

SECTION 5-6 MORE EXERCISES

Perform the following exercises before beginning Section 5-7.

1. The gate to source voltage (V_{GS}) controls the current flow through a MOSFET. When $V_{GS} = 0$ V, some MOSFETs are "normally on" while others are "normally off." Indicate from the devices given if they are "normally on" or "normally off" and what polarity of V_{GS} is needed to increase and decrease current flow through them respectively.

MOSFET Device	*Normal Condition When $V_{GS} = 0$ V*	*V_{GS} Polarity Needed to Increase Current*	*V_{GS} Polarity Needed to Decrease Current*
N-channel depletion type			
P-channel depletion type			
N-channel enhancement type			
P-channel enhancement type			

2. Draw the transfer characteristic curves for the following types of MOSFETs. (Refer to Figures 5-3b and 5-7b, respectively.)

   a. *N*-channel depletion type   b. *N*-channel enhancement type

3. On the following circuits, calculate the indicated values using the formulas $V_D = +V_{DD} - V_{R_D}$, $I_D = V_{R_D}/R_D$, $V_{R_D} = I_D R_D$, $I_G = +V_{DD}/R_A + R_G$, $V_G = I_G R_G$, $V_S = I_S R_S$, and $V_{GS} = V_G - V_S$.

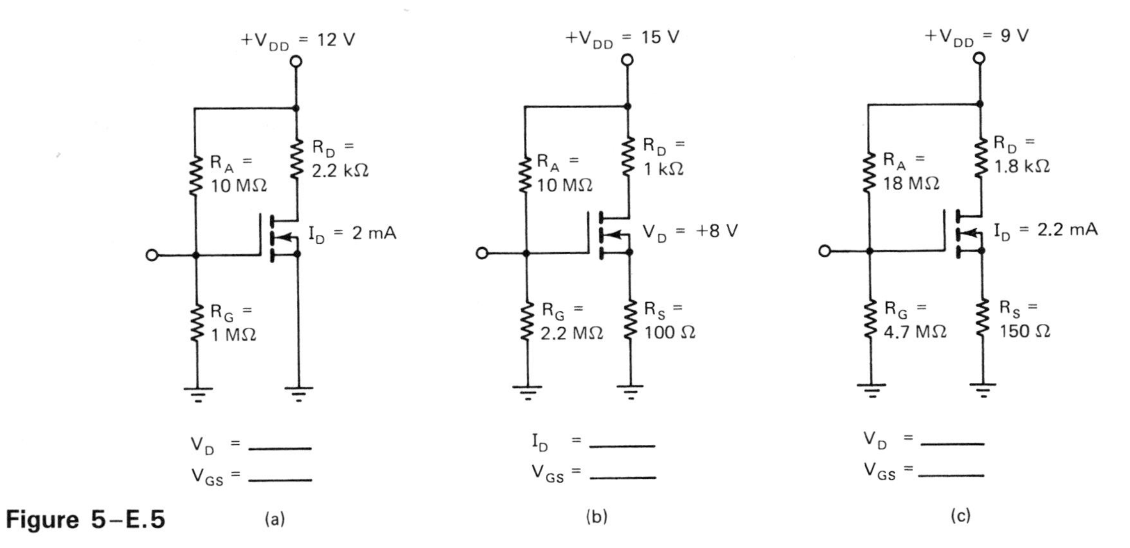

Figure 5–E.5

4. Draw a CMOS circuit with $V_{out} \approx 0$ V and indicate which MOSFET is on and which MOSFET is off. Draw another CMOS circuit with $V_{out} \approx +5$ V and indicate which MOSFET is on and which MOSFET is off. (Refer to Figure 5-10.)

   a. $V_{out} \approx 0$ V   b. $V_{out} \approx +5$ V

5. MOSFETs can be used in the three basic circuit configurations similar to the JFET (see Section 4-6, Exercise 6). Draw and label the circuits for common-source, common-drain, and common-gate amplifiers for the *N*-channel depletion-type MOSFET and the *N*-channel enhancement-type MOSFET (a total of six circuits).

SECTION 5-7 MOSFET INSTANT REVIEW

The MOSFET has a metal gate that is insulated from the channel by a thin layer of silicon dioxide (SiO_2). A bias voltage placed on the gate develops an electrostatic field that cuts into the channel area and controls current flow from source to drain (I_D). This effect indicates that the MOSFET is a voltage-controlled device.

A depletion-type MOSFET is a "normally on" device that operates in the depletion region with reverse bias and in the enhancement region with forward bias.

An enhancement-type MOSFET is a "normally off" device that operates in the enhancement region with forward bias.

SECTION 5-8 SELF-CHECKING QUIZZES

5-8a MOSFET: TRUE-FALSE QUIZ

Place a T for true or an F for false to the left of each statement.

____ **1.** A MOSFET is a bipolar device.

____ **2.** The input impedance of a MOSFET is higher than that of a JFET.

____ **3.** The insulating material of a MOSFET is silicon dioxide.

____ **4.** A depletion-type MOSFET may use forward or reverse bias.

____ **5.** An enhancement-type MOSFET is considered a "normally on" device.

____ **6.** No current flows through a MOSFET when the pinch-off voltage is exceeded.

____ **7.** More care is required in handling MOSFETs than in handling other types of semiconductor devices.

____ **8.** The current, voltage, and lead identification nomenclature is the same for a MOSFET as for a JFET.

____ **9.** For an N-channel MOSFET, when V_{GS} goes negative there will be an increase of electron flow in the channel.

____ **10.** A P-channel MOSFET schematic symbol shows the arrow pointing away from the gate.

5-8b MOSFET: MULTIPLE-CHOICE QUIZ

Circle the correct answers for each question.

1. The primary advantage of MOSFETS over bipolar transistors and JFETs is that they:

a. have a higher gain

b. are physically more rugged

c. have a higher input impedance

d. are larger in size and easier to handle

2. A depletion-type MOSFET can be biased:

a. at zero volts

b. with positive voltage

c. with negative voltage

d. all of the above

e. none of the above

Questions 3 through 5 refer to Figure 5-A.

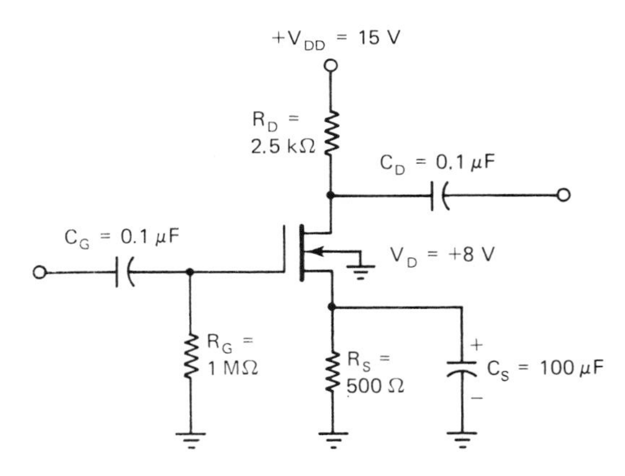

Figure 5-A

3. The circuit shown in Figure 5-A is a:

a. common-gate amplifier

b. common-source amplifier

c. common-drain amplifier

d. none of the above

4. The I_D for this circuit is:

a. 16 mA **b.** 3.2 mA

c. 2.8 mA **d.** 2.3 mA

5. The V_{GS} for this circuit is:

a. +1.6 V **b.** −1.6 V

c. +1.4 V **d.** −1.4 V

6. When V_{GS} goes more positive for an N-channel enhancement-type MOSFET, I_D:

a. decreases **b.** increases

c. remains the same **d.** none of the above

7. Precautions should be used when handling MOSFET, such as:

a. the use of shorting devices and anti-static materials

b. grounding the person's body to the equipment being worked on

c. using a soldering gun

d. only answers a and b are correct

e. answers a, b, and c are correct

Questions 8 through 10 refer to Figure 5-B.

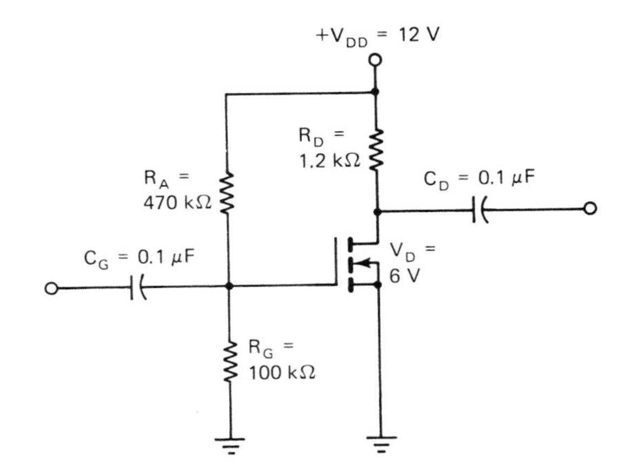

Figure 5-B

8. The I_D for this circuit is:

a. 0 mA **b.** 2 mA

c. 5 mA **d.** 10 mA

9. The V_{GS} for this circuit is:

a. 0 V **b.** +2.1 V

c. +9.9 V **d.** −2.1 V

10. A positive-going signal at the input of this circuit would cause V_D to:

a. increase

b. decrease

c. remain the same

d. not enough information given

ANSWERS TO EXPERIMENTS AND QUIZZES

Experiment 1. **(1)** on **(2)** more, enhancement **(3)** less, depletion **(4)** off

Experiment 2. **(1)** V_{GS}, V_{DS}, I_D (any order) **(2)** I_D, pinch-off **(3)** I_D, V_{DS} **(4)** 4.5 (or your answer) **(5)** voltage

Experiment 3. **(1)** off **(2)** turns on, enhancement **(3)** off **(4)** drain, $+V_{DD}$

Experiment 4. **(1)** V_{GS}, V_{DS}, I_D (any order) **(2)** 5.5 (or your answer) **(3)** channel **(4)** V_{DS}, V_{GS} **(5)** controlled device

Experiment 5. **(1)** gain **(2)** decreases **(3)** increases **(4)** 2.5 (or your answer) **(5)** 2.5 (or your answer)

Experiment 6. **(1)** off, open **(2)** on, closed **(3)** isolation **(4)** infinite, zero

True–False. **(1)** F **(2)** T **(3)** T **(4)** T **(5)** F **(6)** F **(7)** T **(8)** T **(9)** F **(10)** T

Multiple Choice. **(1)** c **(2)** d **(3)** b **(4)** c **(5)** d **(6)** b **(7)** d **(8)** c **(9)** b **(10)** b

Unit 6

Unijunction Transistor

INTRODUCTION The unijunction transistor (UJT) is a device used in triggering and oscillator circuits.

UNIT OBJECTIVES Upon completion of this unit you will be able to:

1. Draw the schematic symbol of a UJT.
2. Describe the structure and operation of the UJT.
3. Define the terms voltage gradients, intrinsic standoff ratio, peak voltage, and valley voltage.
4. Draw a current–voltage characteristic curve for a UJT.
5. Explain what occurs in the negative-resistance region of a UJT.
6. Explain what occurs in the saturation region of a UJT.
7. Show how a UJT functions as an oscillator.
8. Explain the voltage waveforms of a UJT oscillator.
9. Test a UJT with an ohmmeter.
10. Troubleshoot a UJT oscillator circuit.

SECTION 6-1 THEORY OF STRUCTURE AND OPERATION

6-1a STRUCTURE AND SCHEMATIC SYMBOL

The unijunction transistor (UJT) consists of a bar of *N*-type silicon material to which a small amount of *P*-type material has been diffused, as shown in Figure 6-1. Terminals are connected to each end of the *N*-material and labeled base 1 (B_1) and base 2 (B_2). An emitter terminal (E) is connected to the *P* material and forms the single *PN* junction of the UJT. The bar of *N* material between the base terminals has a specific amount of intrinsic resistance, which is termed r_{BB}. The intrinsic resistance from B_1 to E is called r_{B1} and the intrinsic resistance from E to B_2 is labeled r_{B2}. Therefore, $r_{BB} = r_{B_1} + r_{B2}$.

The ratio of r_{B_1} to r_{BB} is called the *intrinsic standoff ratio,* represented by the Greek letter η (eta) and is used to determine the *firing point* (the point when the UJT turns on). It can be expressed mathematically as

$$\eta = \frac{r_{B1}}{r_{B1} + r_{B2}} = \frac{r_{B1}}{r_{BB}}$$

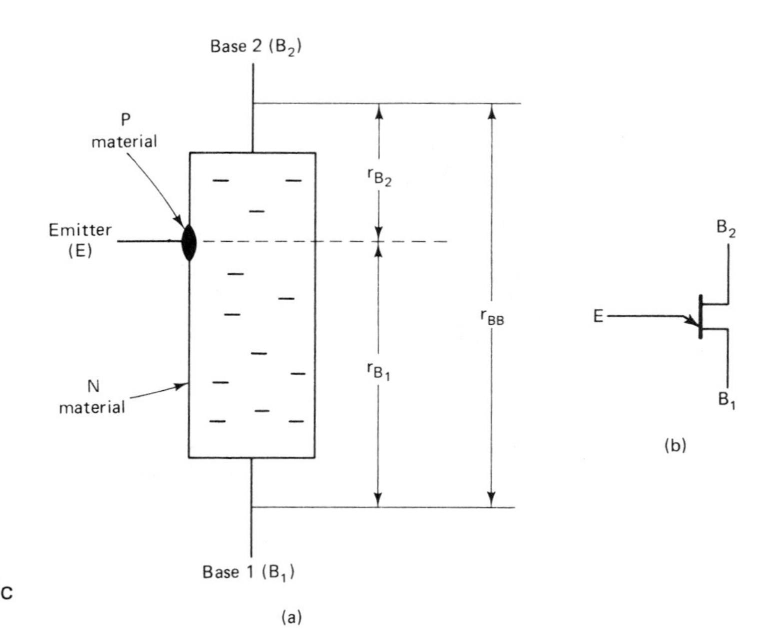

Figure 6–1 Unijunction transistor: (a) structure; (b) schematic symbol.

Typical values of η will range from approximately 0.5 to 0.8. An *N*-type UJT has the arrow pointing toward the schematic symbol. Rarely used are *P*-type UJTs, which have the arrow pointing away from the symbol.

Figure 6–2 UJT operation: (a) voltage gradients; (b) equivalent circuit; (c) before firing; (d) after firing.

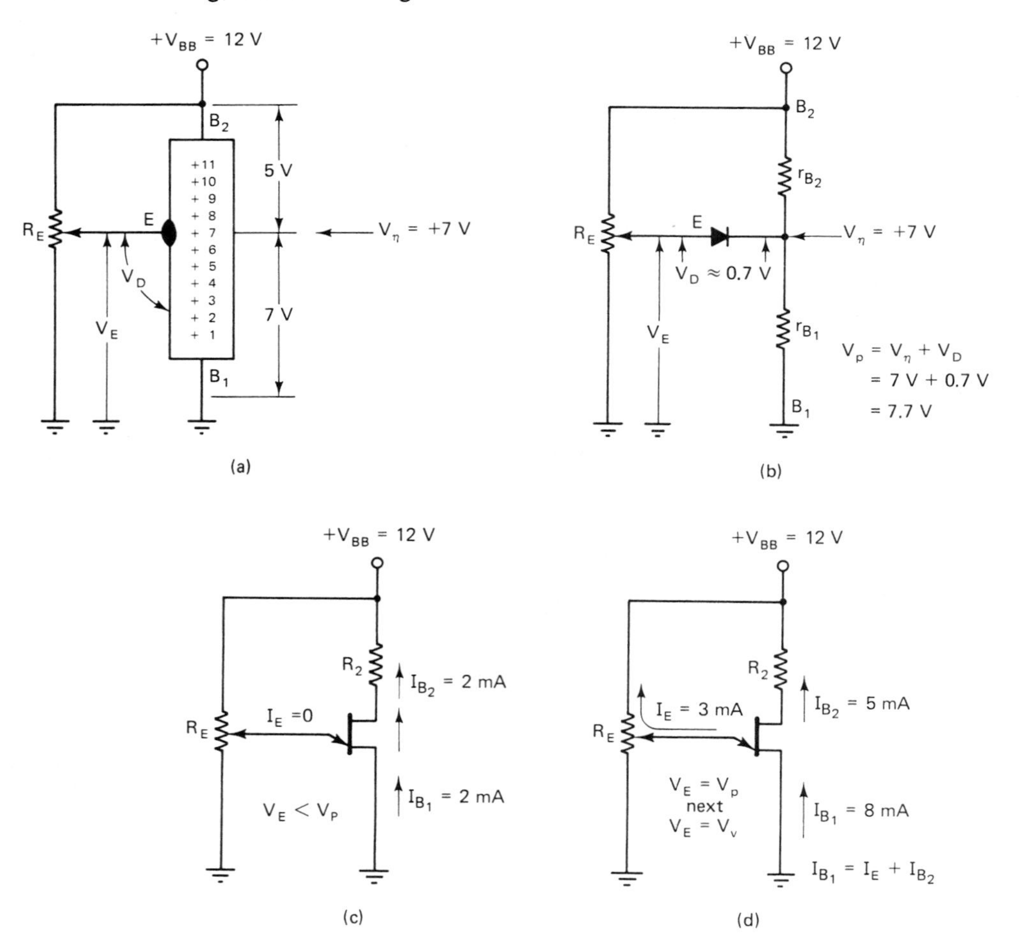

6-1b OPERATION OF A UJT

When a positive voltage is connected to B_2 with the negative or ground connected to B_1, a voltage gradient exists across the *N*-type bar of the UJT, as seen in Figure 6-2. A small current flows from B_1 to B_2. The emitter is connected to the wiper of potentiometer R_E and set at ground. The point opposite the *P* material (V_η) is at +7 V. At this time the *PN* junction is reverse biased and only leakage current flows through it. As the wiper is moved up R_E, the *P* material or anode of the *PN* junction becomes more positive and is referred to as the emitter voltage (V_E). When the V_E exceeds the value of V_η plus the potential barrier (V_D) of the junction, the junction conducts and current flows out the emitter. The current also increases through the *N* material from B_1 to B_2. Unlike its equivalent circuit shown in Figure 6-2b, the r_{B1} of the UJT is not fixed, but decreases when the emitter conducts or fires. This causes V_E to decrease to a minimum value, even though the emitter current (I_E) increases. This condition is opposite to Ohm's law theory, where *I* is proportional to *V*; therefore, the UJT is said to be in the negative-resistance region. As long as the *PN* junction is forward biased, maximum current will flow through the UJT. The UJT can be turned off by reducing V_E below the minimum voltage, which keeps the *PN* junction forward biased.

6-1c CURRENT–VOLTAGE CHARACTERISTICS OF A UJT

The current and voltage relationships of a UJT are shown in the circuit of Figure 6-3a. When V_E reaches the point at which the UJT fires, I_E increases, I_{B1} increases, and I_{B2} increases. The voltage changes are different; V_E decreases because of the negative resistance, V_{B1} increases because I_{B1} increases; and V_{B2} decreases because the total resistance from B_2 to ground decreases.

Figure 6–3 UJT current–voltage characteristics: (a) *I–V* nomenclature; (b) emitter characteristic curve.

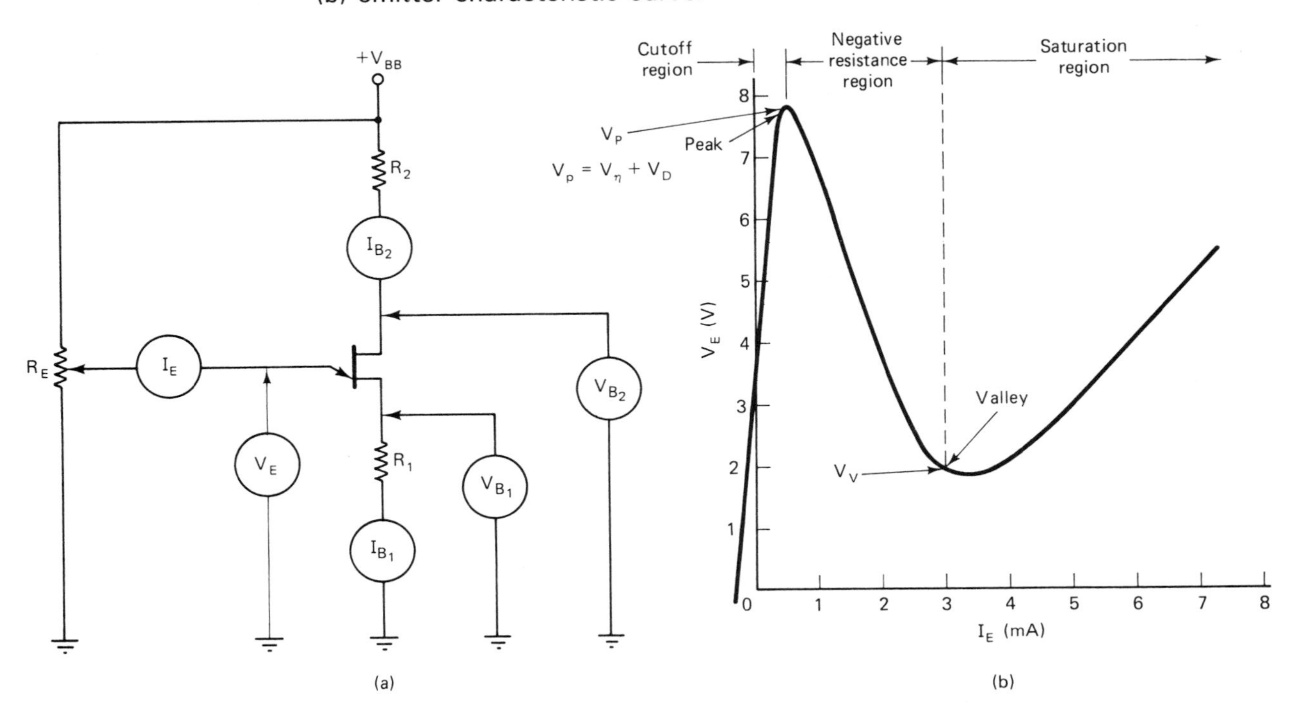

The current-voltage characteristic curve for a UJT is usually based on its emitter operation, as shown in Figure 6-3b. The point just before the UJT fires is referred to as the peak voltage (V_p) of the emitter. The area before this is the cutoff region. The minimum voltage on the emitter after the UJT fires is the valley voltage (V_v). The area between V_p and V_v is the negative-resistance region. If more current is allowed to flow through the emitter after the UJT fires, it is said to be in the saturation region.

6-1d UJT RELAXATION OSCILLATOR

Since the UJT acts as a switch, it is generally used in oscillators and circuits that develop trigger pulses for other circuits. When a series *RC* circuit is connected to a positive voltage, as shown in Figure 6-4a, the capacitor begins to charge up, exponentially, toward the applied voltage. If the switch S_1 is closed before the capacitor is completely charged, as shown in Figure 6-4b, it will discharge back through the switch. The voltage across the capacitor during the charge time is rather long, depending on the time constant of R_E and C_E, whereas the voltage across the capacitor during the discharge time is very short because the switch has negligible resistance. Hence a sawtooth voltage waveform is produced across the capacitor, as shown in Figure 6-4c.

Figure 6–4 UJT relaxation oscillator: (a) and (b) *RC* circuit with switch; (c) voltage waveform across C_E; (d) oscillator with waveform.

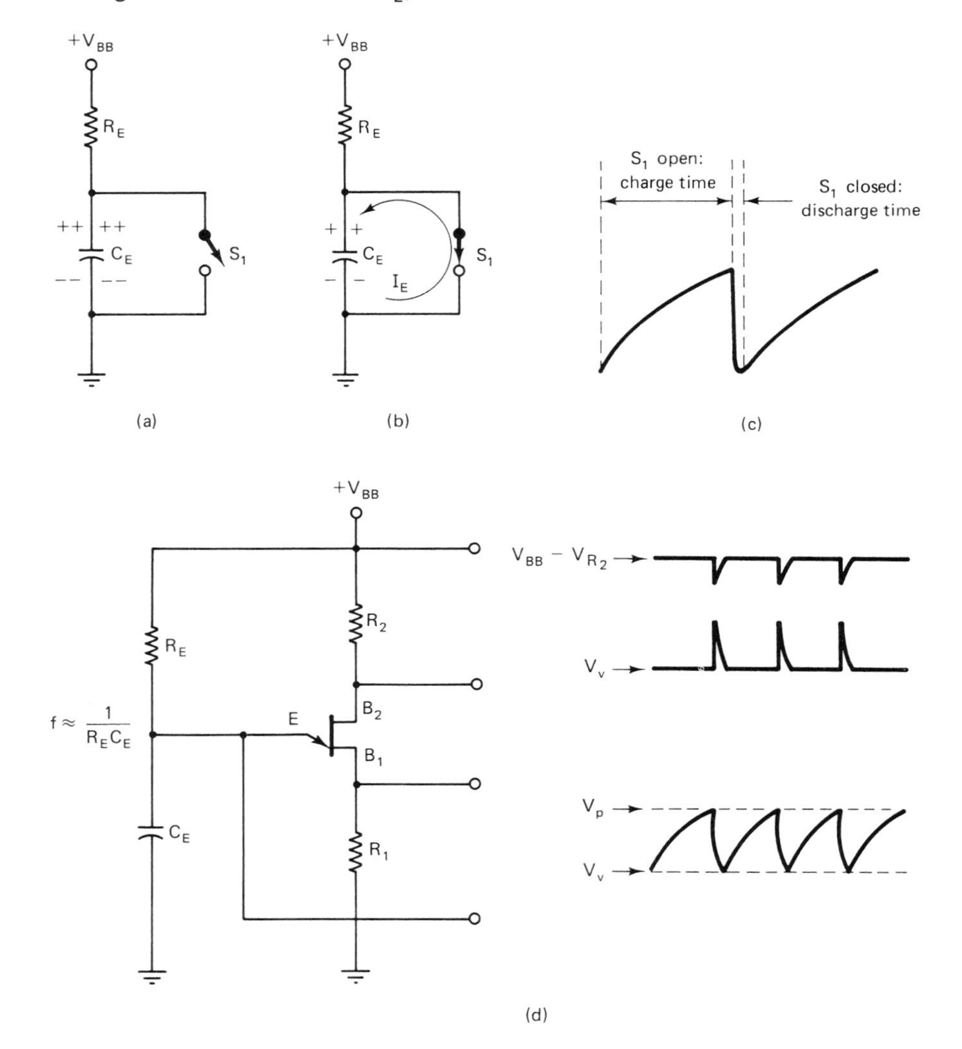

If the switch is replaced by a UJT, a relaxation oscillator is created, as shown in Figure 6-4d. When the supply voltage is applied to the circuit, C_E begins to charge via R_E. The UJT is off, acting like an open switch with only minimum current flowing through the circuit (R_1, UJT, and R_2). Resistances R_1 and R_2 are usually of low value; therefore, the voltage at B_1 will be minimum, but the voltage at B_2 will be maximum ($V_{BB} - V_{R2}$). The voltage on C_E, which is applied to E, continues to increase until V_p is reached. At this time, the UJT fires and acts like a closed switch. Current flows out of E and C_E discharges rapidly back to ground, through R_1 and the UJT. A sharp leading-edge pulse going positive occurs at B_1 because of the discharge current. Also, a sharp leading-edge pulse going negative occurs at B_2, because the resistance of the UJT has decreased. When the voltage at E discharges below V_v, the UJT turns off and C_E begins to charge again to produce the next cycle.

The amount of pulses produced per second is referred to as the pulse repetition rate or generally the frequency of the oscillator. The approximate frequency can be calculated from the formula

$$f \approx \frac{1}{R_E C_E}$$

6-1e UJT DEFINITIONS

V_{BB}	power supply voltage
$V_{B_1B_2}$	voltage across base 1 and base 2 (the interbase voltage)
V_p	peak (maximum) voltage at the emitter before the UJT fires
V_D	forward voltage drop of the *PN* junction
V_v	valley (minimum) voltage at the emitter after the UJT fires
I_{B1}	current flowing into base 1
I_{B2}	current flowing out of base 2
I_E	current flowing out of the emitter
I_{EO}	emitter leakage current when the *PN* junction is reverse biased
I_p	peak (maximum) current at the emitter before the UJT fires
I_V	valley (minimum) current at the emitter after the UJT fires
r_{BB}	interbase resistance between base 1 and base 2
r_{B1}	intrinsic resistance from base 1 to emitter
r_{B2}	intrinsic resistance from base 2 to emitter
η	intrinsic standoff ratio, defined as the base resistance voltage divider ratio, which helps to determine the emitter firing voltage, $\eta = r_{B1}/r_{BB}$

SECTION 6-2 UJT DEFINITION EXERCISE

Refer to the previous sections and write a brief definition of each term.

1. V_{BB} ____

2. $V_{B_1B_2}$ ______

3. V_P ______

4. V_D ______

5. V_V ______

6. I_{B_1} ______

7. I_{B_2} ______

8. I_E ______

9. I_{EO} ______

10. I_P ______

11. I_V ______

12. r_{BB}

13. r_{B1}

14. r_{B2}

15. η (eta)

SECTION 6-3 EXERCISES AND PROBLEMS

Perform the following exercises before beginning Section 6-4.

1. Draw the structure of an N-type UJT. (Label all parts.)

2. Draw the schematic symbol of an N-type UJT. (Label all parts.)

3. Draw a sample V_E versus I_E characteristic curve of an N-type UJT. (Indicate V_P, V_V, cutoff, negative resistance, and saturated regions.)

4. Draw the circuit in Figure 6-2c and indicate with arrows the electron current paths through the UJT before firing. (Label all currents.)

5. Draw the circuit in Figure 6-2d and indicate with arrows the electron current paths through the UJT after firing. (Label all currents.)

6. Draw the circuit shown in Figure 6-3a and write in the voltage and current nomenclature (V_E, V_{B1}, V_{B2}, I_E, I_{B1}, and I_{B2}) indicated by the meters shown in the circuit.

7. Refer to the circuit in Exercise 6 and indicate any change in the voltages and currents after the UJT fires. (Use "increases," "decreases," or "remains the same.")

V_E ________ I_E ________

V_{B1} ________ I_{B1} ________

V_{B2} ________ I_{B2} ________

8. Find the value of the intrinsic standoff ratio for the data given. Use the formula

$$\eta = \frac{r_{B1}}{r_{B1} + r_{B2}} \quad \text{or} \quad \frac{r_{B1}}{r_{BB}}$$

a. $r_{BB} = 9\ \text{k}\Omega$
$r_{B1} = 6.75\ \text{k}\Omega$

$\eta =$ ______

b. $r_{BB} = 20\ \text{k}\Omega$
$r_{B1} = 12\ \text{k}\Omega$

$\eta =$ ______

c. $r_{BB} = 15\ \text{k}\Omega$
$r_{B1} = 8.4\ \text{k}\Omega$

$\eta =$ ______

9. Find the value of V_P for the data given. Use the formula $V_P = \eta V_{BB} + V_D$, where $V_D = +0.7$ V.

a. $\eta = 0.65$
$V_{BB} = +10$ V

$V_P =$ ______

b. $\eta = 0.72$
$V_{BB} = +15$ V

$V_P =$ ______

c. $\eta = 0.63$
$V_{BB} = +12$ V

$V_P =$ ______

10. Find the value of η for the data given. Use the formula $\eta = (V_P - V_D)/V_{BB}$, where $V_D = +0.7$ V.

a. $V_{BB} = +9$ V
$V_P = +6.7$ V

$\eta =$ ______

b. $V_{BB} = +12$ V
$V_P = +9.3$ V

$\eta =$ ______

c. $V_{BB} = +15$ V
$V_P = +10.45$ V

$\eta =$ ______

SECTION 6-4 EXPERIMENTS

EXPERIMENT 1. TESTING A UJT

Objective:

To demonstrate a practical go/no go method of testing a UJT with an ohmmeter. This is called a go/no go test.

Introduction:

The *PN* junction of a UJT can be tested with an ohmmeter similar to testing diodes and bipolar transistors. With the negative lead placed on the emitter and the positive lead placed at either base, the junction is reverse biased and the resistance should be high or infinite. When the positive lead is placed on the emitter and the negative lead is placed at either base, the junction is forward biased and the resistance should be low. There should be a resistance reading of several thousand ohms when the meter is placed across the base leads.

Materials Needed:

A standard or digital ohmmeter
One or several UJTs

Procedure:

1. Set the ohmmeter to the midrange scale.
2. Refer to Figure 6-5a to connect the ohmmeter to the UJT properly for each lead and record the readings in the indicated ohmmeter circles.
3. Refer to Figure 6-5b to connect the ohmmeter to the UJT properly for each lead and record the readings in the indicated ohmmeter circles.

Fill-in Questions:

1. A forward-biased *PN* junction should have ____________ resistance.

2. A reverse-biased *PN* junction should have ____________ resistance.

3. A forward-biased *PN* junction with a high ohmmeter reading indicates that the UJT is ____________ .

4. A reverse-biased *PN* junction with a low ohmmeter reading indicates that the UJT is ____________ .

5. The resistance of a UJT from base to base reads the same regardless of the ____________ of the ohmmeter leads.

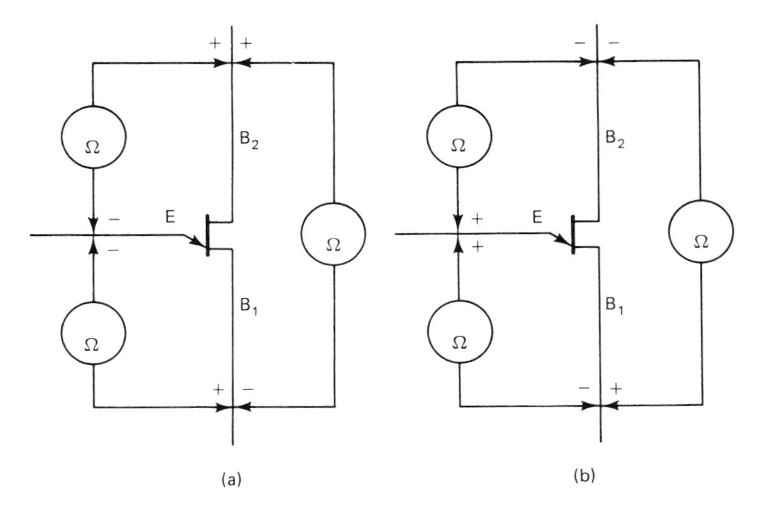

Figure 6–5 Testing a UJT with an ohmmeter.

EXPERIMENT 2. FINDING V_p, V_v AND η OF A UJT

Objective:

To show a practical method for finding V_p and V_v and for determining the intrinsic standoff ratio (η) of a UJT.

Introduction:

The V_p or firing point of a UJT can be found by connecting a voltmeter to the emitter as shown in Figure 6-6a. The voltage on the

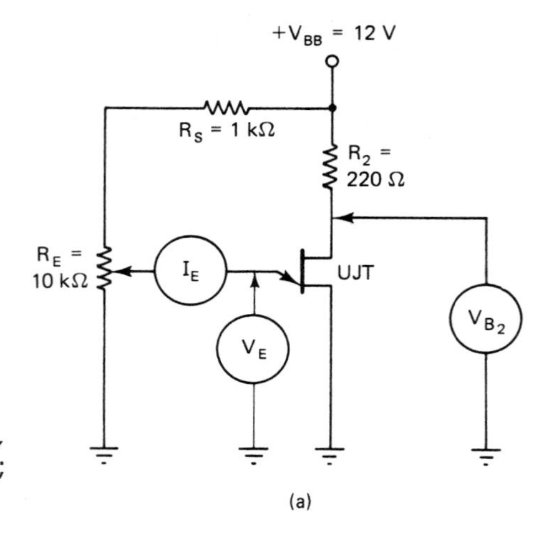

Condition	V_E	V_{B2}	I_E
Before firing	V_p =		
After firing	V_v =		

$$\eta = \frac{V_p - V_D}{V_{BB}} = \frac{V_p - 0.7\text{ V}}{12} = \_\_\_\_\_\_$$

(b)

Figure 6–6 Finding V_p, V_v, and η of a UJT: (a) test circuit; (b) data table and formula.

emitter is slowly increased by adjusting R_E. When the UJT fires, the voltmeter reading will decrease rapidly to the minimum value of V_v. This procedure may be required several times to obtain as accurate a reading as possible. Before firing, I_E should be zero, but after firing, I_E should increase significantly. Resistor R_S is used to limit the current through the emitter, in case R_E is adjusted too close to $+V_{BB}$. Resistor R_2 and the UJT form a resistive voltage divider, with V_{B2} being slightly less than V_{BB} before firing. After firing, the resistance of the UJT decreases and the voltage drop across it, V_{B2}, also decreases.

Since voltage is proportional to resistance, the intrinsic standoff ratio can be found by the formula

$$\eta = \frac{V_p - V_D}{V_{BB}}, \qquad \text{where } V_D = 0.7\text{ V}$$

Materials Needed:

1 Fixed +12-V power supply
1 Standard or digital voltmeter
1 Standard or digital ammeter
1 2N2646 UJT or equivalent
1 220-Ω resistor at 0.5 W (R_2)
1 1-kΩ resistor at 0.5 W (R_S)
1 10-kΩ potentiometer, linear (R_E)
1 Breadboard for constructing circuit

Procedure:

1. Construct the circuit shown in Figure 6-6a.
2. Adjust R_E until V_E reads 0 V.
3. Measure V_{B2} and record in the place in the data table in Figure 6-6b labeled "Before firing."
4. Measure I_E and record next to V_{B2} in the same row.
5. Slowly adjust R_E (with V_E increasing) until the meter reading drops suddenly. Record the reading as close as possible to the V_P before the UJT fires. Repeat this step a few times to obtain accuracy.
6. With the UJT fired, measure V_V and record in its proper place in the data table (second row).
7. Measure V_{B2} and record next to V_V in the data table.
8. Measure I_E and record next to V_{B2} in the row "After firing."
9. Using the value of V_p from the data table, calculate and record the intrinsic standoff ratio (η) in Figure 6-6b.

Fill-in Questions:

1. The voltage on the emitter just before the UJT fires is called ____________ .

2. The voltage on the emitter after the UJT fires is called ____________ .

3. Before the UJT fires, I_E is equal to ____________ .

4. After the UJT fires, I_E ____________ .

5. When the UJT fires, V_{B2} ____________

EXPERIMENT 3. CURRENT–VOLTAGE CHARACTERISTICS OF A UJT CIRCUIT

Objective:

To demonstrate the reaction of voltages and currents when a UJT fires, and to show how to calculate current values from voltage and resistance values.

Introduction:

For the circuit in Figure 6-7a, voltage measurements are taken before and after the UJT fires. The currents are calculated from the formulas shown in the data table of Figure 6-7b. From this table, it can be determined what occurs to each voltage and current after the UJT fires.

Materials Needed:

1 Fixed +12-V power supply
1 Standard or digital voltmeter
1 2N2646 UJT or equivalent
2 100-Ω resistors at 0.5 W (R_1 and R_2)
1 1-kΩ resistor at 0.5 W (R_S)
1 10-kΩ potentiometer, linear (R_E)
1 Breadboard for constructing circuit

Procedure:

1. Construct the circuit shown in Figure 6-7a.
2. Adjust R_E for a V_E reading of 0 V.
3. Measure V_{B1} and V_{B2} and record in the data table (row 1).
4. Calculate I_{B1}, I_{B2}, and I_E and record in the data table (row 1).
5. Adjust R_E until the UJT fires (see Experiment 2 for reference to this indication).
6. Measure V_V, V_{B1}, and V_{B2} and record in the second row of the data table.
7. Calculate I_{B1}, I_{B2}, and I_E and record in the second row of the data table. Compare the values of the data table for differences of voltages and currents before the UJT fires and after the UJT fires.

Fill-in Questions

1. After the UJT fires, V_E ____________, but I_E ____________.

2. After the UJT fires, I_{B1} ____________ and V_{B1} ____________.

3. After the UJT fires, I_{B2} ____________, but V_{B2} ____________.

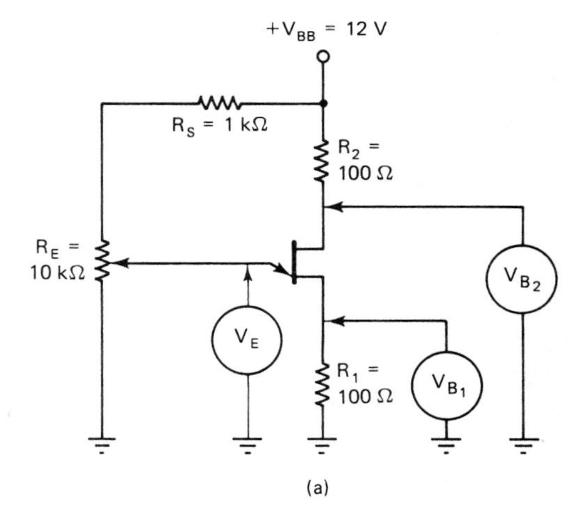

Condition	V_E	V_{B_1}	V_{B_2}	$I_{B_1} = \frac{V_{B_1}}{R_1}$	$I_{B_2} = \frac{+V_{BB} - V_{B_2}}{R_2}$	$I_E = I_{B_1} - I_{B_2}$
Before firing	$< V_p$					
After firing	$V_v =$					

(b)

Figure 6–7 Current–voltage characteristics of a UJT circuit: (a) schematic diagram; (b) data table.

EXPERIMENT 4. UJT RELAXATION OSCILLATOR

Objective:

To show how a UJT is used as a switch in a relaxation oscillator, and to determine the approximate output frequency.

Introduction:

A UJT relaxation oscillator has three available outputs: a positive pulse train at B_1, a negative pulse train at B_2, and an exponential sawtooth waveform at E. Pulses at B_1 and B_2 are normally used to trigger other circuits. Caution is needed when using the output at E, since any circuit placed here is in parallel to C_E, which tends to load it down and changes the frequency of the oscillator. Components R_E and C_E primarily determine the frequency of the oscillator and can be approximated by the formula

$$f \approx \frac{1}{R_E C_E}$$

However, the intrinsic resistance of the UJT and η also play a part in the charge and discharge time of C_E, and the actual frequency will usually be more or less. The power supply voltage has the least effect on the oscillator frequency.

Materials Needed:

1 Fixed +12-V power supply
1 Standard or digital voltmeter
1 Oscilloscope (dual trace preferred)
1 2N2646 UJT or equivalent
2 100-Ω resistors at 0.5 W (R_1 and R_2)
1 10-kΩ resistor at 0.5 W (R_E)
1 22-kΩ resistor at 0.5 W (R_E)
1 4.7-kΩ resistor at 0.5 W (R_E)
1 0.1-μF capacitor at 15 WV dc (C_E)
1 0.2-μF capacitor at 15 WV dc (C_E)
1 0.05-μF capacitor at 15 WV dc (C_E)
1 Breadboard for constructing circuit

Procedure:

1. Construct the circuit shown in Figure 6-8a.
2. Using the voltmeter, measure and record V_E, V_{B1}, and V_{B2} in the blank spaces provided in the figure.
3. Using the oscilloscope, examine the voltage waveforms at E, B_1 and B_2.
4. Draw these voltage waveforms in the spaces provided in the figure and indicate their peak-to-peak values.
5. Calculate the approximate frequency of the oscillator from the values of R_E and C_E and record in the proper place in the data table of Figure 6-8b.
6. Place the oscilloscope at E and measure the actual frequency. Remember that $f = 1/T$, where T is the time period of one cycle. Record this frequency in the proper place in the data table.
7. Change components R_E and C_E as indicated by the data table and repeat steps 5 and 6.

Figure 6–8 UJT relaxation oscillator: (a) schematic diagram; (b) data table.

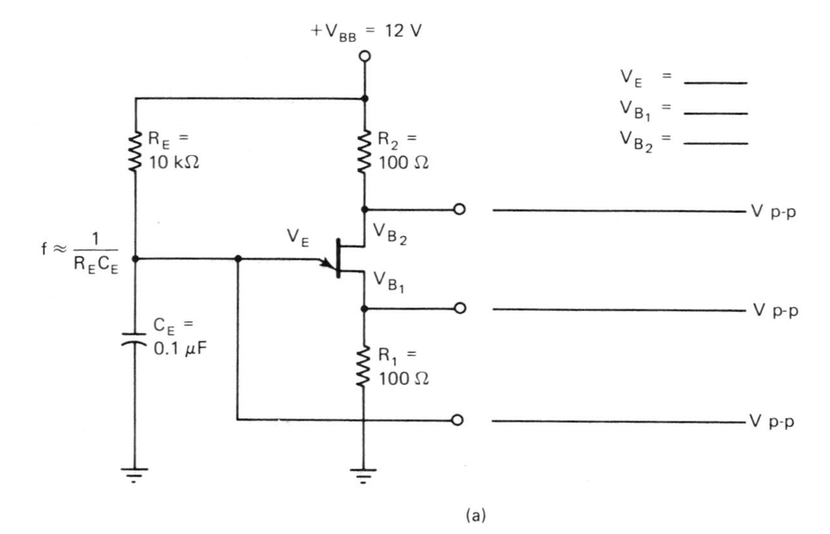

(a)

R_E (kΩ)	C_E (μF)	$f \approx 1/R_E C_E$ hertz	
		Calculated	Measured
10	0.1		
22	0.1		
4.7	0.1		
10	0.2		
10	0.05		

(b)

Fill-in Questions:

1. A ____________ pulse train is found at B_2 of a UJT oscillator.

2. A ____________ pulse train is found at B_1 of a UJT oscillator.

3. The voltage waveform at E of a UJT oscillator is in the form of a __________ .

4. If the value of R_E or C_E increases, the frequency of the oscillator ____________ .

5. If the value of R_E or C_E decreases, the frequency of the oscillator ____________ .

SECTION 6-5 UJT BASIC TROUBLESHOOTING APPLICATION: RELAXATION OSCILLATOR

Construct this circuit using your own values or those of a circuit from a previous section. Open or short the components as listed and record the voltages in the proper place in the table. The abbreviations will help indicate the voltage conditions associated with each problem. All voltages are referenced to ground. When the circuit is inoperative, there will be no voltage waveforms at E, V_{B1}, and V_{B2}.

Figure 6–E.1

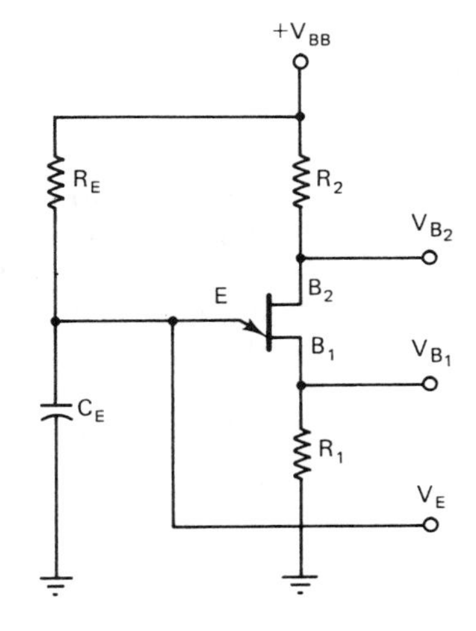

Table Abbreviations:

V_{BB}	Power supply voltage
INC	Increase
DEC	Decrease
NOC	No change
GND	Ground

Condition	V_E	V_{B_1}	V_{B_2}	Comments
Normal				All voltages are proper
R_E open				Open R_E at $+V_{BB}$: capacitor can not charge
	DEC	DEC	INC	
R_2 open				Open R_2 at $+V_{BB}$: no voltage applied to B_2
	DEC	DEC	DEC	
R_1 open				Open R_1 at GND: no current path through UJT
	V_{BB}	V_{BB}	V_{BB}	
C_E open				Open C_E at GND: emitter pulled up toward $+V_{BB}$; UJT saturates
	DEC	INC	DEC	
C_E shorted				Emitter pulled toward GND; UJT cut off
	DEC	DEC	INC	
E open				Open at junction of R_E and C_E: no current path through emitter
	DEC	DEC	NOC	
B_2 open				Open at bottom of R_2 and measure same place: no current through UJT
	INC	INC	V_{BB}	
B_1 open				Open at top of R_1 and measure same place: no current through UJT
	V_{BB}	DEC	V_{BB}	
E-B_1 shorted				R_1 in parallel with C_E
	DEC	NOC	NOC	
E-B_2 shorted				R_2 in parallel with R_E; UJT saturated
	DEC	INC	DEC	

SECTION 6-6 MORE EXERCISES

Perform the following exercises before beginning Section 6-7.

1. Refer to Figure 6-4d and draw a basic UJT relaxation oscillator, indicating the proper voltage waveforms at the respective points in the circuits.

2. Find the approximate operating frequency of the UJT oscillator from the values given for R_E and C_E. Use the formula f = $1/R_EC_E$.

a. R_E = 10 kΩ
C_E = 1μF

f = ______ Hz

b. R_E = 10 kΩ
C_E = 0.2 μF

f = ______ Hz

c. R_E = 3.3 kΩ
C_E = 0.005 μF

f = ______ Hz

3. Find the unknown value to produce the given frequency. Since $f = 1/R_EC_E$, then $R_E = 1/fC_E$ and $C_E = 1/fR_E$.

a. f = 2 kHz
C_E = 0.01 μF

R_E = ______ Ω

b. f = 16 kHz
C_E = 0.5 μF

R_E = ______ Ω

c. f = 30 kHz
R_E = 10 kΩ

C_E = ______ μF

d. f = 50 kHz
R_E = 1 kΩ

C_E = ______ μF

SECTION 6-7 UJT INSTANT REVIEW

A UJT will have some current flowing from base 1 to base 2 before it is turned on (fires). The emitter voltage must become greater than the intrinsic standoff ratio voltage point plus the forward voltage drop of the *PN* junction to turn on (fire) the UJT. After the UJT is turned on (fires), V_E decreases and the currents through the device increase.

SECTION 6-8 SELF-CHECKING QUIZZES

6-8a UJT: TRUE–FALSE QUIZ

Place a T for true or an F for false to the left of each statement.

____ **1.** The UJT is used mainly as a voltage-controlled switch.

____ **2.** Before the UJT fires there is a small current flow from base 1 to base 2.

____ **3.** A positive voltage (with respect to B_1) is used to trigger the UJT.

____ **4.** After the UJT fires, I_E decreases.

____ **5.** The intrinsic standoff ratio of a UJT can be determined by the formula $\eta = r_{BB}/r_{B1} + r_{B2}$.

_____ 6. $I_{B1} = I_E + I_{B2}$

_____ 7. The formula for finding the approximate frequency of a UJT oscillator is $f \approx R_E C_E / 1$.

_____ 8. The negative-resistance region occurs from V_p to V_V.

_____ 9. UJT oscillators are relatively independent of supply voltage variations.

_____ 10. When connected in a circuit, there will be a voltage gradient across the UJT from base to base.

6-8b UJT MULTIPLE-CHOICE QUIZ

Circle the correct answer for each question.

1. In a particular UJT, $r_{B1} = 8\ k\Omega$ and $r_{B2} = 4\ k\Omega$; therefore, its intrinsic standoff ratio is about:

   a. 0.5 b. 0.67
   c. 2.0 d. 1.2

2. A UJT turns on (fires) when:

   a. V_E equals V_{B1} b. V_E exceeds V_{B2}
   c. V_E exceeds V_P d. V_E exceeds V_V

3. When fired, the current through a UJT increases because:

   a. V_E is greater than V_{B2}
   b. the resistance between E and B_2 decreases
   c. the capacitor discharges
   d. the resistance between E and B_1 decreases

4. If V_E is increased after the UJT fires:

   a. cutoff occurs
   b. the negative resistance increases
   c. saturation occurs
   d. nothing happens

5. Negative resistance occurs in a UJT when:

   a. V_E decreases and I_E increases
   b. V_E increases and I_E increases
   c. V_E increases and I_E decreases
   d. none of the above

Questions 6 through 10 refer to Figure 6-A.

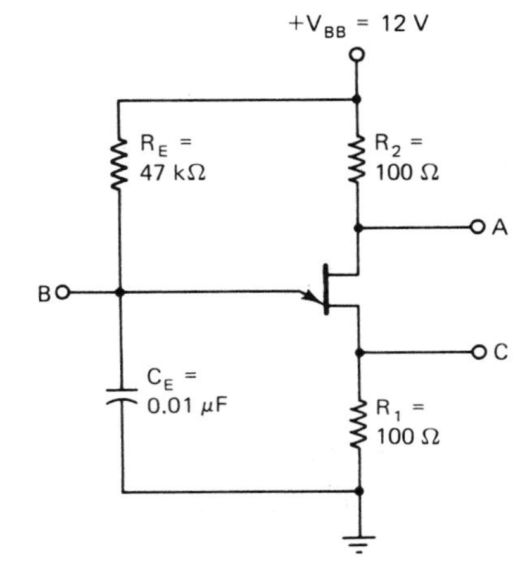

Figure 6–A

6. The circuit shown in Figure 6-A is called:

   a. a linear amplifier
   b. a relaxation oscillator
   c. an emitter follower
   d. an *RC*-coupled amplifier

7. The output frequency of the circuit shown is primarily controlled by:

   a. R_1 and R_2 b. R_1 and C_E
   c. R_E and C_E d. R_2 and C_E

8. The output frequency for this circuit is approximately:

   a. 500 Hz b. 1 kHz
   c. 2.1 kHz d. 10 kHz

9. An oscilloscope connected to terminal A would display:

   a. the out-of-phase amplified input at terminal B

b. a ramp voltage

c. a negative pulse train

d. a positive pulse train

10. An oscilloscope connected to terminal C would display:

a. the in-phase amplified input at terminal B

b. a sawtooth waveform

c. a negative pulse train

d. a positive pulse train

ANSWERS TO EXPERIMENTS AND QUIZZES

Experiment 1. **(1)** low **(2)** high **(3)** open **(4)** shorted **(5)** polarity

Experiment 2. **(1)** V_P **(2)** V_V **(3)** zero **(4)** increases **(5)** decreases

Experiment 3. **(1)** decreases, increases **(2)** increases, increases **(3)** increases, decreases

Experiment 4. **(1)** negative **(2)** positive **(3)** sawtooth **(4)** decreases **(5)** increases

True–False. **(1)** T **(2)** T **(3)** T **(4)** F **(5)** F **(6)** T **(7)** F **(8)** T **(9)** T **(10)** T

Multiple Choice. **(1)** b **(2)** c **(3)** d **(4)** c **(5)** a **(6)** b **(7)** c **(8)** c **(9)** c **(10)** d

Unit 7

Programmable Unijunction Transistor

INTRODUCTION The programmable unijunction transistor (PUT) operates similar to the UJT, but its precise firing point can be determined.

UNIT OBJECTIVES Upon completion of this unit you will be able to:

1. Draw the schematic symbol of the PUT.
2. Explain the operation of the PUT.
3. Describe how the PUT firing point can be calculated and set in a circuit.
4. Show how a PUT is used in an oscillator circuit.
5. Test a PUT with an ohmmeter.
6. Troubleshoot a PUT oscillator circuit.

SECTION 7-1 THEORY OF STRUCTURE AND OPERATION

7-1a STRUCTURE AND SCHEMATIC SYMBOL

The programmable unijunction transistor (PUT) is a member of the thyristor family of semiconductor devices. It consists of four layers of *P* and *N* material as shown in Figure 7-1. Its leads are labeled anode (*A*), gate

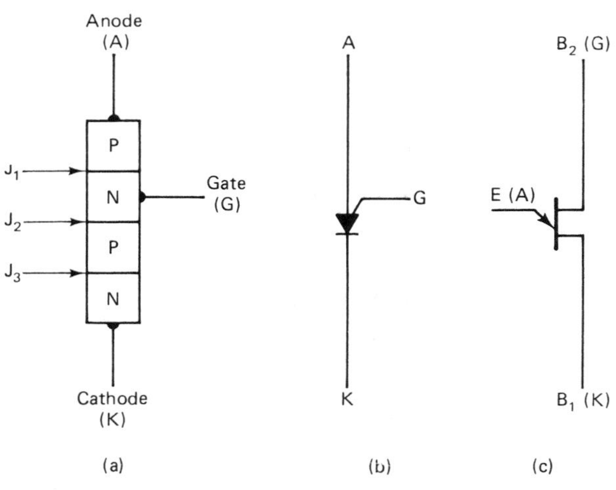

Figure 7–1 Programmable unijunction transistor: (a) structure; (b) schematic symbol; (c) UJT comparison.

(*G*), and cathode (*K*), which have the same function as the UJT leads, emitter, and base 2 and base 1, respectively. The PUT requires a negative-going pulse at its gate to turn it on, since the gate is connected to the *N* material below the anode *P* material.

7-1b OPERATION OF A PUT

The PUT acts like a voltage-controlled diode. When it is connected in a circuit, the anode is positive with respect to the cathode, but no current flows through it unless a voltage pulse occurring at the gate is less than the anode voltage. In the off condition, junctions J_1 and J_3 are forward biased, but junction J_2 is reversed biased until a negative voltage (with respect to anode) is placed on the gate lead.

Figure 7-2 shows how the PUT operates. When the PUT is off, the gate is pulled up to $+V_{CC}$ by resistor R_G and is at the same voltage as the anode. No current is flowing through the PUT. At time T_1, a negative-going pulse at the gate fires the PUT and current flows from *K* to *A* and *G*. Voltage V_A decreases at this time. At time T_2, the pulse is removed from the gate, but current continues to flow from *K* to *A* and *G*. Voltage V_A also remains low, indicating that the PUT is conducting. Therefore, the gate can trigger the PUT on, but then loses control. The PUT can be turned off if the current flowing through it is reduced below the minimum holding current, termed I_H.

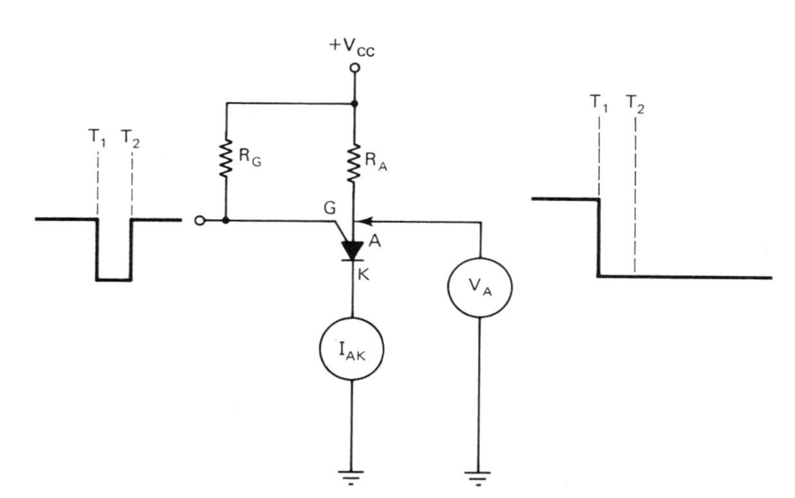

Figure 7–2 Operation of a PUT.

7-1c CURRENT–VOLTAGE CHARACTERISTICS OF A PUT

Similar to the UJT, the characteristic curve used with the PUT is based on the anode voltage (V_A) and anode current (I_A), as shown in Figure 7-3b. The point at which the PUT turns on or fires, V_P, is determined by the gate voltage-divider resistors R_1 and R_2, as shown in Figure 7-3a. These resistors can be selected to form a ratio, which will cause the PUT to fire at a specific percentage of the power supply voltage ($+V_{CC}$). Unlike the intrinsic standoff ratio of the UJT, which is fixed during manufacturing, the firing point of the PUT is said to be programmable through the proper selection of the resistors. The voltage at the gate (V_G) can be determined by the formula

$$V_G = \frac{R_1}{R_1 + R_2}(+V_{CC})$$

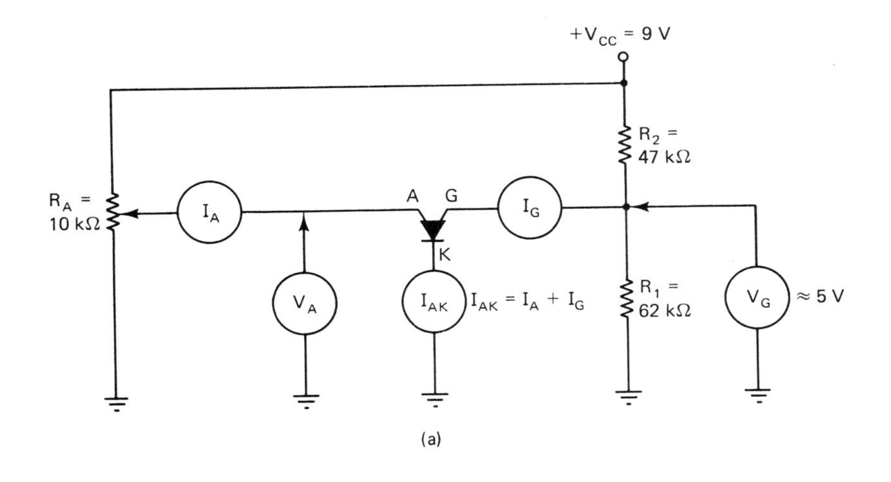

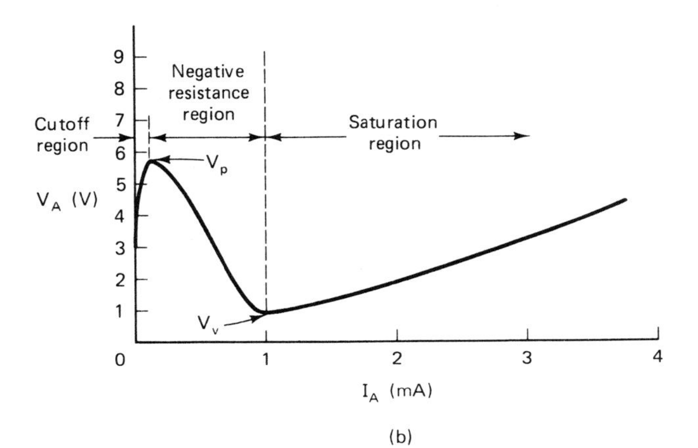

Figure 7–3 PUT V_A and I_A characteristics: (a) test circuit; (b) graph.

The peak voltage (V_P) at which the PUT fires is found by the formula

$$V_P = V_G + V_D$$

where V_D is the forward voltage of the *PN* junction needed to turn on the PUT (≈ 0.7 V).

Before the PUT fires, V_G will be some portion of $+V_{CC}$, and V_A will be less than V_G, which is determined by the setting of the potentiometer R_A. For practical purposes, all currents through the PUT will be zero. When R_A is adjusted to the V_P point, the PUT fires and the following relationships occur:

V_A decreases because of the negative-resistance "action" of the PUT.

V_G also decreases, because of the negative-resistance action.

I_{AK}, I_A, and I_G increase.

Adjusting the anode voltage more positive than the gate voltage has the same effect as applying a negative-going pulse to the gate. The amount of current flowing in the circuit is determined by the resistance of R_A from the anode to $+V_{CC}$ and resistors R_1 and R_2 in the gate circuit. The equivalent resistance of the gate is equal to R_1 and R_2 in parallel and is found by the formula

$$R_G = \frac{R_1 R_2}{R_1 + R_2}$$

Current entering the cathode is equal to the current leaving the anode plus the current leaving the gate ($I_{AK} = I_A + I_G$).

7-1d PUT RELAXATION OSCILLATOR

A PUT relaxation oscillator can produce the same output waveforms as a UJT relaxation oscillator, as shown in Figure 7-4. However, V_P can be programmed by R_1 and R_2 to cause the PUT to fire at exactly 63.2% of $+V_{CC}$, or one time constant. This means that the frequency formula of R_A and C_A will be much more accurate, since its reciprocal is equal to one time constant. The following procedure shows how this can be calculated.

Figure 7–4 Comparison of UJT and PUT relaxation oscillators: (a) UJT; (b) PUT.

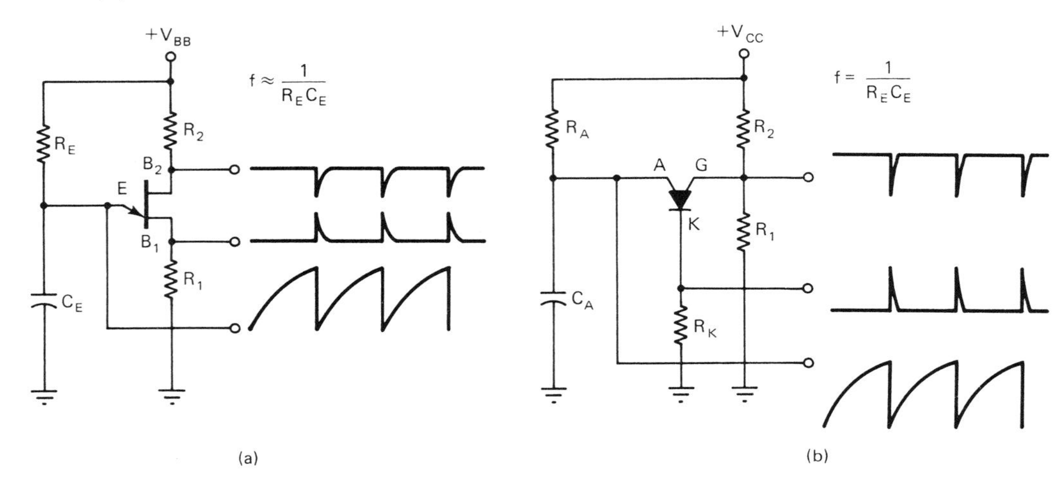

Procedure	Example
1. Select or find $+V_{CC}$	$= +12$ V
2. Find $V_P = 0.632\ (+V_{CC})$	$= 0.632\ (+12\text{ V}) = 7.58$ V
3. Find $V_G = V_P - V_D$	$= 7.58 - 0.7 = 6.88$ V
4. Select total resistance R_T of $R_1 + R_2$	$= 100$ kΩ
5. Find $R_1 = 0.632(R_1 + R_2)$	$= 0.632(100\text{ k}\Omega) = 63.2$ kΩ
6. Find $R_2 = R_T - R_1$	$= 100\text{ k}\Omega - 63.2\text{ k}\Omega = 36.8$ kΩ
[Select commercially available resistors in this range to give V_G close to 6.88 V. For example, let $R_1 = 62$ kΩ and $R_2 = 47$ kΩ; then calculate $V_G = (R_1/R_1 + R_2)(+V_{CC}) = 62\text{ k}\Omega/62\text{ k}\Omega + 47\text{ k}\Omega\ (12) = 6.82$ V.]	
7. Select frequency	$= 1$ kHz
8. Select R_A	$= 47$ kΩ
(*Note:* select this resistor large enough to reduce the current through the PUT below I_H for the given $+V_{CC}$.)	
9. Find $C = \dfrac{1}{Rf}$	$= \dfrac{1}{47\text{ k}\Omega \times 1\text{ k}\Omega} = 0.021\ \mu$F

7-1e PUT DEFINITIONS

The following definitions are some of the more commonly used terms associated with PUTs.

V_{CC}	supply voltage connected to the anode
V_A	voltage from anode to ground
V_G	voltage between R_1 and R_2 to ground
V_D	voltage drop across forward-biased PUT's anode to cathode ($\approx$ 0.7 V)
V_P	peak voltage on anode at which PUT fires
V_V	valley voltage across anode and ground after PUT fires ($\approx$ 0.7 V)
I_{AK}	current flowing into emitter
I_A	current flowing out of anode
I_G	current flowing out of gate
I_P	current flow through PUT before firing
I_V	current flow through PUT after firing

SECTION 7-2
PUT DEFINITION EXERCISE

Refer to the previous sections and write a brief definition of each term.

1. V_{CC} ____________________

2. V_A ____________________

3. V_G ____________________

4. V_D ____________________

5. V_P ____________________

6. V_V ____________________

7. I_{AK} ____

8. I_A ____

9. I_G ____

10. I_P ____

11. I_V ____

SECTION 7-3 EXERCISES AND PROBLEMS

Perform the following exercises before beginning Section 7-4.

1. Draw the structure of a PUT. (Label all parts.)

2. Draw the schematic symbol of a PUT. (Label all leads.)

3. Draw a sample V_A versus I_A characteristic curve of a PUT. (Indicate V_P, V_V, I_P, I_V, cutoff, negative resistance, and saturated regions.)

4. Draw the circuit in Figure 7-3a and write in the voltage and current nomenclature (V_A, V_G, I_A, I_{AK}, and I_G) indicated by the meters shown in the circuit.

5. Refer to the circuit in Exercise 4 and indicate any change in the voltages and currents after the PUT fires. (Use "increases," "decreases," or "remains the same.")

V_A ______________ I_G ______________

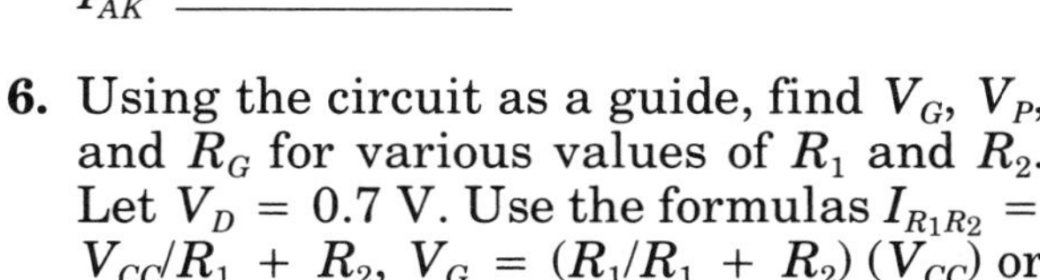

I_A ______________ V_G ______________

I_{AK} ______________

6. Using the circuit as a guide, find V_G, V_P, and R_G for various values of R_1 and R_2. Let $V_D = 0.7$ V. Use the formulas $I_{R1R2} = V_{CC}/R_1 + R_2$, $V_G = (R_1/R_1 + R_2)(V_{CC})$ or $V_G = (R_1/R_1 + R_2)(I_{R1R2})$, $V_P = V_G + V_D$, and $R_G = R_1R_2/R_1 + R_2$.

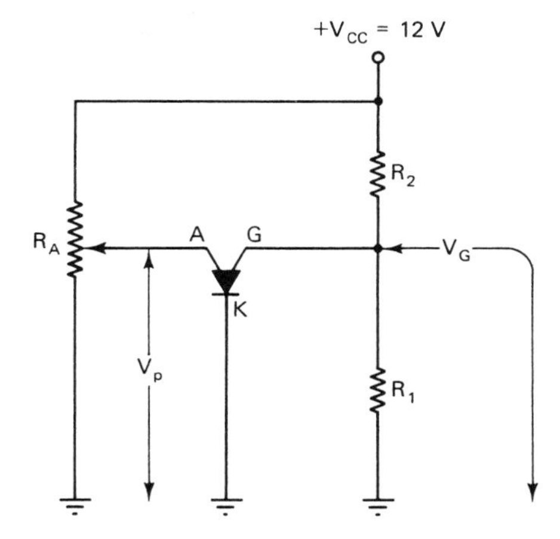

Figure 7–E.1

a. $R_1 = 22$ kΩ
$R_2 = 10$ kΩ

$V_G =$ ________

$V_P =$ ________

$R_G =$ ________

b. $R_1 = 68$ kΩ
$R_2 = 56$ kΩ

$V_G =$ ________

$V_P =$ ________

$R_G =$ ________

c. $R_1 = 47$ kΩ
$R_2 = 33$ kΩ

$V_G =$ ________

$V_P =$ ________

$R_G =$ ________

SECTION 7-4 EXPERIMENTS

EXPERIMENT 1. TESTING A PUT WITH AN OHMMETER

Objective:

To demonstrate a practical go/no go method of testing a PUT with an ohmmeter. This is called a go/no go test.

Introduction:

The *PN* junction from anode to gate of a PUT can be tested with an ohmmeter similar to a regular diode. However, testing from cathode to gate will not indicate a working PUT because one of the *PN* junctions will always be reverse biased. The PUT can be tested with an ohmmeter by placing the positive lead on the anode and the negative lead on the cathode with the gate left open. The meter should indicate high or infinite resistance. Placing a clip lead from the cathode or negative lead of the ohmmeter to the gate will trigger the PUT, and the meter should indicate low resistance. When the clip lead is removed, the meter will continue to indicate low resistance if the power source is sufficient to produce the necessary holding current.

Materials Needed:

A standard or digital ohmmeter
One or several PUTs
A clip lead

Procedure:

1. Set the ohmmeter to the midrange scale.
2. Connect the ohmmeter to the PUT as shown in Figure 7-5b and record the meter reading in the ohmmeter circle.
3. Connect the clip lead as shown in Figure 7-5c and record the reading of the meter in the ohmmeter circle.
4. Remove the clip lead as shown in Figure 7-5d and record the reading of the meter in the ohmmeter circle.

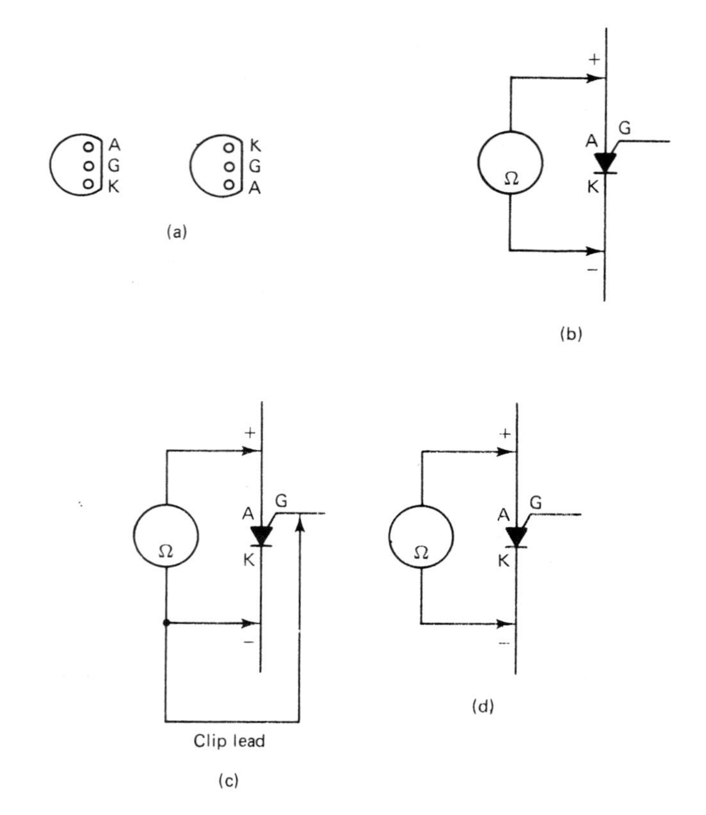

Figure 7–5 Testing a PUT with an ohmmeter: (a) general lead identification; (b) without clip lead; (c) with clip lead; (d) again without clip lead.

Fill-in Questions:

1. A PUT will have ____________ resistance before being triggered.

2. A PUT will have ____________ resistance after being triggered.

3. A PUT is being tested with an ohmmeter. When the clip lead on the gate is removed, the meter indicates high resistance. This does not prove that the PUT is defective, but that the ____________ ____________ of the meter is not sufficient to produce the necessary ________ ____________ through the device.

EXPERIMENT 2. OPERATION OF A PUT

Objective:

To demonstrate the turn on (firing) and turn off (reset) methods for a PUT.

Introduction:

To conduct, the PUT must have its anode more positive than its cathode. When the gate voltage is made more negative than its anode, the PUT turns on or fires and current flows from cathode to anode. When the gate voltage is again made equal to or more positive than the anode, current continues to flow through the PUT. The PUT is turned off or reset by reducing the current through it below its holding current for a specific power supply voltage.

Materials Needed:

1 Fixed +6-V power supply
1 Standard or digital voltmeter

1 Standard or digital ammeter
1 2N6027 PUT or equivalent
1 1-kΩ resistor at 0.5 W (R_L)
1 10-kΩ resistor at 0.5 W (R_G)
2 SPDT switches (S_1 and S_2)
1 Breadboard for constructing circuit

Procedure:

1. Construct the circuit shown in Figure 7-6a.
2. Set switches S_1 and S_2 as indicated and then apply power to the circuit.
3. In the first row of the data table shown in Figure 7-6b, record the values of V_A and I_{AK}.
4. Move S_1 to position B and record the values of V_A and I_{AK} in the second row of the data table.
5. Move S_1 to position A and record the values of V_A and I_{AK} in the third row of the data table.
6. Move S_2 to position B and record the values of V_A and I_{AK} in the fourth row of the data table.
7. Move S_2 to position A and record the values of V_A and I_{AK} in the fifth row of the data table.

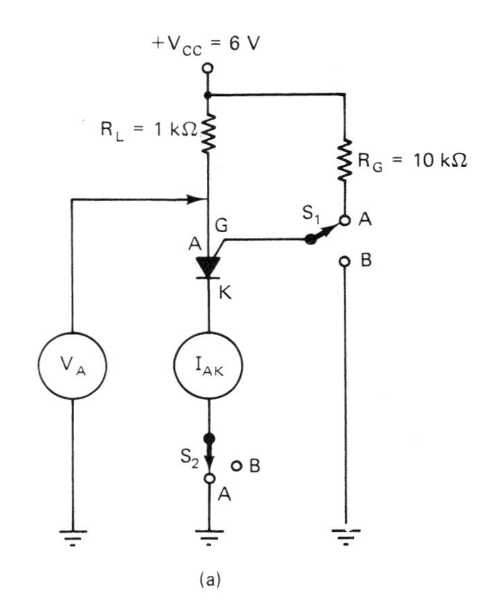

Condition	S_1	S_2	V_A	I_{AK}
Before firing	A	A		
Firing	B	A		
After firing	A	A		
Reset	A	B		
After reset	A	A		

(b)

Figure 7–6 Operation of a PUT: (a) test circuit; (b) data table.

Fill-in Questions:

1. Before firing, the voltage from the anode of the PUT to ground is equal to ______ .

2. When the gate is made more negative, the PUT ____________ and I_{AK} ____________ .

3. Once the PUT is fired, the ____________ ____________ loses control and ____________ continues to flow through the PUT.

4. When the PUT is conducting, V_A is equal to ______ V.

5. The PUT can be reset by reducing the current through it below its ____________ ____________ ____________ .

EXPERIMENT 3. DETERMINING PUT OPERATIONAL FACTORS

Objective:

To show how to calculate and measure V_G and V_P, and how to measure I_{AK} and I_H.

Introduction:

This experiment involves calculating V_G and V_P and then measuring for these voltages. It will be shown how the values of R_1 and R_2 influence the total current I_{AK} and the holding current (I_H) through the PUT.

Materials Needed:

1 Fixed +9-V power supply
1 Standard or digital voltmeter
1 Standard or digital ammeter
1 2N6027 PUT or equivalent
1 10-kΩ potentiometer (R_A)
1 6.8-kΩ resistor at 0.5 W ⎫
1 47-kΩ resistor at 0.5 W ⎬ R_1
1 68-kΩ resistor at 0.5 W ⎭
1 5.6-kΩ resistor at 0.5 W ⎫
1 33-kΩ resistor at 0.5 W ⎬ R_2
1 56-kΩ resistor at 0.5 W ⎭
1 Breadboard for constructing circuit

Procedure:

1. Construct the circuit shown in Figure 7-7a using the values of R_1 and R_2 from the first row of the data table in Figure 7-7b.
2. Adjust R_A until V_A reads 0 V.
3. Calculate V_G and record on the data table.
4. Measure V_G and record in the data table.
5. Calculate V_P and record in the data table.
6. By adjusting R_A, slowly increase V_A until the meter indicates a sudden decrease. The point just before the meter reading decreases is V_P. Slowly perform this step several times to obtain an accurate measurement for V_P.
7. Record V_P in the data table.
8. With the PUT fired (at the V_P point), measure I_{AK} and record in the data table.
9. With the PUT fired, slowly adjust R_A so that I_{AK} decreases. The meter reading will suddenly decrease to zero. The point just before this sudden decrease is the PUT's holding current. Slowly perform this step several times to obtain an accurate measurement for I_H.
10. Record I_H in the data table.
11. With the PUT fired, note the meter readings of V_A and V_G. The PUT is acting like a switch.
12. Repeat steps 2 through 11 for the other values of R_1 and R_2 given in the data table.

Figure 7–7 Determining PUT operational factors: (a) test circuit; (b) data table.

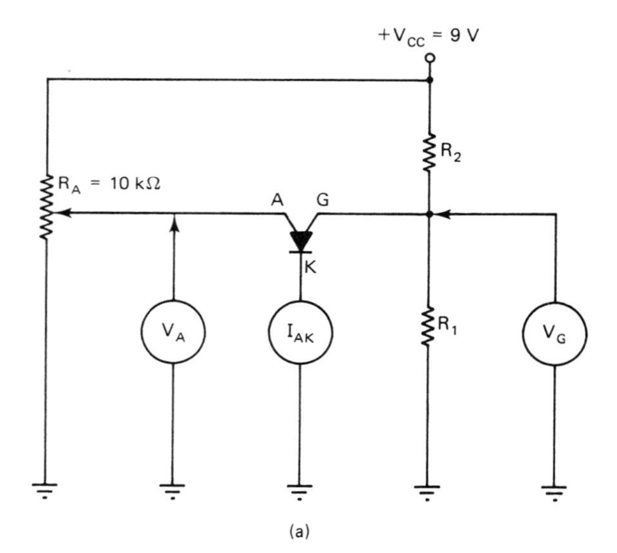

R_1 (kΩ)	R_2 (kΩ)	$V_G = \frac{R_1}{R_1 + R_2}(+V_{CC})$	V_G	$V_P = V_G + V_D$ (0.7 V)	V_P (V_A)	I_{AK} (mA)	I_H (mA)
6.8	5.6						
47	33						
68	56						

(b)

Fill-in Questions:

1. V_P is approximately ______ V greater than V_G.
2. When the PUT fires, V_A ____________ .
3. When the PUT fires, V_G ____________ .
4. When the PUT fires, I_{AK} ____________ .
5. When the values of R_1 and R_2 increase, I_H ____________ .

EXPERIMENT 4. PUT RELAXATION OSCILLATOR

Objective:

To demonstrate how a PUT is used as a switch in a relaxation oscillator, and to determine the output frequency.

Introduction:

The PUT oscillator is similar to the UJT oscillator, except that the trigger pulses at the gate and cathode appear to be sharper. Resistors R_1 and R_2 should be selected to produce a V_G that when added to V_D will equal 63.2% of $+V_{CC}$ (the value of V_P). Resistor R_A should be large enough to limit current below the holding current when the PUT is fired, or the PUT will never turn off. Therefore, no oscillations will be present. Components R_A and C_A determine the frequency of the oscillator and can be found with the formula

$$f = \frac{1}{R_A C_A}$$

Materials Needed:

1 Fixed +12-V power supply
1 Standard or digital voltmeter
1 Oscilloscope (dual trace preferred)
1 2N6027 PUT or equivalent
1 100-Ω resistor at 0.5 W (R_K)
1 5.6-kΩ resistor at 0.5 W (R_2)
1 6.8-kΩ resistor at 0.5 W (R_1)
1 47-kΩ resistor at 0.5 W ⎫
1 100-kΩ resistor at 0.5 W ⎬ R_A
1 220-kΩ resistor at 0.5 W ⎭
1 0.01-μF capacitor ⎫
1 0.02-μF capacitor ⎬ C_A
1 0.1-μF capacitor ⎭
1 Breadboard for constructing circuit

Procedure:

1. Construct the circuit shown in Figure 7-8a.
2. Using the voltmeter, measure and record the operating voltages V_A, V_G, and V_K in the blank spaces provided in the figure.
3. Using the oscilloscope, examine the voltage waveforms at *G, A,* and *K.*
4. Draw these voltage waveforms in the spaces provided in the figure and indicate their peak-to-peak values.

Figure 7–8 PUT relaxation oscillator: (a) schematic diagram; (b) data table.

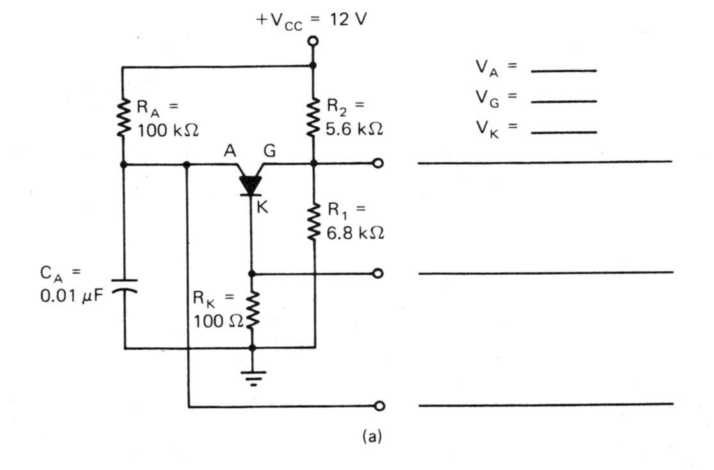

(a)

R_A (kΩ)	C_A (μF)	$f = 1/R_A C_A$ (Hz) Calculated	Measured
100	0.01		
100	0.1		
100	0.02		
47	0.01		
220	0.01		

(b)

5. Calculate the frequency of the oscillator from the values of R_A and C_A and record in the first row of the data table of Figure 7-8b.
6. Place the oscilloscope at A and measure the actual frequency. Remember that $f = 1/T$, where T is the time period of one cycle. Record this frequency in the proper place of the data table.
7. Change components R_A and C_A as indicated by the data table and repeat steps 5 and 6.

Fill-in Questions:

1. A ____________ pulse train is found at the gate of a PUT oscillator.

2. A ____________ pulse train is found at the cathode of a PUT oscillator.
3. The voltage waveform at the anode of a PUT oscillator is in the form of a ____________ .
4. If the value of R_A or C_A increases, the frequency of the oscillator __________ .
5. If the value of R_A or C_A decreases, the frequency of the oscillator __________ .

SECTION 7-5
PUT BASIC TROUBLESHOOTING APPLICATION: RELAXATION OSCILLATOR

Construct this circuit using your own values or those of a circuit from a previous section. Open or short the components as listed and record the voltages in the proper place in the table. The abbreviations will help indicate the voltage conditions associated with each problem. All voltages are referenced to ground. When the circuit is inoperative, there will be no voltage waveforms at V_G, V_K, and V_A.

Figure 7–E.2

Table Abbreviations

$+V_{CC}$	Power supply voltage
INC	Increase
DEC	Decrease
NOC	No change
GND	Ground

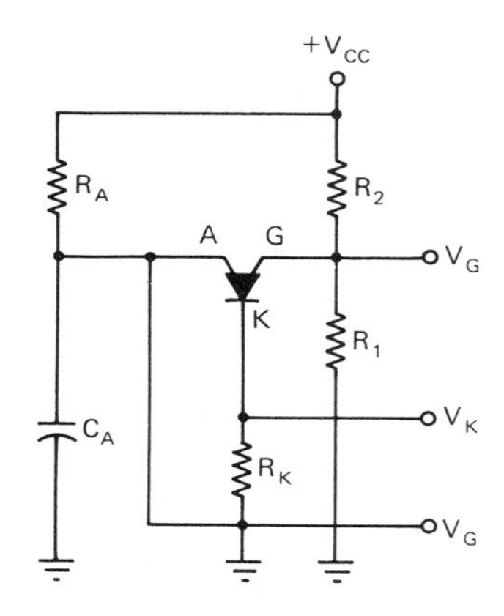

Condition	V_A	V_G	V_K	Comments
Normal				All voltages proper
R_A open	DEC	INC	NOC	Open R_A at $+V_{CC}$: no voltage applied to RC charging circuit
C_A open	NOC	NOC	NOC	Open C_A at GND: circuit may oscillate from stray capacitance; look for higher frequency
C_A shorted	DEC	INC	NOC	Anode pulled to GND; PUT cutoff
R_K open	INC	INC	INC	Open R_K at GND: no discharge path for C_A; anode/gate forward biased
R_1 open	INC	$+V_{CC}$	NOC	Open R_1 at GND: no reference voltage V_G, for PUT
R_2 open	DEC	DEC	NOC	Open R_2 at $+V_{CC}$: PUT saturates

A open	$+V_{CC}$	INC	NOC	Open A and measure between R_A and C_A: no current through PUT
K open	INC	INC	NOC	Open K at R_K and measure top of R_K: anode/gate forward biased
G open	DEC	INC	NOC	Open G and measure between R_1 and R_2: PUT is on
A-K shorted	DEC	INC	NOC	PUT shorted similar to C_A shorted
A-G shorted	INC	INC	NOC	PUT not conducting
G-K shorted	DEC	DEC	INC	PUT constantly on

Figure 7–E.2 (con't)

SECTION 7-6 MORE EXERCISES

Perform the following exercises before beginning Section 7-7.

1. Refer to Figure 7-4b and draw a basic PUT relaxation oscillator indicating the proper voltage waveforms at the respective points in the circuit.

2. Find the approximate value of the unknown resistor for the oscillator from the values given for $V_P = 0.632V_{CC}$. Use the formulas $V_G = V_P - V_D$, $V_{R1} = V_G$, and $V_{R2} = V_{CC} - V_{R1}$. Since $V_{R2}/V_{R1} = R_2/R_1$, then $R_1 = V_{R1}R_2/V_{R2}$ and $R_2 = V_{R2}R_1/V_{R1}$.

a. $R_2 = 10\ \text{k}\Omega$

$R_1 =$ ______

b. $R_2 = 22\ \text{k}\Omega$

$R_1 =$ ______

c. $R_1 = 47\ \text{k}\Omega$

$R_2 =$ ______

d. $R_1 = 68\ \text{k}\Omega$

$R_2 =$ ______

3. Find the operating frequency of the PUT oscillator from the values given for R_A and C_A. Use the formula $f = 1/R_AC_A$.

a. $R_A = 47\ \text{k}\Omega$
$C_A = 0.01\ \mu\text{F}$

$f =$ ______

b $R_A = 22\ \text{k}\Omega$
$C_A = 0.1\ \mu\text{F}$

$f =$ ______

c. $R_A = 68\ \text{k}\Omega$
$C_A = 0.002\ \mu\text{F}$

$f =$ ______

SECTION 7-7 PUT INSTANT REVIEW

A PUT is very similar to a silicon-controlled rectifier. It must be operated in the forward-biased condition where the anode is more positive than the cathode, except that no current flows through it until a negative voltage (with respect to the anode) is placed on the gate. When the negative voltage on the gate is removed, current continues to flow through the PUT, showing that the gate can turn it on, but then loses control. The PUT will continue to stay on until the current through it decreases below its holding current.

SECTION 7-8 SELF-CHECKING QUIZZES

7-8a PUT: TRUE–FALSE QUIZ

Place a T for true or an F for false to the left of each statement.

____ **1.** The PUT can be turned on with a negative pulse applied to the gate.

____ **2.** The forward voltage drop of a PUT from anode to cathode is about 2.1 V.

____ **3.** The PUT can replace an SCR in a circuit.

____ **4.** A PUT can be turned on if its anode voltage is made more positive than its gate voltage.

____ **5.** When the gate trigger pulse is removed, the PUT stops conducting.

____ **6.** The PUT is programmable by selection of the gate resistors, which can determine the exact firing point.

____ **7.** Unlike a UJT, the PUT does not have a negative-resistance region.

____ **8.** The resistor in the anode circuit of a PUT oscillator must be large enough to reduce the current through it below the holding current or it will never turn off.

____ **9.** A PUT oscillator can produce the same voltage waveforms as a UJT oscillator.

____ **10.** A PUT circuit can produce a sharp pulse to trigger other circuits.

7-8b PUT: MULTIPLE-CHOICE QUIZ

Circle the correct answer for each question. All five questions refer to Figure 7-A.

1. The V_G of the circuit is about:

a. 4.5 V **b.** 5.0 V
c. 5.7 V **d.** 6.0 V

2. The V_P of the circuit is about:

a. 4.5 V **b.** 5.0 V
c. 5.7 V **d.** 6.0 V

3. The V_V of the circuit is about:

a. 0.2 V **b.** 0.7 V
c. 4.0 V **d.** 3.3 V

4. The approximate frequency of the circuit is:

a. 50 Hz **b.** 500 Hz
c. 2.5 kHz **d.** 5.0 kHz

5. A sawtooth voltage waveform can be seen with an oscilloscope at point:

a. A **b.** B
c. C **d.** none of the above

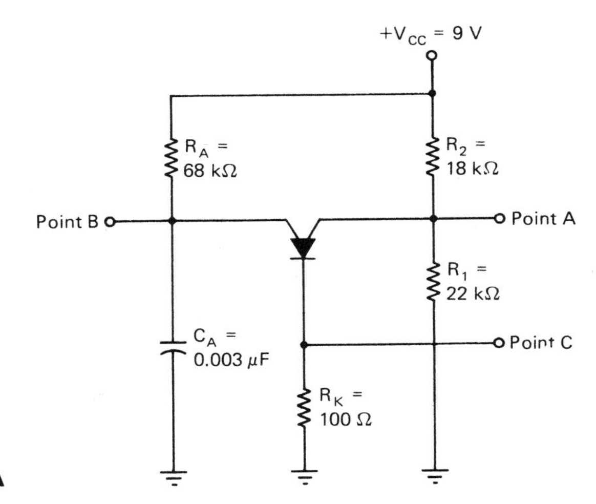

Figure 7-A

ANSWERS TO EXPERIMENTS AND QUIZZES

Experiment 1.	**(1)** high **(2)** low **(3)** power source, holding current
Experiment 2.	**(1)** $+V_{CC}$ **(2)** fires, increases **(3)** gate, current **(4)** 0.7 **(5)** holding current
Experiment 3.	**(1)** 0.7 **(2)** decreases **(3)** decreases **(4)** increases **(5)** decreases
Experiment 4.	**(1)** negative **(2)** positive **(3)** sawtooth **(4)** decreases **(5)** increases
True–False.	**(1)** T **(2)** F **(3)** F **(4)** T **(5)** F **(6)** T **(7)** F **(8)** T **(9)** T **(10)** T
Multiple Choice.	**(1)** b **(2)** c **(3)** b **(4)** d **(5)** b

Unit 8

Silicon-Controlled Rectifier

INTRODUCTION The silicon-controlled rectifier (SCR) is used in rectifying, triggering, and protection circuits.

UNIT OBJECTIVES Upon completion of this unit you will be able to:

1. Draw the SCR schematic symbol.
2. Describe four-layer *pn* bipolar semiconductor devices.
3. Explain the operation of the SCR.
4. Draw the current–voltage characteristic curve for an SCR.
5. Show how the phase angle can be varied for an SCR with ac circuits.
6. Identify SCR voltage waveforms in a circuit.
7. Define the terms breakover voltage, holding current, trigger time, and conduction time.
8. Test an SCR with an ohmmeter.
9. Demonstrate how an SCR can be used as an overvoltage protection device.
10. Troubleshoot a basic SCR trigger circuit.

SECTION 8-1 THEORY OF STRUCTURE AND OPERATION

8-1a STRUCTURE AND SCHEMATIC SYMBOL

The silicon-controlled rectifier (SCR) is a four-layer *PN* bipolar semiconductor device with three terminals, as shown in Figure 8-1. The SCR belongs to the thyristor family of electronic devices, which operates on the principle of current conduction when the breakover voltage is reached. These devices may also have a control element, such as a gate that can trigger the device into conduction below the breakover voltage point. An SCR has an anode, a cathode, and gate terminals. It operates similar to a normal diode, where current flows only in the forward-biased condition, but must be triggered into conduction by the gate terminal. Once the gate has triggered the SCR into conduction, it acts like a latched switch, and the gate no longer has control of the current flow.

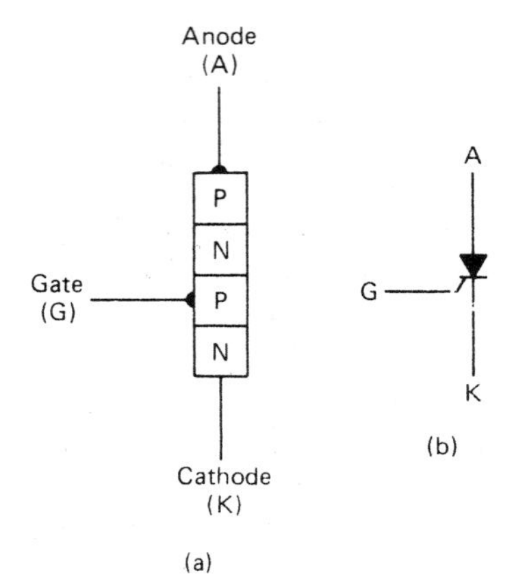

Figure 8–1 Silicon-controlled rectifier: (a) structure; (b) schematic symbol.

8-1b OPERATION OF AN SCR

A better understanding of the operation of an SCR is given in Figure 8-2. The anode is connected through a series-limiting resistor (R_L) to a positive voltage. The cathode is connected to ground via switch S_2 and the gate is connected to switch S_1, which is set to ground. In this condition (Figure 8-2a), junctions 1 and 3 (J_1 and J_3) are forward biased, but junction 2 (J_2) is reverse biased, which prevents any appreciable current from flowing through the SCR. When S_1 is moved up to the bottom side of R_A (Figure 8-2b), a small gate electron current flows toward the supply voltage ($+V_{AA}$). This introduces holes into the P-type gate region, which allows a larger current flow from cathode to anode through the SCR. In turn, J_2 is now forward biased. When the gate is set back to ground via S_1 (Figure 8-2c), the large current flowing through the SCR tends to keep J_2 internally forward biased, and the SCR is on or latched. The SCR can only be turned off if this main current flowing from cathode to anode is reduced below its minimum holding current (I_H). This can be accomplished by momentarily opening switch S_2 in the cathode lead of the circuit. Junction J_2 is again reverse biased, and the SCR can be considered reset or off. The SCR can be turned on again by triggering with the gate lead.

Figure 8–2 Operation of an SCR: (a) off condition; (b) triggering on; (c) on condition without triggering.

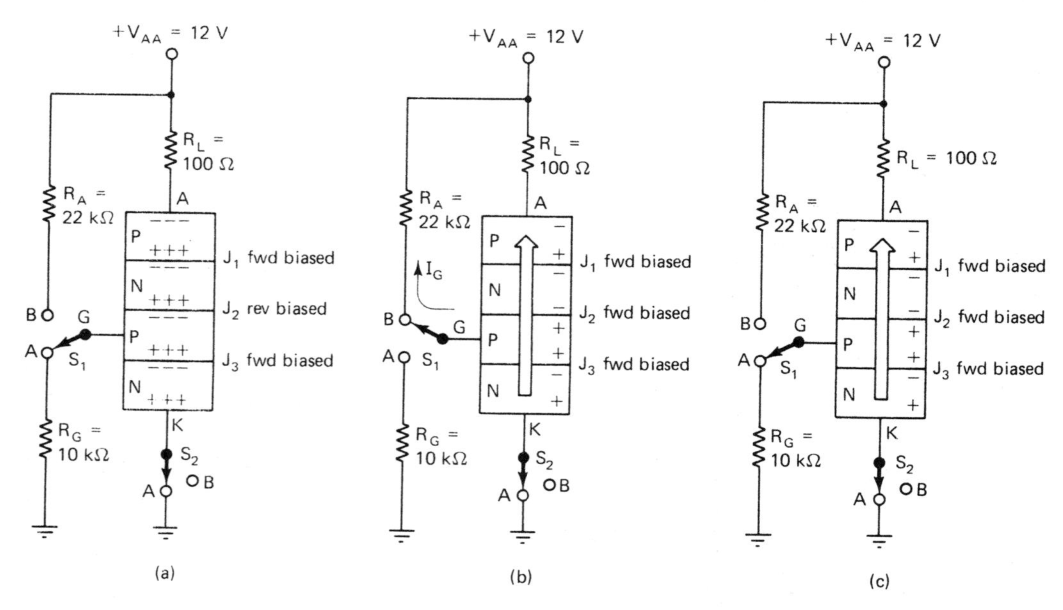

8-1c CURRENT–VOLTAGE CHARACTERISTICS OF AN SCR

The SCR operates similar to normal diodes when in the reverse-biased condition, as shown in Figure 8-3a. The SCR exhibits a very high internal impedance, with perhaps a slight reverse blocking current. However, if the reverse breakdown voltage is exceeded, the reverse current rapidly increases to a large value and may destroy the SCR.

In the forward-biased condition (with the gate connected to ground), the internal impedance of the SCR is very high with a small current flowing, called the *forward blocking current.* When the forward voltage ($+V_F$) is increased beyond the forward breakover voltage point, an avalanche breakdown occurs and the current from cathode to anode increases rapidly. A regenerative action occurs with the conduction of the *PN* junctions, and the internal impedance of the SCR decreases.

Figure 8–3 Current–voltage characteristics for an SCR: (a) *I–V* curves; (b) test for gate current; (c) gate current curves.

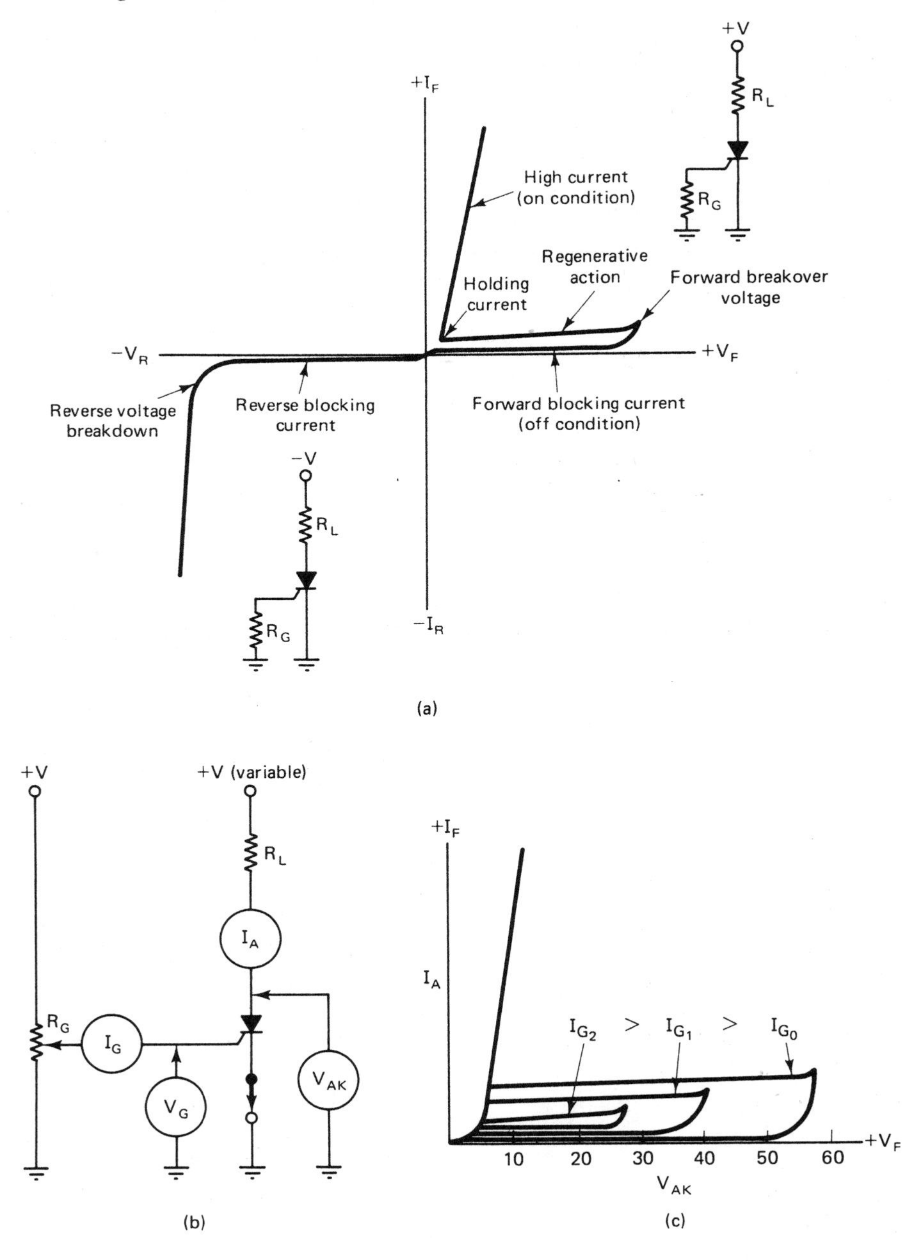

This results in a decrease in voltage across the anode and cathode, as verified by Ohm's law, where $E = IR$. When R is small, so is the voltage drop across it. The forward current through the SCR is limited primarily by the impedance of the external circuit, and the SCR will remain on as long as this current does not fall below the holding current.

If gate current is allowed to flow as shown in Figure 8-3b, the forward breakover voltage point will be less. The more gate current flowing, the lower the point at which forward breakover will occur, as illustrated in Figure 8-3c. Normally, SCRs are operated with applied voltages lower than the forward breakover voltage point (with no gate current flowing) and the gate triggering current is made sufficiently large to assure complete turn on.

8-1d AC TRIGGERING OF AN SCR

The conduction time of an SCR can be controlled for a portion of a sine wave, as shown in Figure 8-4a. Potentiometer R_G is sufficiently large so as to limit and control the gate current that fires the SCR. Resistor R_A prevents the gate from being shorted to the anode when the wiper of R_G is placed all the way to the top. As the voltage rises across the circuit with the positive alternation of the sine wave, the voltage across the SCR also rises and the gate begins to draw current. At this time the SCR is off and the applied voltage is across its A–K terminals. Since no current is flowing through the SCR, there is no IR drop across R_L and its voltage is zero. When sufficient gate triggering current is developed (depending on the setting of R_G), the SCR turns on. The voltage across its A–K terminals decreases. With current flowing through R_L, a voltage drop is developed across R_L for the remaining portion of the positive

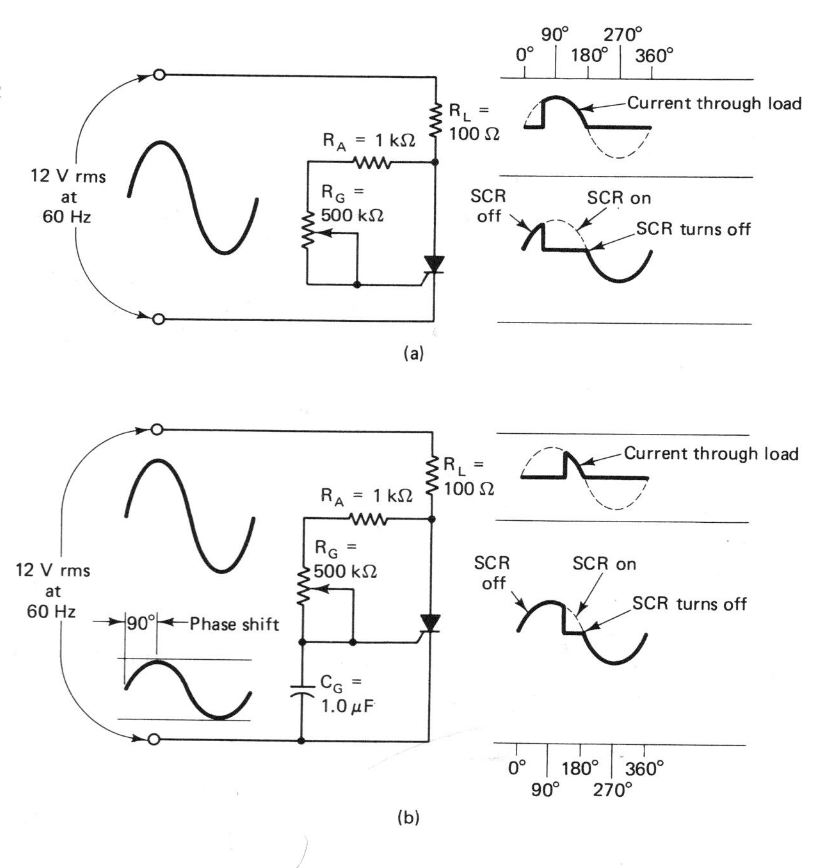

Figure 8–4 AC triggering of an SCR: (a) resistive triggering time ≈ 0° to 90°; (b) *RC* triggering time ≈ 0° to 180°.

alternation. When the negative alternation of the sine wave appears, the SCR turns off and voltage is again present across its A–K terminals. With no current flowing through R_L, its voltage drop is zero. Since the SCR turns on only with a positive-going signal on the gate, the variable firing time is limited to the first 90° of a sine wave. This variable firing time is less than a full 90°, because a few degrees must lapse to provide sufficient voltage across the SCR to create the gate current to fire the SCR. The next positive alternation again causes the SCR to turn on at the same point.

The trigger time of the SCR can be extended to 180° of the sine wave with the addition of a capacitor, as shown in Figure 8-4b. The voltage across the capacitor in a series RC circuit lags the applied voltage by 90°. Therefore, the voltage across C_G lags the voltage applied across the A–K terminals of the SCR, and it takes longer to reach the triggering point of the gate. Potentiometer R_G is adjusted for the desired triggering angle, but notice that, beyond the 90° firing point, current through the SCR and R_L flows for less time.

8-1e SCR DEFINITIONS

The following definitions are some of the more commonly used terms associated with SCRs and thyristors in general.

V_{AA}	supply voltage connected to the anode
V_A	voltage from anode to ground
V_G	voltage from gate to ground
V_{AK}	voltage from anode to cathode
V_F	forward voltage across anode and cathode when the thyristor is conducting
V_{BO}	forward breakover voltage with the gate open; also indicates maximum allowable voltage from cathode to anode for a specific thyristor
V_{DROM}	repetitive peak off-state voltage with the gate open; similar to V_{BO}, but includes only repetitive transient voltages
PRV	peak reverse voltage; maximum allowable reverse voltage from cathode to anode of the thyristor
PIV	peak inverse voltage, the same as PRV
V_{RROM}	repetitive peak reverse voltage with the gate open; similar to PRV, but includes only repetitive transient voltages
I_F	forward current flowing from cathode to anode when the thyristor is on
$I_{F(\text{off})}$	forward blocking current from cathode to anode when the thyristor is off
I_H	holding current; the minimum current from cathode to anode required to keep the thyristor on
I_R	reverse current flowing from anode to cathode when the thyristor is reverse biased
$I_{T(\text{rms})}$	effective (rms) forward current when the thyristor is conducting
I_{TSM}	allowable short-time duration current (nonrepetitive) when the thyristor is conducting
I_{GT}	minimum gate current required to switch the thyristor from the off condition to the on condition

SECTION 8-2
SCR DEFINITION EXERCISE

Refer to the previous sections and write a brief definition for each term.

1. V_{AA} supply voltage connected to anode.

2. V_A voltage from anode to ground

3. V_G voltage from gate to ground

4. V_{AK} voltage from anode to cathode

5. V_F forward voltage across anode & cathode when thyristor is conducting.

6. V_{BO} forward breakover voltage with the gate open.

7. V_{DROM} repetitive peak-off state voltage with gate open

8. PRV Peak reverse voltage; maximum allowable reverse voltage from cathode to anode of the thyristor

9. PIV Peak inverse voltage

10. V_{RROM} repetitive peak- reverse voltage with the gate open.

11. I_F forward current flowing from cathode to anode when the thyristor is on

12 $I_{F\,(off)}$ forward blocking current from cathode to anode when the thyristor is off.

13. I_H holding current, minimum current from cathode to anode required to keep thyristor on

14. I_R reverse current flowing from anode to cathode when the thyristor is reversed-biased.

15. $I_{T\,(rms)}$ effective (rms) forward current when the thyristor is conducting

16. I_{TSM} allowable short-time duration current when the thyristor is conducting

17. I_{GT} minimum gate current required to switch thyristor from the off condition to the on condition.

SECTION 8-3
EXERCISES AND PROBLEMS

Perform the following exercises before beginning Section 8-4.

1. Draw the structure of an SCR. (Label all parts.)

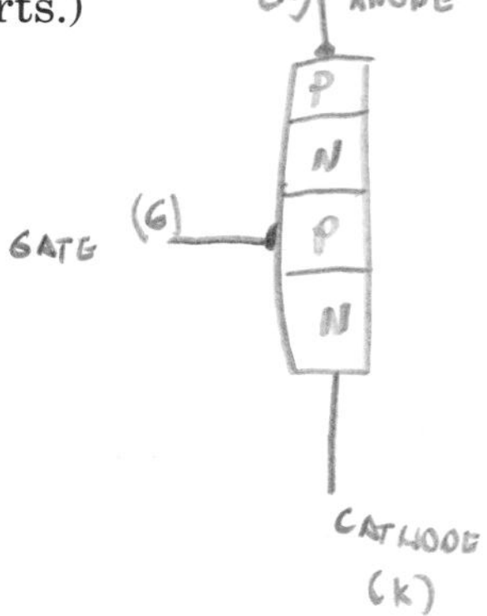

2. Draw the schematic symbol of an SCR. (Label the leads.)

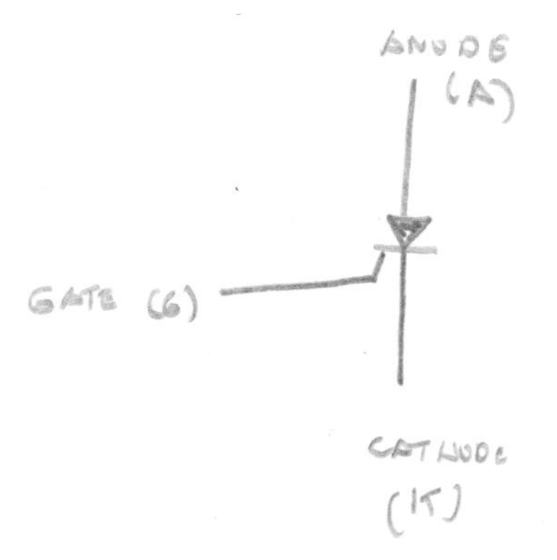

3. Refer to Figure 8-2 and draw the operation of an SCR.

4. Draw a simple characteristic curve for an SCR. (Label the forward and reverse voltages and currents.)

5. Draw the circuit in Figure 8-3b and write in the voltage and current nomenclature (V_G, V_{AK}, I_A, and I_G) indicated by the meters shown in the circuit.

6. Circle the correct statements that apply to an SCR.
   a. The anode of an SCR must be connected to a more positive voltage than its cathode for proper operation.
   b. A negative-going pulse applied to the gate will turn on an SCR.
   c. The SCR is a latching device.
   d. Once fired, the gate loses control of the main current flowing from cathode to anode of an SCR.
   e. A postive-going pulse applied to the gate will turn on an SCR.
   f. An SCR conducts only in one direction.
   g. An SCR continues to conduct until the current through it falls below its minimum holding current level.

SECTION 8-4 EXPERIMENTS

EXPERIMENT 1. TESTING AN SCR WITH AN OHMMETER

Objective:

To demonstrate a practical go/no go method of testing an SCR with an ohmmeter. This is called a go/no go test.

Introduction:

The *PN* junction from gate to cathode of an SCR can be tested with an ohmmeter similar to a regular diode. However, testing from anode to gate will not indicate if an SCR is working properly, because one of the *PN* junctions is always reverse biased. The SCR can be tested with an ohmmeter by placing the positive lead on the anode and the negative lead on the cathode with the gate left open. The meter should read high or infinite resistance. Placing a clip lead from the anode or positive lead of the ohmmeter to the gate triggers the SCR and the meter should indicate low resistance. When the clip lead is removed, the meter continues to indicate low resistance if the power source is sufficient to produce the required holding current.

Materials Needed:

A standard or digital ohmmeter

One or several SCRs

A clip lead

Procedure:

1. Set the ohmmeter to the midrange scale.
2. Connect the ohmmeter to the SCR as shown in Figure 8-5b and record the meter reading in the ohmmeter circle.
3. Connect the clip lead as shown in Figure 8-5c and record the reading of the meter in the ohmmeter circle.

Figure 8–5 Testing an SCR with an ohmmeter: (a) general lead identification; (b) without clip lead; (c) with clip lead; (d) again without clip lead.

4. Remove the clip lead as shown in Figure 8-5d and record the reading of the meter in the ohmmeter circle.

Fill-in Questions:

1. An SCR will have ___high___ resistance before being triggered.
2. An SCR will have ___Low___ resistance after being triggered.
3. The ___GATE___-to-___CATHODE___ resistance of an SCR can be checked like a normal diode.
4. An SCR is being tested with an ohmmeter. When the clip lead on the gate is removed, the meter indicates high resistance. This does not prove that the SCR is defective, but that the power source of the meter is not sufficient to produce the necessary ___holding___ ___current___ through the device.

EXPERIMENT 2. OPERATION OF AN SCR

Objective:

To show the turn-on (fire) and turn-off (reset) methods for an SCR.

Introduction:

To conduct, the SCR must have its anode more positive than its cathode. When the gate voltage is made more positive than its cathode, the SCR turns on or fires and current flows from cathode to anode. When the gate voltage is again made equal to or more negative than the cathode, current continues to flow through the SCR. The SCR is turned off or reset by reducing the current through it below its holding current.

Materials Needed:

1 Fixed +12-V power supply
1 Standard or digital voltmeter
1 C106Y1 SCR or equivalent
1 100-Ω resistor at 0.5 W (R_L)
1 10-kΩ resistor at 0.5 W (R_G)
1 22-kΩ resistor at 0.5 W (R_A)
2 DPST switches (S_1 and S_2)
1 Breadboard for constructing circuit

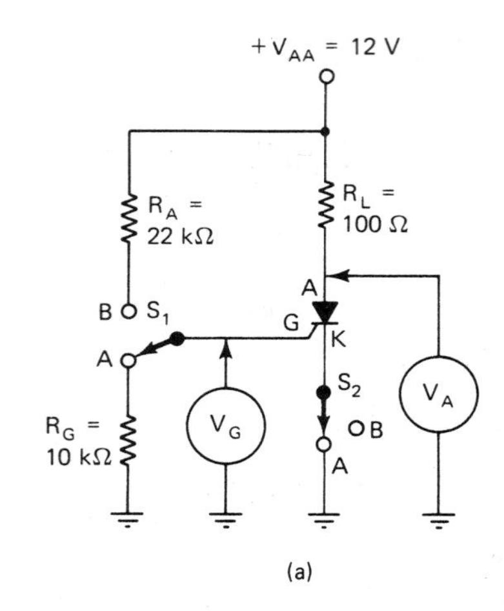

S_1 condition	S_2 condition	V_G	V_A	Condition of SCR (on or off)
A	A	0	12	off
B	A	0.7	0.7	ON
A	A	0.7	0.7	ON
A	B	0	12v	off
A	A	0	12v	off

(b)

Figure 8–6 Operation of an SCR: (a) test circuit; (b) data table.

Procedure:

1. Construct the circuit shown in Figure 8-6a.
2. Set switches S_1 and S_2 as indicated and then apply power to the circuit.
3. In the first row of the data table shown in Figure 8-6b, record the values of V_G and V_A.
4. Move S_1 to position B and record the values of V_G and V_A in the second row of the data table.
5. Move S_1 to position A and record the values of V_G and V_A in the third row of the data table.
6. Move S_2 to position B and record the values of V_G and V_A in the fourth row of the data table.
7. Move S_2 to position A and record the values of V_G and V_A in the fourth row of the data table.

Fill-in Questions:

1. Before firing, the voltage from anode to ground of the SCR is equal to +VAA supply voltage.

2. When the gate is made more POSITIVE, the SCR fires and I_{AK} INCREASES.

3. Once the SCR fires, the gate Loses control and the current CONTINUES to flow through the SCR.

4. When the SCR is conducting, the voltage from anode to ground is equal to ≈ 0.7V.

5. The SCR can be reset by reducing the current through it below its holding current.

EXPERIMENT 3. CURRENT CONTROL OF AN SCR

Objective:

To demonstrate the effect that gate current has to turn on an SCR, and to determine the minimum holding current to keep the SCR conducting.

Introduction:

This experiment shows that sufficient gate current must flow in order to turn on the SCR and that the minimum holding current can be found with the addition of a large-value potentiometer in the anode circuit.

Materials Needed:

1 Fixed +12-V power supply
1 Standard or digital voltmeter
1 Standard or digital ammeter
1 C106Y1 SCR or equivalent
1 100-Ω resistor at 0.5 W (R_L)
1 22-kΩ resistor at 0.5 W (R_B)
1 100-kΩ resistor at 0.5 W (R_A)
1 50-kΩ potentiometer (R_H)
2 DPST switches (S_1 and S_2)
1 Breadboard for constructing circuit

Procedure:

1. Construct the circuit shown in Figure 8-7a.
2. Set both switches as indicated and then apply power to the circuit.
3. Calculate the gate current, I_{RA}, flowing through R_A and record in the place indicated.
4. Measure V_{AK} and record in the place indicated next to I_{RA}. Is the SCR on or off? ______
5. Move S_1 to position *B*.
6. Calculate the gate current I_{RB} flowing through R_B and record in the place indicated.
7. Measure V_{AK} and record in the place indicated next to I_{RB}. Is the SCR on or off? ______
8. Remove the power supply voltage from the circuit.
9. Modify the circuit by adding the ammeter and 50-kΩ potentiometer (R_H) as shown in Figure 8-7b.
10. Set the wiper of R_H so that the resistance is completely "shorted out."
11. Make sure that S_1 and S_2 are set as indicated and then apply power to the circuit.
12. Momentarily move S_1 from position A to position B and back again.
13. Record the reading of V_{AK} and I_A here:

    V_{AK} = ______

    I_A = ______

14. Slowly adjust R_H so that the current I_A begins to decrease.
15. Remember the reading of I_A when V_{AK} increases to $+V_{AA}$. Record this value in the place indicated for the minimum holding current of the SCR. (Perform steps 10 through 15 a few times for a more accurate reading.)

Figure 8–7 Current control of an SCR: (a) gate current control; (b) holding current control.

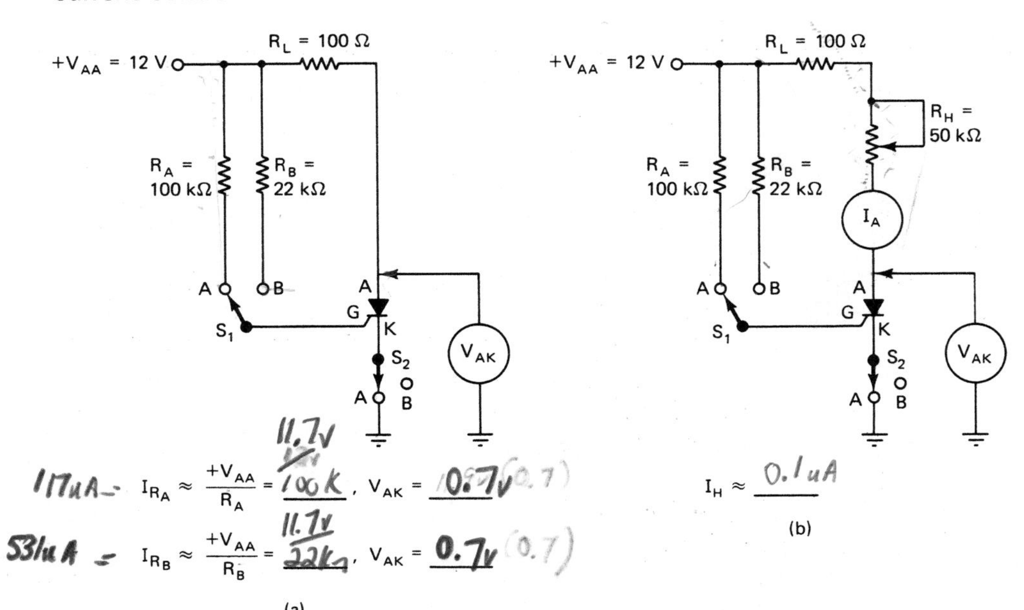

Fill-in Questions:

1. If the gate resistor is too large, not enough gate current will flow to trigger the SCR.

2. When the gate resistor is sufficient, the proper value of gate current will flow to trigger the SCR.

3. Sufficient holding current is required to keep the SCR conducting.

4. If the load resistance in series with the anode is too large, not enough current flows from cathode to anode and the SCR will turn off.

EXPERIMENT 4. SCR OVERVOLTAGE PROTECTION CIRCUIT

Objective:

To demonstrate how an SCR can be used to protect a given load from transient increases in the power supply voltage.

Introduction:

Some electronic components, such as integrated circuits, cannot withstand a voltage greater than their rating. The circuit shown in Figure 8-8 allows the SCR to shunt current away from the load (R_L) when overvoltages occur in the power supply. Resistors R_S, R_Z, and R_G are used for current limiting to protect the zener diode and SCR. The zener diode establishes a reference voltage for the output. Capacitor C_Z allows a delay in the rising voltage at the gate of the SCR when the power supply is turned on, thus preventing any false triggering. Diodes D_1 and D_2 have a total voltage drop of 1.4 V. Switch S_1 in parallel with these diodes is used to simulate an overvoltage condition. When S_1 is momentarily closed, the voltage to the protection circuit increases, and the zener diode draws more current, which triggers the SCR into conduction. Most of the circuit current now flows through the SCR instead of the 1-kΩ load resistor. Switch S_2 is used to reset or restore the voltage across the load after the overvoltage condition has been corrected or cleared. This type of circuit is referred to as a "crowbar" since, in effect, it pries the load loose from the power supply.

Materials Needed:

1 Variable 0- to 20-V power supply
1 Standard or digital voltmeter
1 MCR 106-2 SCR or equivalent with 2 PRV at 60 V
1 1N5231 5.1-V zener diode or equivalent
2 1N4001 diodes or equivalent (D_1 and D_2)
1 10-Ω resistor at 0.5 W (R_G)
1 100-Ω resistor at 0.5 W (R_Z)
1 100-Ω resistor at 2 W (R_S)
1 1-kΩ resistor at 0.5 W (R_L)
1 0.1-μF capacitor at 12 VW dc (C_Z)
2 DPST switches (S_1 and S_2)
1 Breadboard for constructing circuit

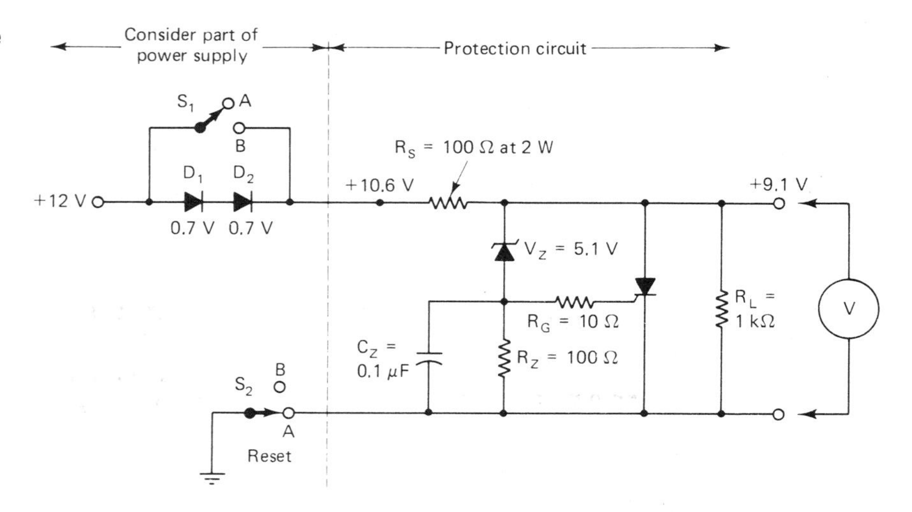

Figure 8–8 Overvoltage protection circuit.

Procedure:

1. Construct the circuit shown in Figure 8-8.
2. Place the voltmeter across R_L. It should indicate about +9 V. (If it does not, make sure that S_1 is in position A and then momentarily open and close S_2. If the SCR continues to turn on, the +12-V supply may need to be adjusted slightly lower.)
3. Momentarily move S_1 to position B and then back to position A.
4. Measure the voltage across R_L and record the value here: 5.1 . always
5. Momentarily move S_2 to position B and then back to position A. (This turns off the SCR and restores the power supply voltage to R_L.) Measure and record the voltage across R_L here: OPEN .
6. Repeat steps 3 through 5 a few times to understand fully the operation of the circuit.

Lose reference
no ground

Fill-in Questions:

1. An overvoltage protection circuit protects the load against increases in power supply voltage.
2. When an overvoltage occurs, the SCR conducts and in effect "shorts out" the LOAD .
3. When the SCR is conducting, most of the circuit current will flow through the SCR .
4. When the protection circuit is reset, the SCR is off and power supply voltage is restored to the LOAD ______ .

EXPERIMENT 5. AC TRIGGERING OF AN SCR

Objective:

To show how ac current to a load can be controlled by an SCR, depending on what portion of the positive alternation of a sine wave the SCR turns on.

Introduction:

The first part of this experiment uses only a variable resistance to vary the trigger time from 0° to 90°. An oscilloscope is used to view the voltage waveforms across the SCR and R_L.

A capacitor and diode are added to the original circuit to complete the second part of the experiment. The capacitor extends the trigger time to nearly 180°, and the diode produces a sharper current pulse when it conducts, to provide more trigger control.

Materials Needed:

1 12-V rms transformer or ac source
1 Oscilloscope (use only one channel)
1 C106Y1 SCR or equivalent
1 1N4001 diode or equivalent
1 100-Ω resistor at 0.5 W (R_L)
1 1-kΩ resistor at 0.5 W (R_A)
1 500-kΩ potentiometer (R_G)
1 0.2-μF capacitor at 25 WV dc (C_G)
1 Breadboard for constructing circuit

Procedure:

1. Construct the circuit shown in Figure 8-9a.
2. Place the oscilloscope across the SCR.
3. Vary R_G back and forth and view the voltage waveform across the SCR.
4. Adjust R_G so that the SCR triggers about halfway between 0° and 90°.
5. Draw the voltage waveform across the SCR in the space provided, making sure to align it with the proper degrees for one cycle. (Indicate peak-to-peak voltage.)
6. Place the oscilloscope across R_L.
7. Draw the voltage waveform across R_L in the space provided, making sure to align it with the proper degrees for one cycle. (Indicate peak-to-peak voltage.)
8. Modify the circuit as shown in Figure 8-9b by adding the capacitor and diode.
9. Place the oscilloscope across the SCR.

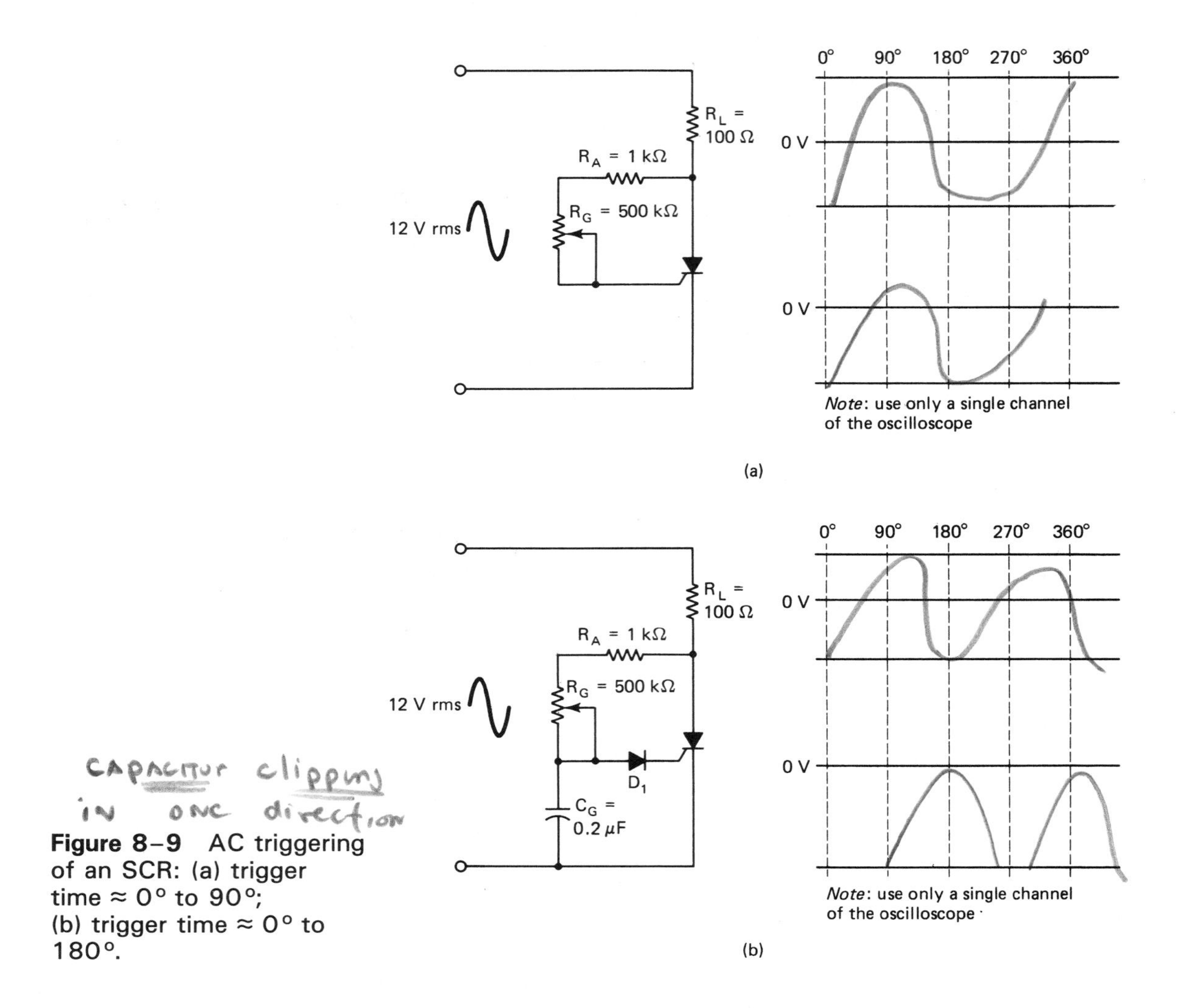

Figure 8–9 AC triggering of an SCR: (a) trigger time ≈ 0° to 90°; (b) trigger time ≈ 0° to 180°.

10. Vary R_G back and forth and view the voltage waveform across the SCR.
11. Adjust R_G so that the SCR triggers past 90°, but not at the 180° point.
12. Draw the voltage waveform across the SCR in the space provided, making sure to align it with the proper degrees for one cycle. (Indicate peak-to-peak voltage.)
13. Place the oscilloscope across R_L.
14. Draw the voltage waveform across R_L in the space provided, making sure to align it with the proper degrees for one cycle. (Indicate peak-to-peak voltage.)

Fill-in Questions:

1. Using only a potentiometer, the ac trigger time of an SCR can be varied from about ___0___ to ___90___ degrees.
2. Using a potentiometer and a capacitor, the ac trigger time of an SCR can be varied from about ___0___ to ___180___ degrees.
3. When the SCR conducts, the voltage across R_L is about equal to the ___applied___ ___voltage___.
4. When the SCR conducts, the voltage across its *A–K* terminals is about ___0.7___ V.
5. The voltage across R_L when the SCR conducts is the result of ___current___ times ___resistance___.

SECTION 8-5
SCR BASIC TROUBLESHOOTING APPLICATION: BASIC TRIGGER CIRCUIT

Construct the circuit shown. Open or short the components as listed and record the voltages in the proper place in the table. The first half of this troubleshooting application is when the SCR is off or before triggering; the last half is with the SCR on or triggered (note the position of S_1). The abbreviations will help indicate the voltage conditions associated with each problem. All voltages are referenced to ground.

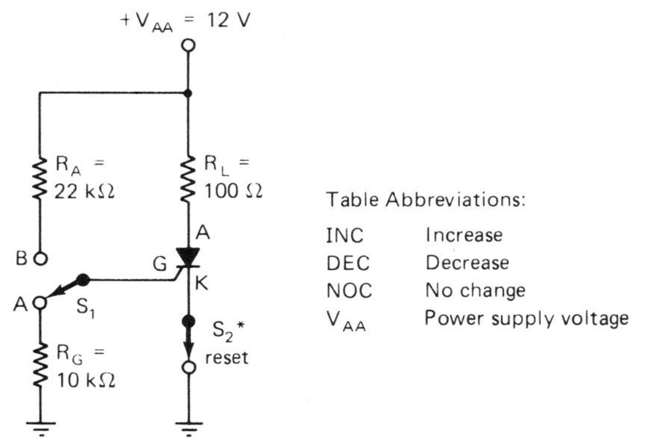

*Reset S_2 after each trouble condition.

Position of S_1	Condition	V_{AK}	V_G	Comments
A	Normal			All voltages proper (SCR off)
A	R_L open	DEC	NOC	No voltage applied to SCR
A	R_G open	V_{AA}	NOC	Little or no change unless SCR triggers because of static charge on gate
A	SCR open	V_{AA}	NOC	Open at K lead: no indications before SCR is triggered on
A	SCR shorted	DEC	NOC	Place short A to K: SCR conducts current before being triggered on
A	Gate open	V_{AA}	NOC	Open at S_1: no indications before SCR is triggered on
B	Normal			All voltages proper (SCR on)
B	R_A open	V_{AA}	DEC	No gate current flows
B	R_L open	DEC	DEC	No voltage applied to SCR
B	R_G open	NOC	NOC	Little or no change unless erratic operation is visible
B	SCR open	V_{AA}	DEC	Open at K lead: no current flow through SCR
B	SCR shorted	DEC	DEC	Place short A to K: current flow continuous through SCR
B	Gate open	V_{AA}	V_{AA}	Open at S_1: no gate current flows

Figure 8–E.1

SECTION 8-6 MORE EXERCISES

Perform the following exercises before beginning Section 8-7.

1. Draw the circuits shown in Figure 8-5 and indicate in the meters the ohmic readings (high or low) for a normally operating SCR.

2. Referring to Figure 8-6b, indicate which SCRs are conducting (On) or not conducting (Off) from the voltage readings given.

a. $V_{AA} = +12$ V
$V_{AK} = +0.5$ V

Conclusion: ______

b. $V_{AA} = +12$ V
$V_{AK} = +12$ V

Conclusion: ______

c. $V_{AA} = +15$ V
$V_{AK} = +15$ V

Conclusion: ______

3. On the following circuits, calculate the indicated values using the formulas $V_{AK} = +V_{DD} - V_{RL}$, $V_{RL} = +V_{DD} - V_{AK}$, $V_{RL} = I_A R_L$, $I_A = +V_{DD} - V_{AK}/R_L$, and $R_L = V_{RL}/I_A$.

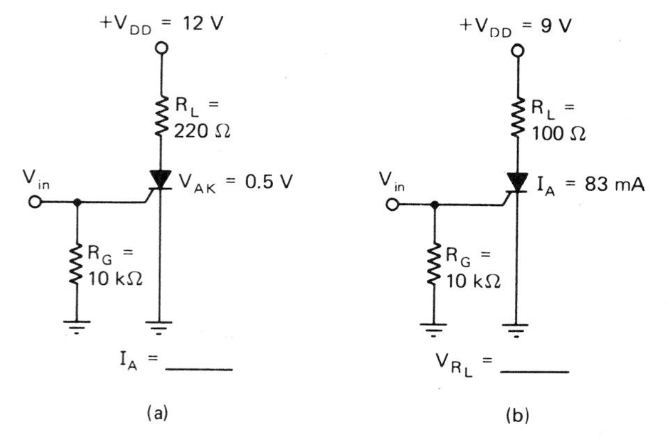

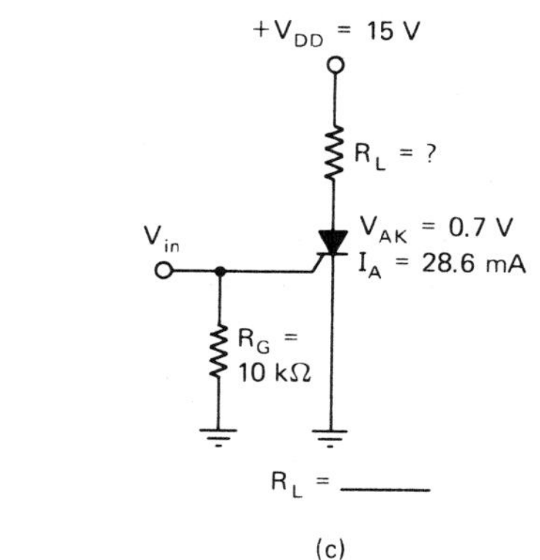

Figure 8-E.2

SECTION 8-7 SCR INSTANT REVIEW

An SCR is a bipolar device that is normally forward biased with a positive voltage applied to the anode and a negative voltage applied to the cathode. However, the SCR will not conduct unless a positive voltage is applied to the gate lead. When this gate voltage is removed, the SCR continues conducting and the gate loses control over the current flowing through the SCR. Sufficient gate current must flow to turn on the SCR, and the minimum holding current is what keeps the SCR conducting.

SECTION 8-8 SELF-CHECKING QUIZZES

8-8a SCR: TRUE–FALSE QUIZ

Place a T for true or an F for false to the left of each statement.

__F__ **1.** The SCR is a unipolar device. (bi)

__T__ **2.** Once the SCR is "triggered on," the gate loses control over the anode current.

__F__ **3.** An SCR is triggered on with a negative pulse to the gate. (positive

__F__ **4.** An SCR has three semiconductor layers.

__T__ **5.** The minimum value of forward current required to keep an SCR conducting is called holding current.

__T__ **6.** An SCR can be turned on by exceeding its forward breakover voltage or with a positive pulse to the gate.

__T__ **7.** An SCR's forward breakover voltage will decrease when its gate current is increased.

__T__ **8.** In most cases an SCR can be tested with an ohmmeter.

__T__ **9.** The SCR belongs to the thyristor family of semiconductor devices.

__F__ **10.** Once triggered on, the SCR will conduct for the entire cycle of an ac voltage.

8-8b SCR: MULTIPLE-CHOICE QUIZ

Circle the correct answer for each question.

1. The SCR is used primarily as an:

a. amplifier
b. oscillator
c. electronic switch
d. none of the above

2. Once triggered on, an SCR can be turned off in a dc circuit when the:

a. positive voltage is removed from the gate
b. negative voltage is removed from the gate
c. holding current falls below the minimum point
d. none of the above

3. The trigger time of an SCR circuit using a gate capacitor is approximately:

a. 0° to 45° **b.** 0° to 90°
c. 90° to 180° **d.** 0° to 180°

4. The correct waveform that might be seen across the SCR in Figure 8-A when it is conducting is:

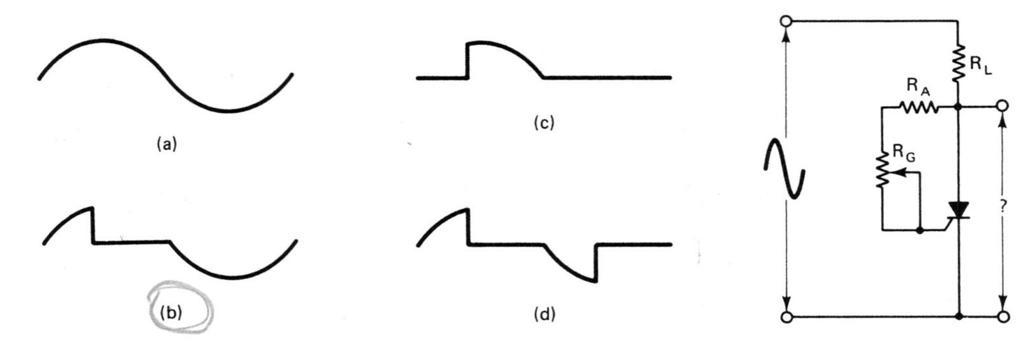

Figure 8–A

5. When an SCR is used with ac voltage and is "triggered" into operation, it conducts:
   a. for the entire ac cycle
   b. for only the positive alternation of the cycle
   c. for only the negative alternation of the cycle
   d. on neither alternation of the ac cycle, because it is used only with dc circuits

ANSWERS TO EXPERIMENTS AND QUIZZES

Experiment 1. (1) high (2) low (3) gate, cathode (4) holding current

Experiment 2. (1) $+V_{AA}$ (2) positive, increases (3) loses, continues (4) ≈ 0.7 V (5) reset

Experiment 3. (1) trigger (2) sufficient (3) holding current (4) off

Experiment 4. (1) power supply (2) load (3) SCR (4) SCR, load

Experiment 5. (1) 0, 90 (2) 0, 180 (3) applied voltage (4) 0.7 (5) current, resistance

True–False. (1) F (2) T (3) F (4) F (5) T (6) T (7) T (8) T (9) T (10) F

Multiple Choice. (1) c (2) c (3) d (4) b (5) b

Unit 9

DIAC/TRIAC (Thyristors)

INTRODUCTION Thyristors are used in trigger circuits and as protection devices in electronic circuits. This unit presents various other breakover voltage devices.

UNIT OBJECTIVES Upon completion of this unit you will be able to:

1. Draw the schematic symbol for a DIAC and TRIAC.
2. Describe the operation of a DIAC and TRIAC.
3. Draw the current–voltage characteristic curve for a DIAC and TRIAC.
4. Explain the phase angle triggering and conducting time for a DIAC–TRIAC circuit.
5. Identify DIAC and TRIAC voltage waveforms associated with ac circuits.
6. Describe the operation of miscellaneous thyristor devices.
7. Test DIACs and TRIACs.
8. Construct circuits using DIACs and TRIACs.
9. Troubleshoot a basic TRIAC trigger circuit.

SECTION 9-1 THEORY OF STRUCTURE AND OPERATION

9-1a DIAC STRUCTURE AND SCHEMATIC SYMBOL

The DIAC is a three-layer (two-junction) *PN* device, as shown in Figure 9-1a. Its official name is "bidirectional diode thyristor," which means that it is bilateral or will conduct in both directions. Its schematic symbol (Figure 9-1b) is drawn to indicate this. The DIAC operates similar to two zener diodes placed face to face or back to back in series (Figure 9-1c).

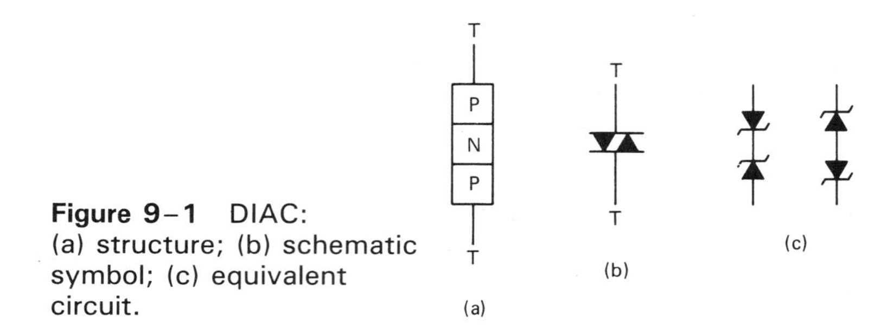

Figure 9–1 DIAC: (a) structure; (b) schematic symbol; (c) equivalent circuit.

9-1b DIAC OPERATION

The DIAC conducts when the voltage across its terminals exceeds the breakover point, as shown in Figure 9-2. When a positive voltage is applied to the DIAC (Figure 9-2a), junction J_1 is forward biased, while junction J_2 is reverse biased. As the applied voltage is increased, the forward voltage breakover ($+V_{BO}$) point is reached (Figure 9-2b), where junction J_2 becomes forward biased as a result of avalanche breakdown, and an appreciable amount of current flows through the DIAC. If the DIAC is turned around or a negative voltage is used (Figure 9-2c), junction J_1 is reverse biased and J_2 is forward biased. As the applied voltage increases more negatively, the reverse voltage breakover ($-V_{BO}$) point is reached (Figure 9-2d), where junction J_1 becomes forward biased as a result of avalanche breakdown, and again an appreciable amount of current flows, but in the opposite direction.

Figure 9–2 DIAC operation: (a) no current before $+V_{BO}$; (b) current after $+V_{BO}$; (c) no current before $-V_{BO}$; (d) current after $-V_{BO}$.

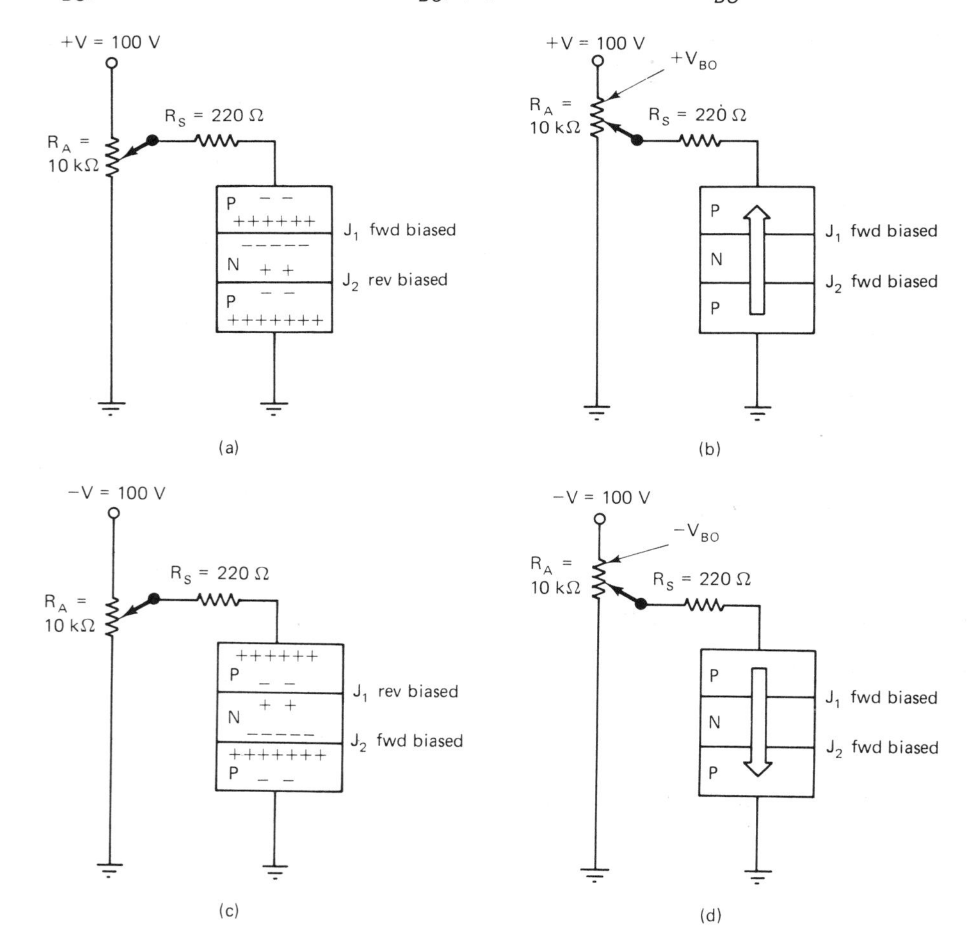

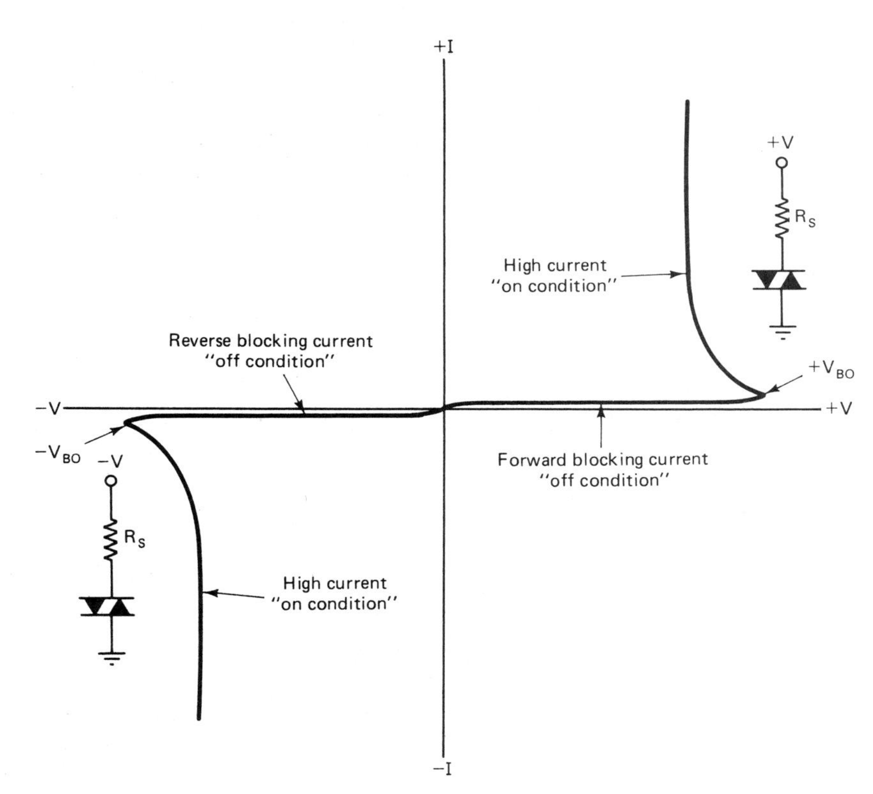

Figure 9–3 Current–voltage characteristics for a DIAC.

9-1c CURRENT–VOLTAGE CHARACTERISTICS OF A DIAC

The current–voltage characteristics of a DIAC are shown in Figure 9-3. Before the voltage breakover point is reached, a very small blocking current flows through the device, but the DIAC is considered to be in the "off condition." When the DIAC fires, a high current flows through it and it is considered to be in the "on condition." The current in the circuit is limited by the series resistor R_S. Notice that when it is turned on the voltage across the DIAC is less than the voltage breakover point. This is a result of negative resistance or regeneration of the device.

9-1d DIAC USED FOR AC PEAK LIMITING

A DIAC is used primarily to produce a sharp trigger current pulse to turn on other types of thyristors. It can be used in a simple protection circuit as shown in Figure 9-4. In this circuit the DIAC acts like a voltage limiter. The peak voltage of one alternation is 50V; however, the V_{BO} of the DIAC is 40 V. As the voltage increases, the DIAC is off or open, and the applied voltage is seen across its terminals. When 40 V is reached (in either direction), the DIAC conducts and the voltage is clamped at about 35 V across the DIAC.

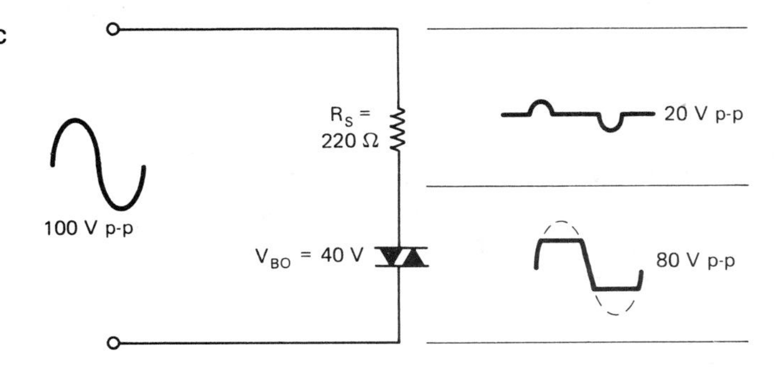

Figure 9–4 Limiting ac peaks with a DIAC.

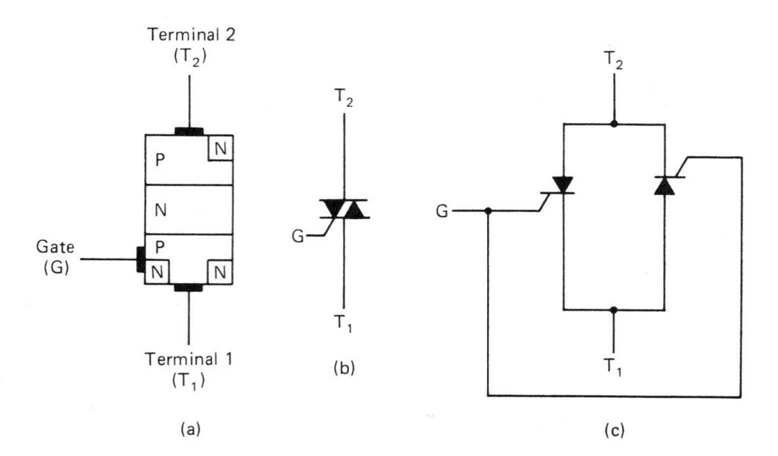

Figure 9–5 TRIAC: (a) structure; (b) schematic diagram; (c) equivalent circuit.

9-1e TRIAC STRUCTURE AND SCHEMATIC SYMBOL

The TRIAC is basically a three-layer *PN* device with added *N* regions, as shown in Figure 9-5a. These added *N* regions permit the leads to contact both *P*- and *N*-type material, so the TRIAC can be triggered into conduction in either direction, making it a bilateral device. This is indicated by its schematic symbol (Figure 9-5b). The official name of the TRIAC is "bidirectional triode thyristor." It is similar to two SCRs connected in inverse parallel with their gates tied together (Figure 9-5c).

9-1f TRIAC OPERATION

The TRIAC conducts when the voltage across its main terminals (T_1 and T_2) exceeds the breakover point or if it is triggered by a positive or negative voltage pulse applied to the gate lead, as shown in Figure 9-6. It operates similar to an SCR or DIAC, where one of the internal junctions is reverse biased, until conduction occurs. Once it is triggered on, the gate loses control of the current flow, and the holding current will keep it conducting. Switch S_1 can be used to interrupt the holding current and turn the TRIAC off.

There are four modes of triggering for a TRIAC. When T_2 is positive and a positive pulse is applied to the gate, the TRIAC turns on and current tends to flow through it in one area or portion (Figure 9-6a). A negative pulse will also turn it on, but the current tends to flow in the

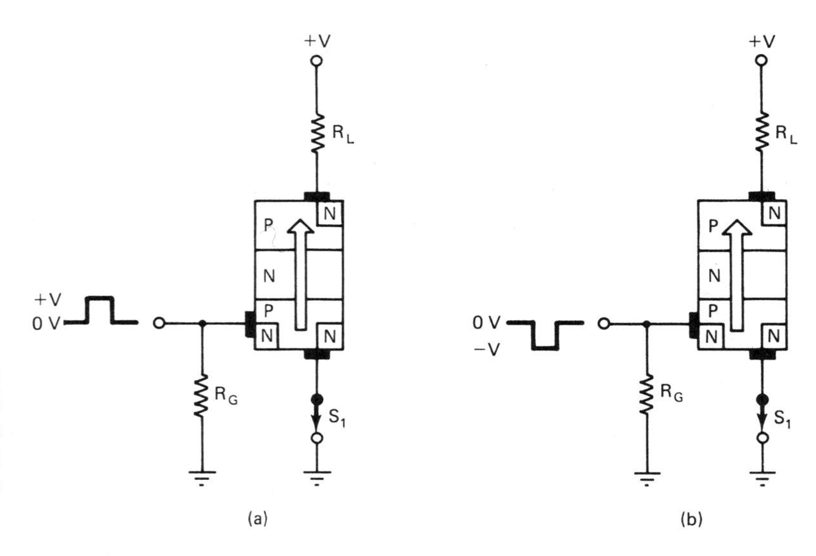

Figure 9–6 Operation of a TRIAC: (a) positive terminal voltage with positive gate voltage; (b) positive terminal voltage with negative gate voltage;

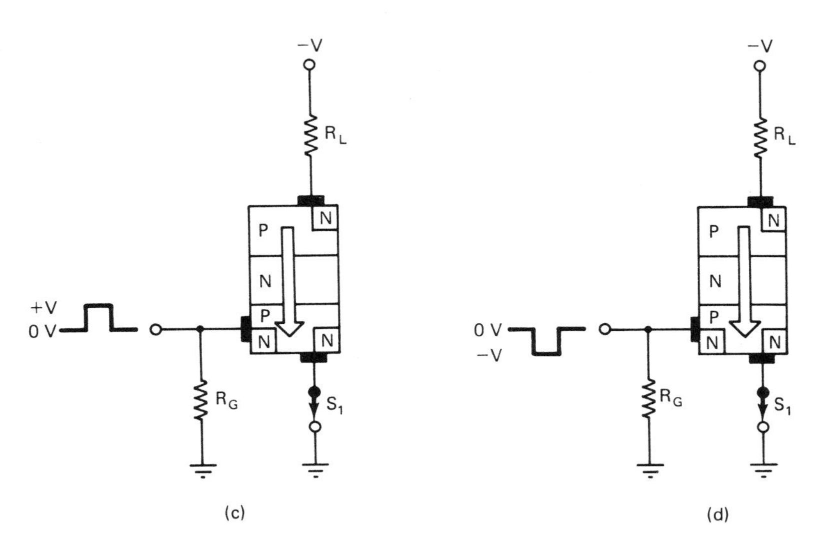

Figure 9–6 (c) negative terminal voltage with positive gate voltage; (d) negative terminal voltage with negative gate voltage.

opposite area or portion of the device (Figure 9-6b). When T_2 is negative and a positive pulse is applied to the gate, the TRIAC turns on and the current tends to flow in one area, but in the opposite direction (Figure 9-6c). A negative pulse will also turn it on, but the current tends to flow in the other portion of the device (Figure 9-6d).

9-1g CURRENT–VOLTAGE CHARACTERISTICS OF A TRIAC

The current–voltage characteristics of a TRIAC, shown in Figure 9-7, are similar to those of an SCR, except that conduction occurs in the reverse-biased condition also. Severe regeneration develops in both directions when the TRIAC turns on. The current through the device increases to a high level, while the resistance of the junctions decreases,

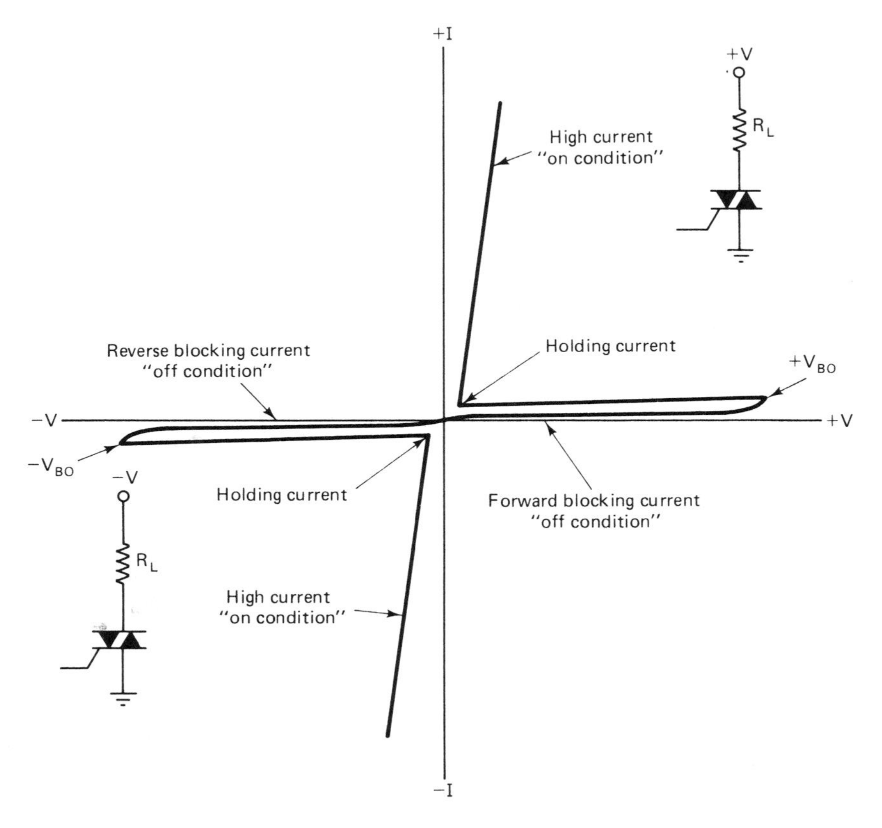

Figure 9–7 Current–voltage characteristics for a TRIAC.

which results in a low-voltage drop across its main terminal. This curve indicates that the breakover voltages occur with the gate open; however, lower breakover voltage points are directly proportional to the amount of gate current. The holding current keeps the TRIAC "latched" in either direction.

9-1h AC CONTROL USING A TRIAC

The TRIAC is truly an ac control device, since it can control the amount of current applied to a load for the positive and negative alternations of a cycle, as shown in Figure 9-8. A TRIAC usually will not trigger at the same point for each alternation of a cycle and is referred to as being unsymmetrical. Very often a triggering device, such as a DIAC, is placed in series with the gate lead of a TRIAC to make its triggering time more symmetrical (that is, occur at the same place for each alternation of a cycle).

With a simple resistive trigger circuit (Figure 9-8a), the triggering time of the TRIAC can be controlled for approximately 0° to 90° and 180° to 270° of a sine wave. The conduction time would be about 0° to 180° for the positive alternation and 180° to 360° for the negative alternation. When the TRIAC is off, the applied voltage is present across its main terminals, and little or no current flows through it and the load (R_L). The voltage drop across the load at this time is zero. When the TRIAC is on, the voltage drop across its terminals is low, but current now flows through it and the load, which develops a voltage drop across the load.

The addition of a capacitor to the gate circuit, as shown in Figure 9-8b, delays the voltage applied to the gate, thereby extending the trigger time of the TRIAC for nearly the entire cycle. Thus the conduction

Figure 9-8 AC triggering of a TRIAC: (a) resistive triggering time ≈ 0° to 90° and 180° to 270°; (b) RC triggering time ≈ entire cycle.

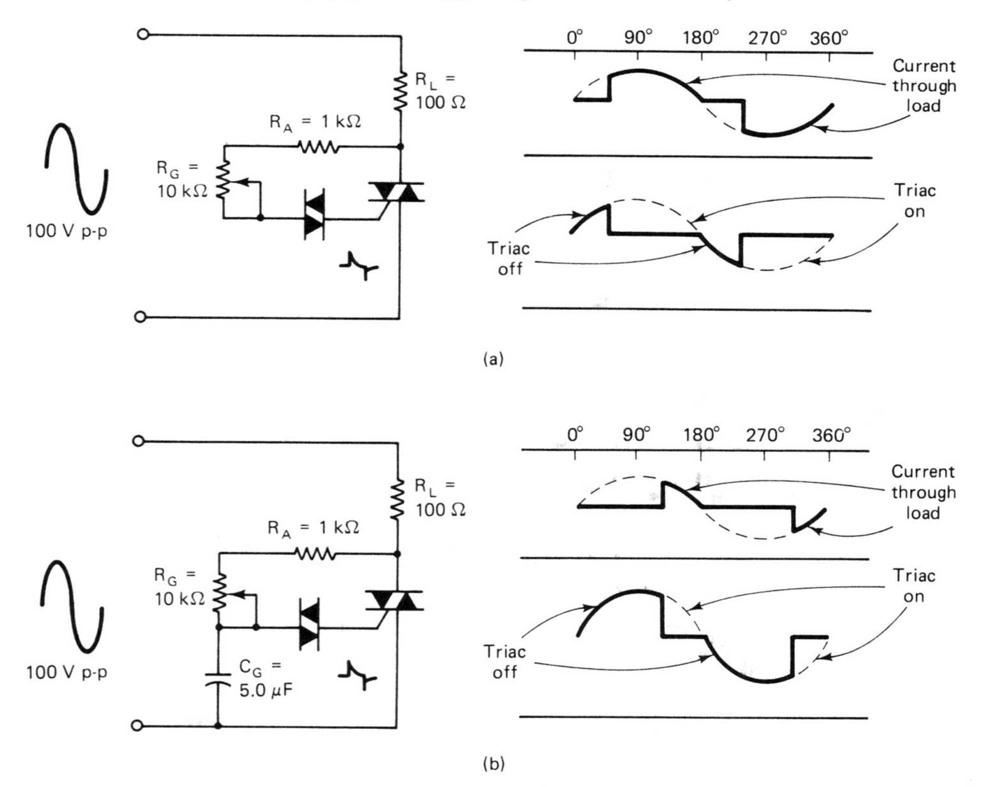

time is more accurately controlled, and small amounts and/or large amounts of current can be allowed to flow through the load.

A circuit very similar to this one is used in light-dimmer control circuits in houses. The incandescent light replaces R_L, and control potentiometer R_G can be adjusted to make the light glow brighter or dimmer.

9-1i THYRISTOR DEFINITIONS

Since the SCR is also a thyristor, most of the definitions given in Section 8-1e apply also to the DIAC and TRIAC. The following terms apply more specifically to these bilateral devices.

$+V$	forward voltage or positive alternation applied to main terminals of device
$-V$	reverse voltage or negative alternation applied to main terminals of device
$+V_{BO}$	forward breakover voltage
$-V_{BO}$	reverse breakover voltage
V_{T_1}	voltage from terminal 1 to ground (TRIAC)
V_{T_2}	voltage from terminal 2 to ground (TRIAC)
$+I$	forward current when the thyristor is on
$-I$	reverse current when the thyristor is off

9-1j MISCELLANEOUS THYRISTORS

A few other thyristor devices are used in various triggering applications. Figure 9-9 shows the schematic symbol and voltage–current characteristic curve for these thyristors.

9-1j.1 Shockley Diode

The *Shockley diode* (Figure 9-9a), also referred to as a reverse-blocking diode thyristor, is a four-layer device that operates similarly to a DIAC. It is triggered into conduction in the forward-biased mode by exceeding the anode breakover voltage.

9-1j.2 Silicon-controlled Switch

The *silicon-controlled switch* (SCS) (Figure 9-9b), also called a reverse-blocking tetrode thyristor, operates like an SCR, but can be triggered on by either gate lead. A positive pulse on the cathode gate lead turns it on, or a negative pulse to the anode gate lead accomplishes the same thing.

9-1j.3 Silicon Unilateral Switch

The *silicon unilateral switch* (SUS) (Figure 9-9c) has a zener junction added to the gate, which determines the positive voltage trigger level. The SUS can be triggered by normal breakover voltage or a positive or negative pulse applied to the gate.

9-1j.4 Silicon Bilateral Switch

The *silicon bilateral switch* (SBS) (Figure 9-9d) operates like a SUS, but will conduct in either direction. It can be triggered into conduction by

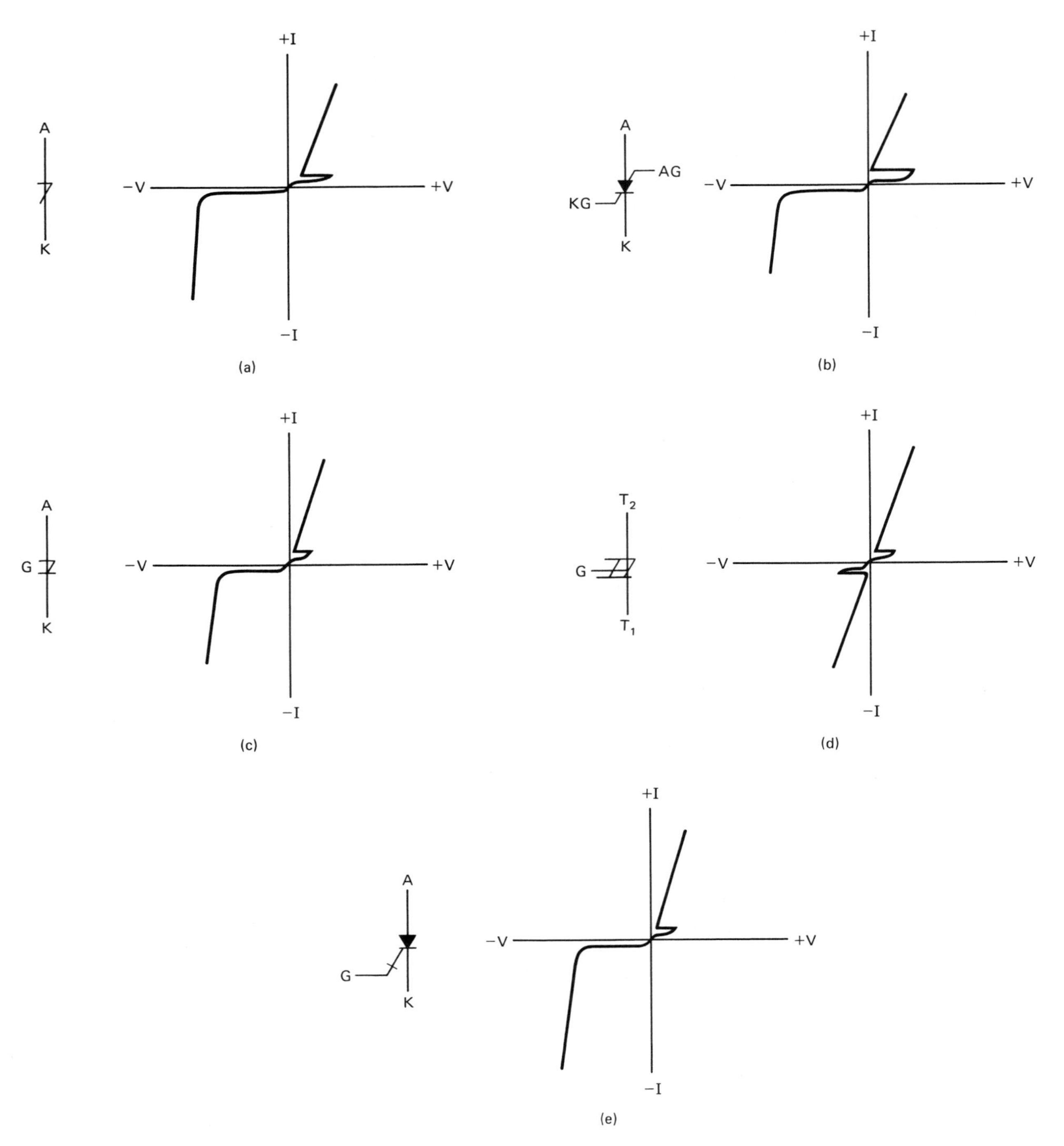

Figure 9–9 Miscellaneous thyristors: (a) Shockley diode; (b) SCS; (c) SUS; (d) SBS; (e) GCS.

exceeding the breakover voltage in either direction or by a positive or negative pulse applied to the gate. The SBS can be considered a low-voltage TRIAC.

9-1j.5 Gate-controlled Switch

The *gate-controlled switch* (GCS) (Figure 9-9e) is called a turnoff thyristor and is similar to an SCR. A positive pulse to the gate triggers it into conduction, while a negative pulse to the gate turns it off.

Notice that the Shockley diode, SCS, SUS, and GCS have a low forward breakover voltage, whereas, the reverse breakover voltage is greater and has a zener effect.

SECTION 9-2
THYRISTOR DEFINITION EXERCISE

Refer to the previous sections and write a brief description of each term.

1. $+V$ ______________________________

____

____

2. $-V$ ______________________________

____

____

3. $+V_{BO}$ ______________________________

____

____

4. $-V_{BO}$ ______________________________

____

____

5. V_{T_1} ______________________________

____

____

6. V_{T_2} ______________________________

____

____

7. $+I$ ______________________________

____

____

8. $-I$ ______________________________

____

____

SECTION 9-3 EXERCISES AND PROBLEMS

Perform the following exercises before beginning Section 9-4.

1. Draw the structure of a DIAC. (Label all parts.)

2. Draw the schematic diagram of a DIAC. (Label the leads.)

3. Draw a sample characteristic curve for a DIAC. (Label the forward and reverse voltages and currents.)

4. Draw the circuit in Figure 9-4, but use an input voltage of 120 V p-p and draw the output waveform indicating peak-to-peak voltage.

5. Draw the structure of a TRIAC. (Label all parts.)

6. Draw the schematic diagram of a TRIAC. (Label the leads.)

7. Draw a sample characteristic curve for a TRIAC. (Label the forward and reverse voltages and currents.)

8. Draw the circuit in Figure 9-12a and write in the voltage nomenclature (V_G and V_{T_2}).

SECTION 9-4 EXPERIMENTS

EXPERIMENT 1. TESTING A DIAC

Objective:

To demonstrate how a DIAC can be tested with a power supply voltage.

Introduction:

A DIAC can be tested with a power supply voltage that is at least a few volts greater than its V_{BO}. The DIAC is connected in series with a small-value limiting resistor, which in turn is connected to a potentiometer or variable power supply voltage. The voltage across the circuit is then increased until V_{BO} is reached. The power supply is then set to zero volts. The DIAC is reversed in the circuit, and the same procedure is performed to test the reverse direction.

Materials Needed:

1 100-V power supply
1 Standard or digital voltmeter
1 R.S. 276-1050 DIAC or equivalent with $V_{BO} = 40$ V
1 470-Ω resistor at 0.5 W (R_S)
1 10-kΩ potentiometer (R_A)
1 Breadboard for constructing circuit

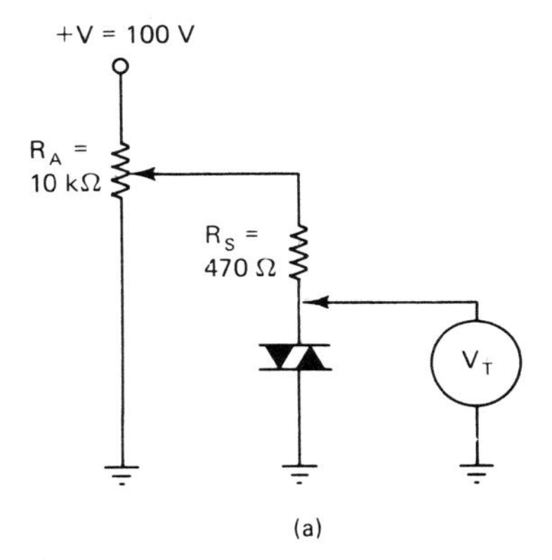

Circuit condition	V_T	Condition of Diac (on, off)
Before firing	0 to V_{BO}	
After firing		
Reverse Diac in circuit		
Before firing	0 to V_{BO}	
After firing		

(b)

Figure 9–10 Testing a DIAC: (a) test circuit; (b) data table.

Procedure:

1. Construct the circuit shown in Figure 9-10.
2. Place the wiper of R_A to ground.
3. Place the voltmeter across the DIAC.
4. Record the condition of the DIAC in the first line of the data table of Figure 9-10b.
5. Adjust R_A so that the voltage indicated by the meter begins to increase.
6. Continue to increase the voltage until there is a slight decrease in the voltage reading.

7. Record the condition of the DIAC on the second line of the data table.
8. Place the wiper of R_A back to ground.
9. Reverse the DIAC in the circuit.
10. Record the condition of the DIAC on the third line of the data table.
11. Repeat steps 5 and 6.
12. Record the condition of the DIAC on the fourth line on the data table.

Fill-in Questions:

1. The DIAC turns on when its __________ ____________ is reached in either direction.

2. When the DIAC turns on in either direction, the voltage across its terminals ____________ slightly, indicating a ____________-resistance action.

3. The voltage across the DIAC decreases when it turns on, but the current through it ____________ .

4. The DIAC is similar to two ____________ ____________ face to face or back to back in series.

EXPERIMENT 2. TESTING A TRIAC WITH AN OHMMETER

Objective:

To show how to test a TRIAC for conduction in both directions.

Introduction:

A TRIAC can be tested with an ohmmeter similar to testing an SCR or PUT. The positive lead of the ohmmeter is placed on T_2 and the negative lead is placed on T_1. The meter should read infinite resistance. A clip lead is placed from the positive lead to the gate, which should trigger on the TRIAC. The meter should now indicate low resistance. When the clip lead is removed, the meter continues to indicate low resistance if the power source is sufficient to produce the required holding current. The meter leads are reversed on the main terminals of the TRIAC, and a clip lead is placed from the negative lead to the gate to test for conduction in the reverse direction. This is a go/no go test.

Materials Needed:

A standard or digital voltmeter
One or several TRIACS
A clip lead

Procedure:

1. Set the ohmmeter to the low-range scale.
2. Connect the ohmmeter to the TRIAC as shown in Figure 9-11b and record the meter reading in the ohmmeter circle.
3. Connect the clip lead as shown in Figure 9-11c and record the meter reading in the ohmmeter circle.
4. Remove the clip lead as shown in Figure 9-11d and record the reading of the meter in the ohmmeter circle.
5. Connect the ohmmeter to the TRIAC as shown in Figure 9-11e and record the meter reading in the ohmmeter circle.
6. Connect the clip lead as shown in Figure 9-11f and record the reading of the meter in the ohmmeter circle.
7. Remove the clip lead as shown in Figure 9-11g and record the reading of the meter in the ohmmeter circle.

Fill-in Questions:

1. A TRIAC will have ____________ resistance in either direction before being triggered.

2. A TRIAC will have ____________ resistance in either direction after being triggered.

3. A TRIAC is being tested with an ohmmeter. When the clip lead is removed, the meter indicates high resistance. This does not prove that the TRIAC is defective, but that the power source of

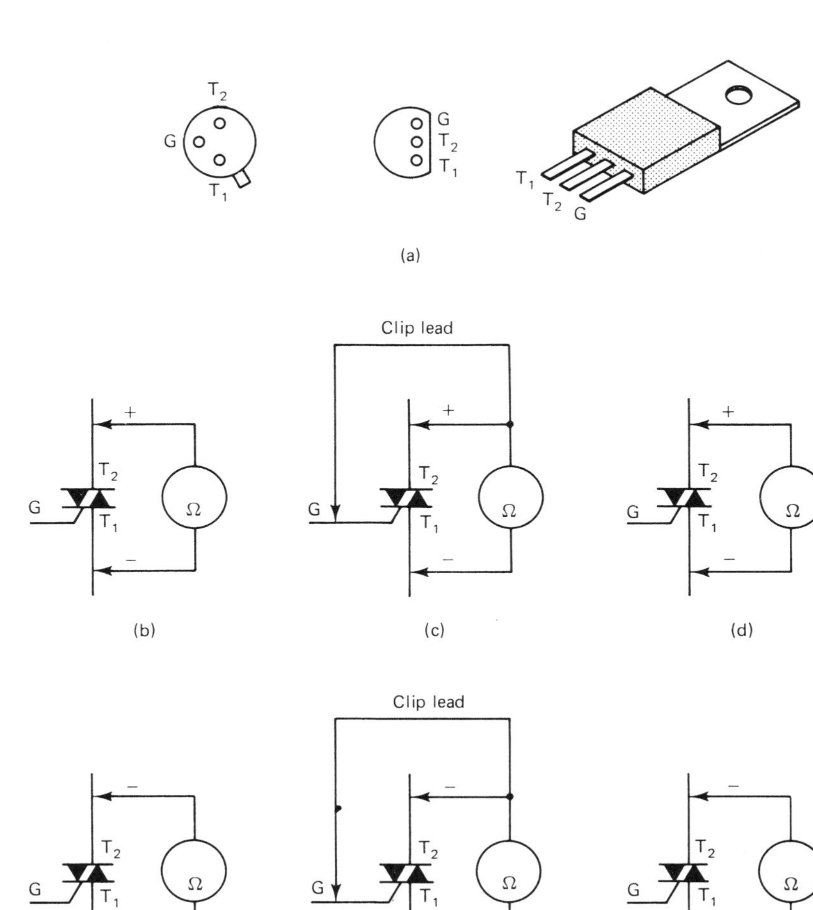

Figure 9–11 Testing a TRIAC with an ohmmeter: (a) general lead identification; (b) without clip lead; (c) with clip lead; (d) again without clip lead; (e) without clip lead; (f) with clip lead; (g) again without clip lead.

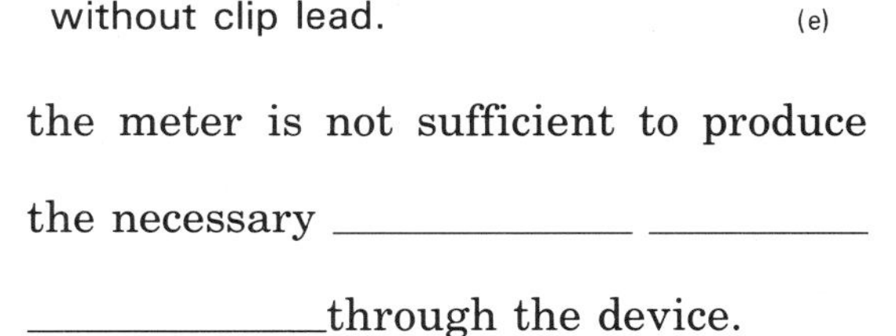

the meter is not sufficient to produce the necessary ____________ __________ ____________through the device.

4. If the ohmmeter shows low resistance before the TRIAC is triggered, this indicates that the TRIAC is ____________ .

5. If the ohmmeter shows infinite resistance after the TRIAC is triggered, this indicates that the TRIAC is ____________ .

EXPERIMENT 3. TRIGGERING MODES OF A TRIAC

Objective:

To demonstrate how the TRIAC conducts in both directions, and how it can be triggered with positive and negative voltages.

Introduction:

There are four modes of triggering a TRIAC:

1. Positive terminal voltage with positive trigger voltage
2. Positive terminal voltage with negative trigger voltage
3. Negative terminal voltage with positive trigger voltage
4. Negative terminal voltage with negative trigger voltage

Materials Needed:

1 15-V dual power supply
1 Standard or digital voltmeter

1 2N5754 TRIAC or equivalent
1 100-Ω resistor at 0.5 W (R_L)
2 1-kΩ resistors at 0.5 W (R_A and R_B)
1 10-kΩ resistor at 0.5 W (R_G)
1 TPST switch (S_1) (a single wire may be used)
1 DPST switch (S_2)
1 Breadboard for constructing circuit

Procedure:

1. Construct the circuit shown in Figure 9-12a.
2. Open and close S_2 to make sure that the TRIAC is off.
3. Measure and record V_G and V_{T_2} in the first line of the data table in Figure 9-12b.
4. Indicate on the same line of the data table if the TRIAC is on or off.
5. Move S_1 to position B. Measure and record the data in the table on the second line as done in steps 3 and 4.
6. Move S_1 to position A and again measure and record the data on the third line of the data table.
7. Move S_2 to position B and then back to position A. Measure and record the data on the fourth line of the data table.
8. Move S_1 to position C. Measure and record the data on the fifth line of the data table.

Figure 9–12 Triggering modes of a TRIAC: (a) test circuit with +V; (b) data table; (c) test circuit with −V; (d) data table.

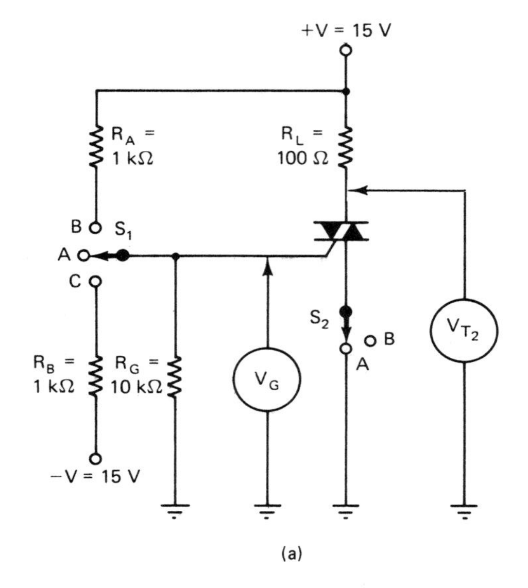

(a)

S_1 condition	S_2 condition	V_G	V_{T_2}	Condition of Triac (on or off)
A	A			
B	A			
A	A			
A	A → B → A			
C	A			
A	A			
A	A → B → A			

(b)

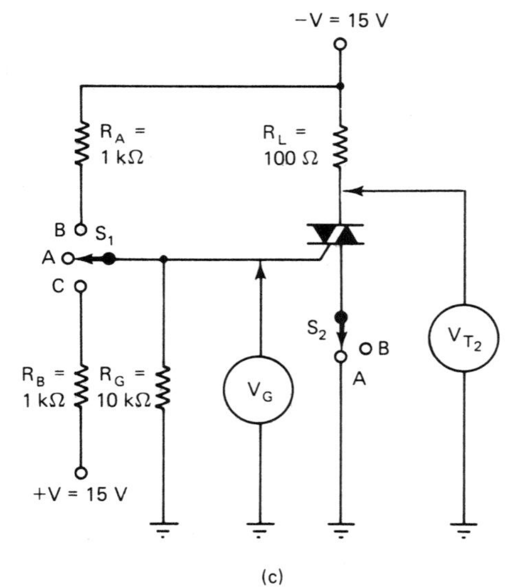

(c)

S_1 condition	S_2 condition	V_G	V_{T_2}	Condition of Triac (on or off)
A	A			
B	A			
A	A			
A	A → B → A			
C	A			
A	A			
A	A → B → A			

(d)

9. Move S_1 to position A. Measure and record the data on the sixth line of the data table.
10. Move S_2 to position B and then back to position A. Measure and record the data on the seventh line of the data table.
11. Reverse the power supply voltages as shown in Figure 9-12c to test the TRIAC for conduction in the other direction.
12. Repeat steps 1 through 10, but record the data in the data table of Figure 9-12d.

Fill-in Questions:

1. Before the TRIAC is triggered on, the voltage across its main terminals is equal to the ________ ________ voltage.
2. When the TRIAC is triggered on, the voltage across its main terminals is about ______ V.
3. Once the TRIAC is conducting, the gate ________ control and the current ________ to flow through the TRIAC.
4. The TRIAC can be turned off by reducing the current through it below its ________ .
5. The TRIAC can have ________ or ________ terminal voltage and be triggered on by ________ or ________ voltage applied to the gate.

EXPERIMENT 4. PEAK VOLTAGE LIMITING WITH A DIAC

Objective:

To show how a DIAC can be used to limit voltage peaks and serve as a circuit-protection device.

Introduction:

This experiment is similar to Experiment 5 in Section 2-4. The input voltage will be clipped at slightly less than the $+V_{BO}$ for each alternation of the sine wave. Observing the waveforms across the DIAC and resistor R_S will show when current is not flowing and flowing in the circuit. Remember that the voltage drop across R_S is the result of the current flow through it, times its resistance.

Materials Needed:

1 100-V p-p ac power source
1 Oscilloscope (use only one channel)
1 R.S. 276-1050 DIAC or equivalent with V_{BO} = 40 V
1 1-kΩ resistor at 0.5 W (R_S)
1 Breadboard for constructing circuit

Procedure:

1. Construct the circuit shown in Figure 9-13.
2. Place the oscilloscope across the DIAC and draw the waveform displayed in the proper place. Indicate proper phase

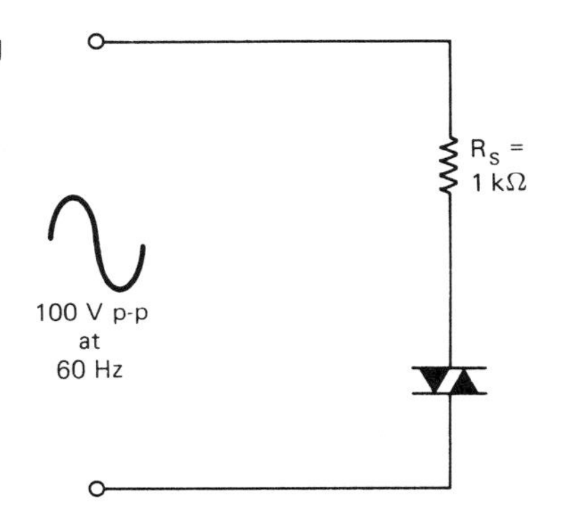

Figure 9–13 Peak limiting with a DIAC.

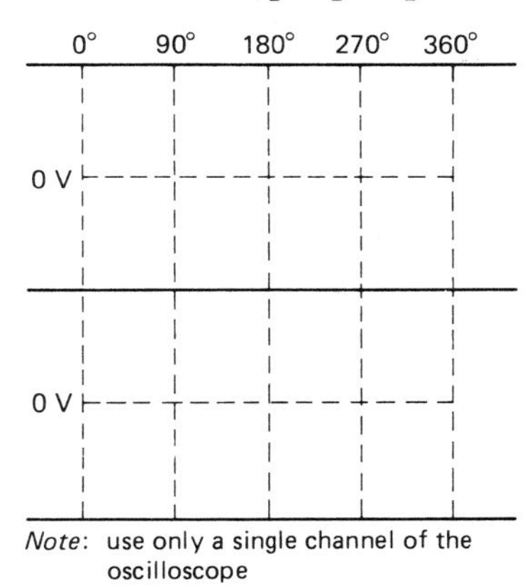

and peak-to-peak voltage measurements.

3. Place the oscilloscope across R_S and draw the waveform displayed in the place shown. Indicate proper phase and peak-to-peak voltage measurements.

Fill-in Questions:

1. When the applied voltage reaches $\pm V_{BO}$, the DIAC ______________ .

2. When the DIAC is conducting, ______________ flows through R_S and a voltage waveform is present ______________ ______________ it.

3. The DIAC is used primarily to trigger other ______________ , but it can be used as a ______________ device for other circuits.

EXPERIMENT 5. AC TRIGGERING OF A TRIAC

Objective:

To demonstrate how ac current to a load can be controlled by a TRIAC.

Introduction:

The first part of this experiment uses only a variable resistor to vary the trigger time from 0° to 90° and 180° to 270°. An oscilloscope is used to view the voltage waveforms across the TRIAC and R_L.

In the second part of this experiment, a capacitor is added to the original circuit to extend the trigger time for nearly the entire cycle. The zener diodes simulate a DIAC, which makes the trigger time more symmetrical.

Materials Needed:

1 24-V rms transformer or ac source

1 Oscilloscope (use only one channel)

Figure 9–14 AC triggering of a TRIAC: (a) resistive triggering time ≈ 0° to 90° and 180° to 270°; (b) RC triggering time ≈ the entire cycle.

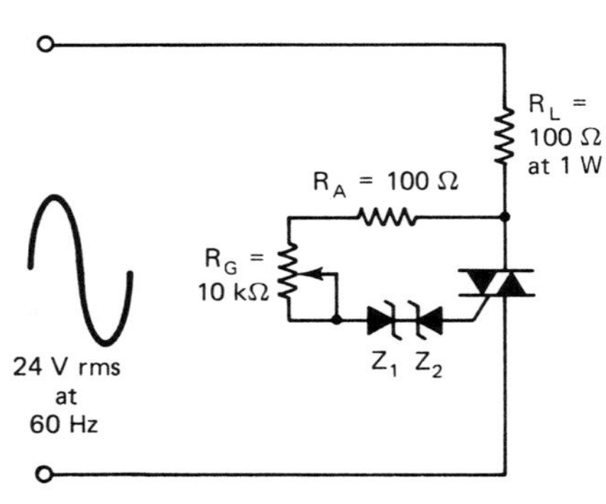

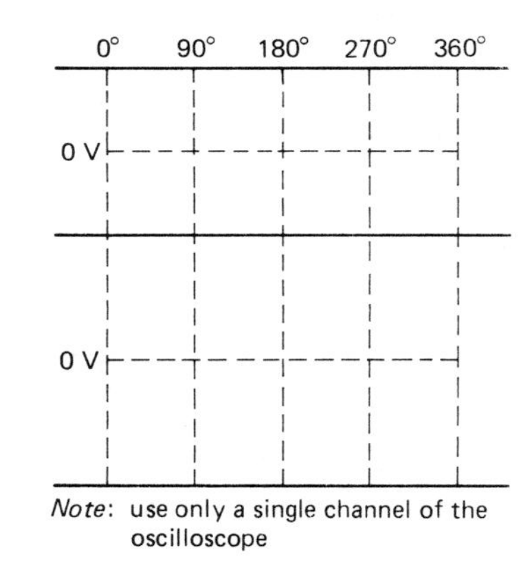

(a)

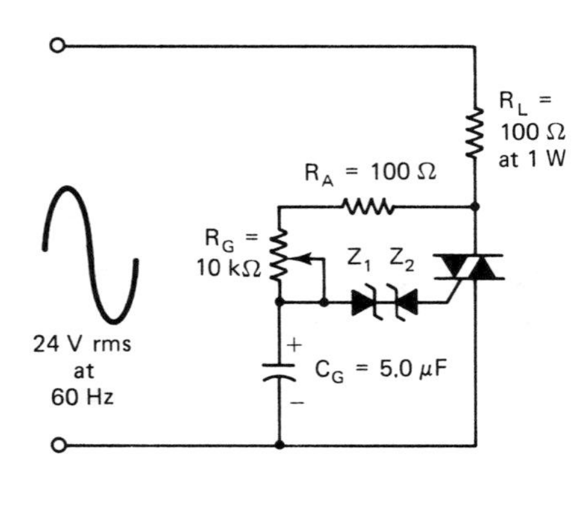

0° 90° 180° 270° 360°

0 V

0 V

Note: use only a single channel of the oscilloscope

(b)

1 2N5754 TRIAC or equivalent
2 1N5231 zener diodes or equivalent (Z_1, Z_2)
2 100-Ω resistors at 0.5 W (R_L and R_A)
1 10-kΩ potentiometer (R_G)
1 5.0-μF capacitor at 50 WV dc (C_G)
1 Breadboard for constructing circuit

Procedure:

1. Construct the circuit shown in Figure 9-14a.
2. Place the oscilloscope across the TRIAC.
3. Vary R_G back and forth and view the voltage waveform across the TRIAC.
4. Adjust R_G so that the TRIAC triggers about halfway between 0° and 90° and 180° and 270°.
5. Draw the voltage waveform across the TRIAC in the space provided, making sure to align it with the proper degrees for one cycle (indicate peak-to-peak voltage).
6. Place the oscilloscope across R_L.
7. Draw the voltage waveform across R_L in the space provided, making sure to align it with the proper degrees for one cycle. (Indicate peak-to-peak voltage.)
8. Modify the circuit as shown in Figure 9-14b by adding the capacitor.
9. Place the oscilloscope across the TRIAC.
10. Vary R_G back and forth and view the voltage waveform across the TRIAC.
11. Adjust R_G so that the TRIAC triggers past 90° and 270°.
12. Draw the voltage waveform across the TRIAC in the space provided, making sure to align it with the proper degrees for one cycle (indicate peak-to-peak voltage).
13. Place the oscilloscope across R_L.
14. Draw the voltage waveform across R_L in the space provided, making sure to align it with the proper degrees for one cycle (indicate peak-to-peak voltage).

Fill-in Questions:

1. Using only a potentiometer, the ac trigger time of a TRIAC can be varied about ______ to ______ degrees and ______ to ______ degrees.

2. Using a potentiometer and a capacitor, the ac trigger time of a TRIAC can be varied over nearly the __________ cycle.

3. When the TRIAC conducts, the voltage across R_L is about equal to the __________ __________.

4. When the TRIAC conducts, the voltage across its main terminals is about ______ V.

5. When the TRIAC is not conducting, the voltage across its main terminals is equal to the __________ __________.

SECTION 9-5 TRIAC BASIC TROUBLESHOOTING APPLICATION: BASIC AC CONTROL CIRCUIT

Construct the circuit shown. Open or short the components as listed and record the voltages in the proper place in the table. The oscilloscope will be used to measure the voltage waveforms. All voltages are referenced to ground. The abbreviations will help indicate the voltage conditions associated with each problem.

Table Abbreviations:

INC	Increase
DEC	Decrease
NOC	No change

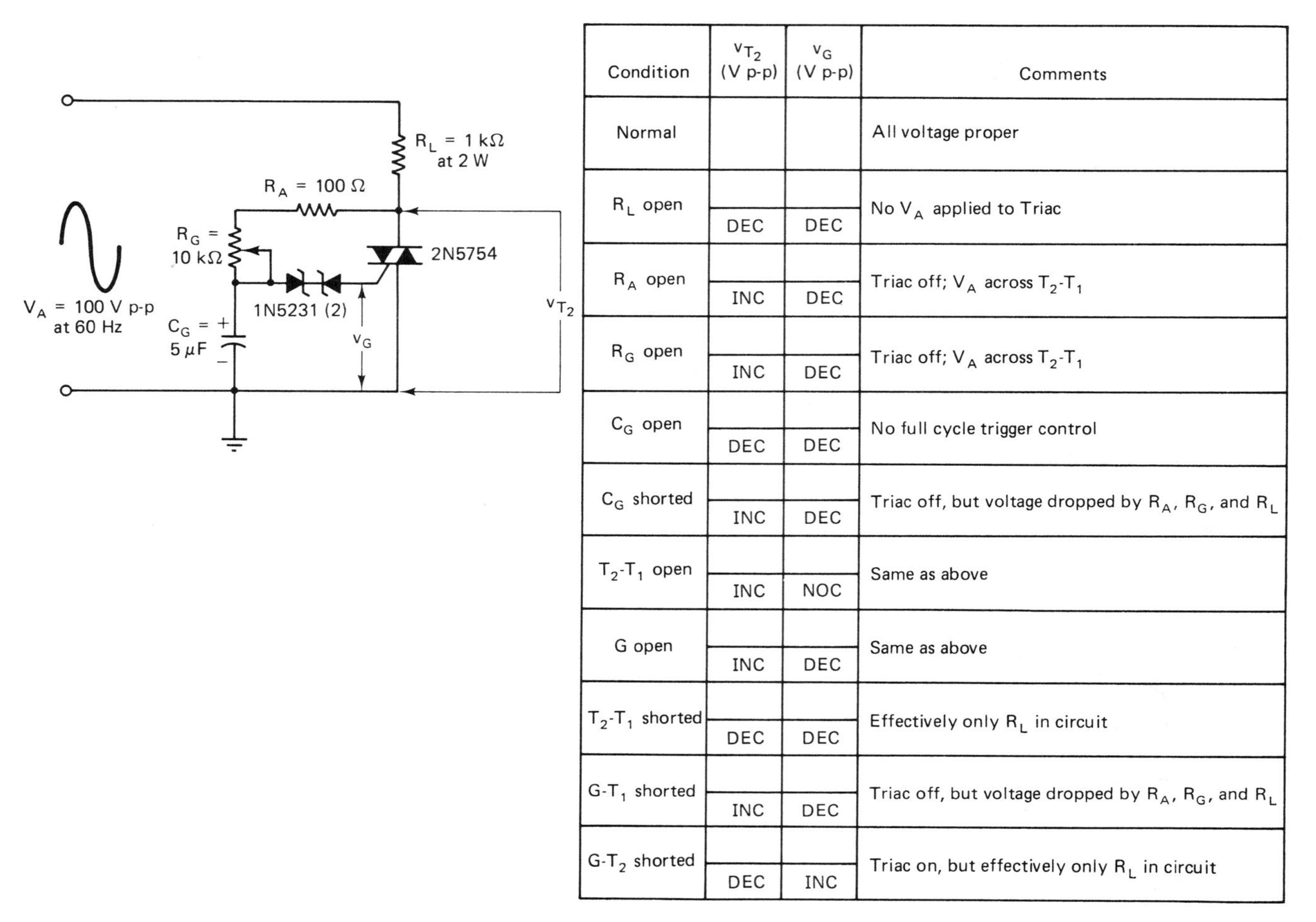

Condition	V_{T_2} (V p-p)	V_G (V p-p)	Comments
Normal			All voltage proper
R_L open	DEC	DEC	No V_A applied to Triac
R_A open	INC	DEC	Triac off; V_A across T_2-T_1
R_G open	INC	DEC	Triac off; V_A across T_2-T_1
C_G open	DEC	DEC	No full cycle trigger control
C_G shorted	INC	DEC	Triac off, but voltage dropped by R_A, R_G, and R_L
T_2-T_1 open	INC	NOC	Same as above
G open	INC	DEC	Same as above
T_2-T_1 shorted	DEC	DEC	Effectively only R_L in circuit
G-T_1 shorted	INC	DEC	Triac off, but voltage dropped by R_A, R_G, and R_L
G-T_2 shorted	DEC	INC	Triac on, but effectively only R_L in circuit

Figure 9–E.1

SECTION 9-6 MORE EXERCISES

Perform the following exercises before beginning Section 9-7.

1. Match the correct statements for a DIAC and TRIAC (some statements may apply to both). *Choices:* 1. DIAC; 2. TRIAC.

_______ **a.** conducts in both directions.

_______ **b.** is usually triggered into operation by the gate lead.

_______ **c.** has a low internal voltage drop.

_______ **d.** is turned off by reducing the current through it below its minimum holding current.

_______ **e.** often used as a peak voltage limiter.

_______ **f.** often used to trigger a TRIAC.

_______ **g** primarily used in ac circuits.

_______ **h.** has no gate lead.

2. Referring to Figure 9-11, draw the methods of testing a TRIAC with an ohmmeter and give proper ohmic readings of the meter.

3. On the following circuits, calculate the indicated values using the formulas $V_{T_2} = +V - V_{RL}$, $V_{RL} = +V - V_{T_2}$, $V_{RL} = I_T R_L$, $I_T = +V - V_{T_2}/R_L$, and $R_L = V_{RL}/I_T$.

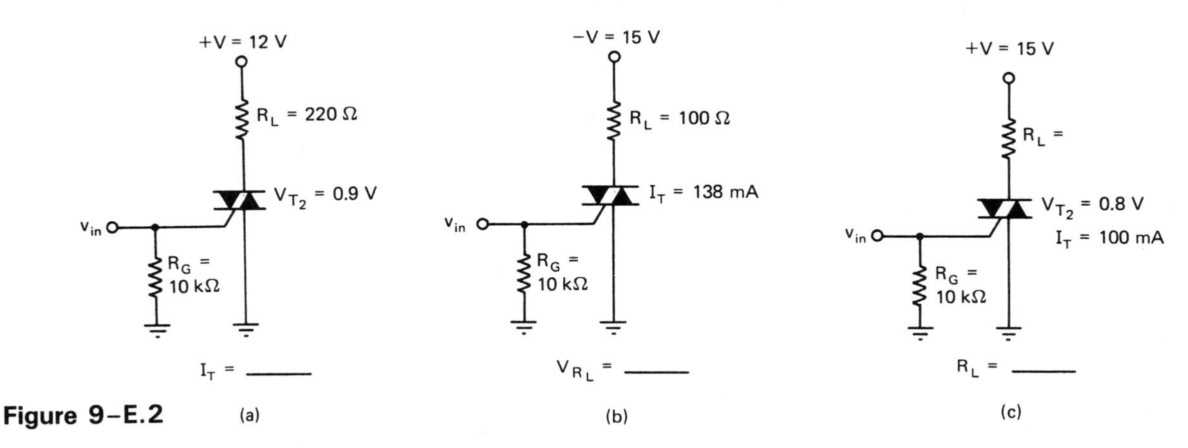

Figure 9–E.2

SECTION 9-7
THYRISTOR INSTANT REVIEW

DIAC: The DIAC is a bilateral (conducts in both directions) solid-state switch. Conduction occurs when the voltage across the terminals exceeds the rated breakover voltages $+V_{BO}$ and $-V_{BO}$. The DIAC is used to provide a sharp trigger current pulse to control other thyristors, such as the SCR and TRIAC.

TRIAC: The TRIAC is a bilateral (conducts in both directions) solid-state switch whose conduction can be controlled by the gate lead. There are four modes of operation, where the main terminals may be either positive or negative voltage and a positive or negative voltage applied to the gate lead will turn on the device. Similar to an SCR, the gate loses control over the current flowing through the TRIAC and the holding current keeps the device conducting.

SECTION 9-8
SELF-CHECKING QUIZZES

9-8a DIAC/TRIAC: TRUE–FALSE QUIZ

Place a T for true or an F for false to the left of each statement.

____ **1.** The DIAC is a three-layer *PN* semiconductor device with two main terminal leads and a gate lead.

____ **2.** The DIAC and TRIAC are bilateral solid-state switches.

____ **3.** The DIAC has a forward breakover voltage point and a reverse breakover voltage point.

____ **4.** The gate lead of a TRIAC is used to turn the device on and off.

____ **5.** A DIAC is similar to two zener diodes connected face to face or back to back in series.

____ **6.** A TRIAC can be tested with an ohmmeter.

____ **7.** A DIAC is used primarily to trigger other thyristors.

____ **8.** The TRIAC is similar to two SCRs connected in inverse parallel with the gates tied together.

____ **9.** A DIAC can be tested with an ohmmeter.

____ **10.** A TRIAC can be triggered on the positive and negative alternations of a sine wave.

9-8b DIAC/TRIAC: MULTIPLE-CHOICE QUIZ

Circle the correct answer for each question.

1. Thyristors are primarily used as:

- **a.** amplifiers
- **b.** power supply rectifiers
- **c.** impedance matching devices
- **d.** solid-state switches

2. The minimum holding current of a thyristor:

- **a.** is from one main terminal (or cathode) to the other main terminal (or anode)
- **b.** keeps the gate in a state of readiness
- **c.** is the current needed to keep the device on
- **d.** answers a and c are correct
- **e.** answers b and c are correct

3. The correct output voltage waveform for Figure 9-A is:

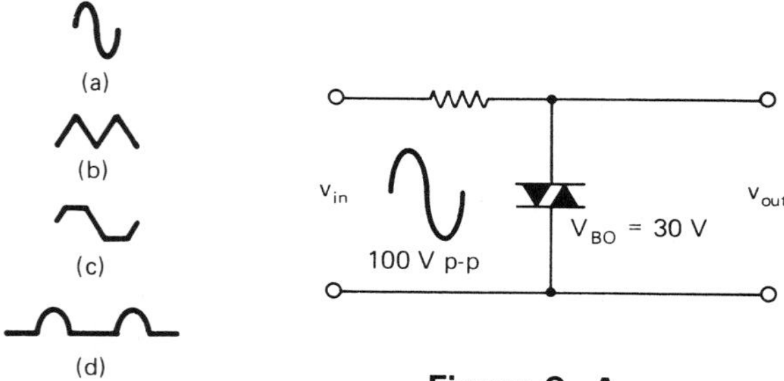

Figure 9–A

4. A TRIAC can be triggered into conduction when the voltage on:

- **a.** its T_2 terminal is positive and the gate is positive
- **b.** its T_2 terminal is positive and the gate is negative
- **c.** its T_2 terminal is negative and the gate is positive
- **d.** its T_2 terminal is negative and the gate is negative
- **e.** all of the above

5. The device that has a low-voltage drop across its main terminal in either direction when conducting is a:

- **a.** SCR
- **b.** DIAC
- **c.** TRIAC
- **d.** none of the above

6. The approximate trigger times for a TRIAC circuit without a capacitor are:

- **a.** 0° to 45° and 180° to 270°

b. 0° to 90° and 270° to 360°
c. 0° to 90° and 180° to 270°
d. 0° to 180° and 180° to 360°

7. The best device to use as a light dimmer in an ac circuit is the:
   a. bipolar transistor
   b. TRIAC
   c. SCR
   d. rheostat

8. A DIAC is mainly used to:
   a. limit voltage peaks
   b. trigger other thyristors
   c. rectify ac voltage
   d. limit current in a current

9. The waveform shown in Figure 9-B is that of a:

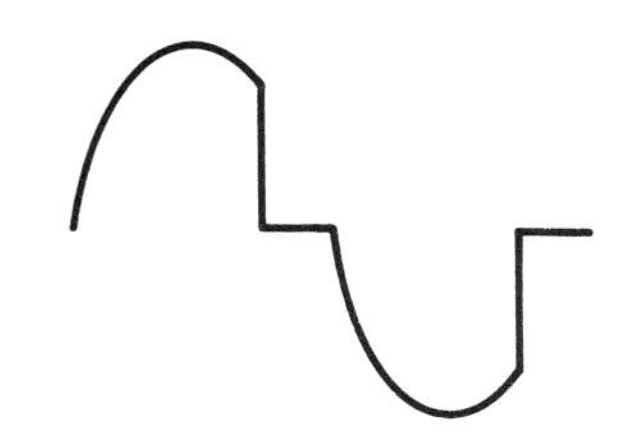

Figure 9–B

   a. SCR
   b. DIAC
   c. TRIAC
   d. JFET

10. The trigger time of the waveform shown in Figure 9-B can be:
   a. 0° to 90° and 180° to 270°
   b. 0° to 90° and 270° to 360°
   c. 0° to 180°
   d. nearly the entire 360°

ANSWERS TO EXPERIMENTS AND QUIZZES

Experiment 1: **(1)** V_{BO} **(2)** decreases, negative **(3)** increases **(4)** zener diodes

Experiment 2. **(1)** high **(2)** low **(3)** holding current **(4)** shorted **(5)** open

Experiment 3. **(1)** power supply **(2)** 1 (or your measurement) **(3)** loses, continues **(4)** holding current **(5)** positive, negative, positive, negative

Experiment 4. **(1)** conducts **(2)** current, across **(3)** thyristors, protection

Experiment 5. **(1)** 0, 90, 180, 270 **(2)** entire **(3)** applied voltage **(4)** 1 (or your measurement) **(5)** applied voltage

True–False. **(1)** F **(2)** T **(3)** T **(4)** F **(5)** T **(6)** T **(7)** T **(8)** T **(9)** F **(10)** T

Multiple Choice. **(1)** d **(2)** d **(3)** c **(4)** e **(5)** c **(6)** c **(7)** b **(8)** b **(9)** c **(10)** d

Unit 10

Optoelectronic Devices

INTRODUCTION Optoelectronics uses light energy in connection with electronic devices, particularly semiconductor devices, and has become one of the most used branches of sensing and control in the electronics industry.

UNIT OBJECTIVES Upon completion of this unit you will be able to:

1. Identify light-source and light-sensing devices.
2. Define the terms photon, LED, photodetector, photodiode, phototransistor, photo-Darlington, photofet, LASCR, photoresistor, solar cell, optoisolator, and other terms associated with optoelectronics.
3. Describe the operation of an LED and various photodetectors.
4. Calculate the value of the series resistor required by an LED.
5. Test an LED and photodetectors with an ohmmeter.
6. Show how light is used to control electronic circuits.
7. Connect an optoisolator circuit to show how electronics produces light, which in turn controls other electronic circuits.
8. Troubleshoot a light-controlled electronic circuit.

SECTION 10-1 THEORY OF STRUCTURE AND OPERATION

10-1a PROPERTIES OF LIGHT

Light is identified as electromagnetic radiation of specific frequencies that can be perceived by the human eye. The band of light frequencies is from 300 to 300,000,000 GHz or 3×10^{11} to 3×10^{17} Hz; however, only a narrow band of frequencies is visible to the eye, which extends from about 400,000 to 750,000 GHz. Invisible light below this narrow band is called infrared light and invisible light above this band is called ultraviolet light. The total band of light frequencies is above radio frequencies used for communication and below X-ray frequencies used in medical equipment.

Light is also conceived of as vibrating minute particles known as photons. The photon theory can perhaps best describe the operation of optoelectronic devices. Optoelectronic devices combine the technologies

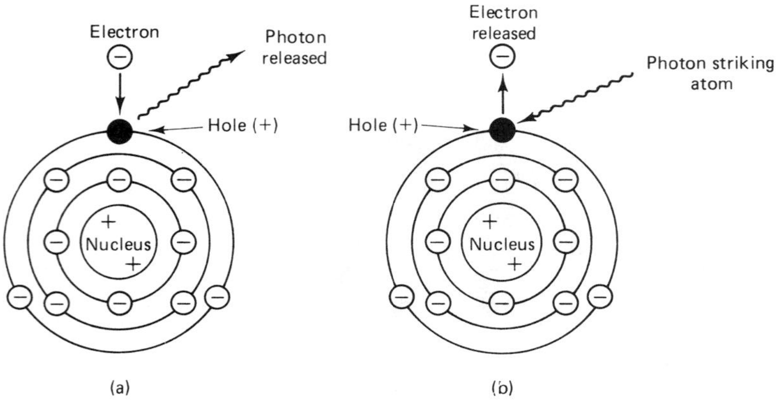

Figure 10–1 Light energy versus electrical energy: (a) valence electron falling to a lower energy level releases photon; (b) photon striking valence electron raises its energy level so that it can move to another atom.

of optics and electronics, commonly using semiconductor materials. Light emission occurs in semiconductor material when electrons combine with holes, as shown in Figure 10-1a. The electrons fall to a lower energy level as they travel from one atom to another and during the process photons are released. Light detection occurs in semiconductor material when photons strike the material and release valence electrons, as shown in Figure 10-1b.

The most widely used semiconductor materials are germanium (Ge) and silicon (Si), which are also responsive to light, which is why normal solid-state devices are contained in light-tight packages. Materials that are better suited for optoelectronic devices are cadmium sulfide (CdS), gallium phosphide (GaP), cadmium selenide (CdSe), and gallium arsenide (GaAs).

Converting electrical energy into light energy is known as light emitting, light source, or photo source. Converting light energy into electrical energy is called light detecting, light sensing, or photodetecting. Optoelectronics refers to these devices as photo source or photodetector. Photodetector devices respond to visible light and invisible infrared light.

10-1b LIGHT-EMITTING DIODE

A light-emitting diode (LED) is a photo-source device consisting of a *PN* junction made of GaP or GaAsP materials. The LED is biased the same as a regular diode, except that in the forward-biased condition it emits light when electrons cross the junction and combine with holes, as shown in Figure 10-2a. The schematic symbol is the same as for a regular diode, but with one or more arrows pointing away from it, as shown in Figure 10-2b.

Actually, little or no current flows or light is produced until the forward-biased voltage is equal to or greater than the inherent forward voltage drop (V_F) of the LED, which is typically about 2 V. In the reverse-biased condition, little or no current flows and no light is produced.

Figure 10-2c shows how the plastic package covering the LED is used to focus the light and serves as a clear or diffused lens to give more visible light. LEDs come in various sizes and have clear and colored lenses (red, yellow, orange, green, and blue). The cathode is next to the notch or flat side of the package, as shown in Figure 10-2d. An infrared LED (IRLED) is made of GaAs material and produces invisible light when forward biased.

In most applications, the LED must be protected with a series current-limiting resistor (R_S) as shown in Figure 10-3. The value of this resistor is easily calculated using Ohm's law. The V_F of the LED is

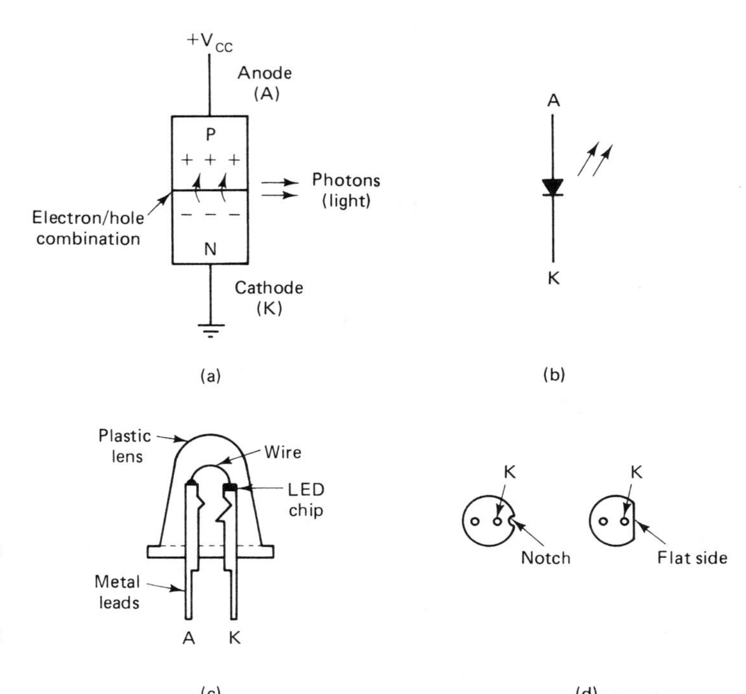

Figure 10–2 Light-emitting diode: (a) structure and operation; (b) schematic; (c) construction (side view); (d) lead identification (bottom view).

fairly constant; therefore, the voltage across R, V_{R_S}, is the difference between the applied voltage, V_{CC}, and V_F: $V_{R_S} = V_{CC} - V_F$. A safe forward current (I_F) is chosen for the LED that produces sufficient light. This current, which also flows through R_S, is divided into V_{R_S} to find the value of R_S: $R_S = V_{R_S}/I_F$.

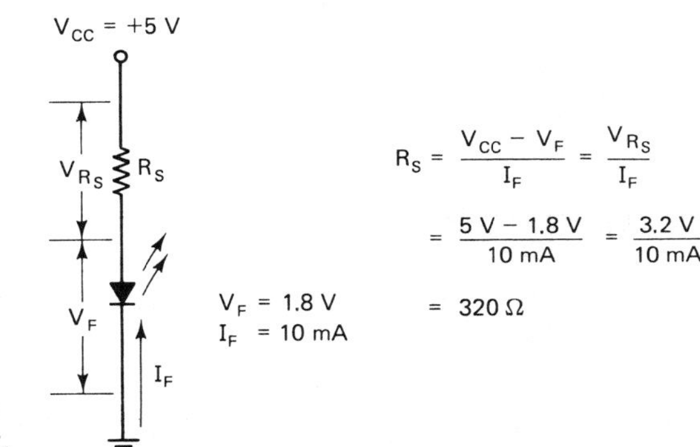

Figure 10–3 Calculation of current-limiting resistor.

10-1c PHOTODETECTORS

A photodiode, shown in Figure 10-4a and b, is a photodetector device similar to a normal diode except with a small glass window in the package, which allows light to strike the *PN* junction. In normal operation, the photodiode is reverse biased, which causes the depletion region to increase. With no light striking the diode, very little current flows; this is referred to as dark current (I_D). The diode acts like an open circuit. When light strikes the diode, electron–hole pairs are generated, which cross the depletion region and combine to produce a larger reverse current known as light current (I_L). This reverse current is proportional to the amount of light striking the diode. The schematic symbol of a photodiode is shown in Figure 10-4a. Notice that the light-indicating arrow points toward the diode.

In the circuit shown in Figure 10-4b, when light is blocked from the diode, it acts like an extremely high resistance or open circuit. For

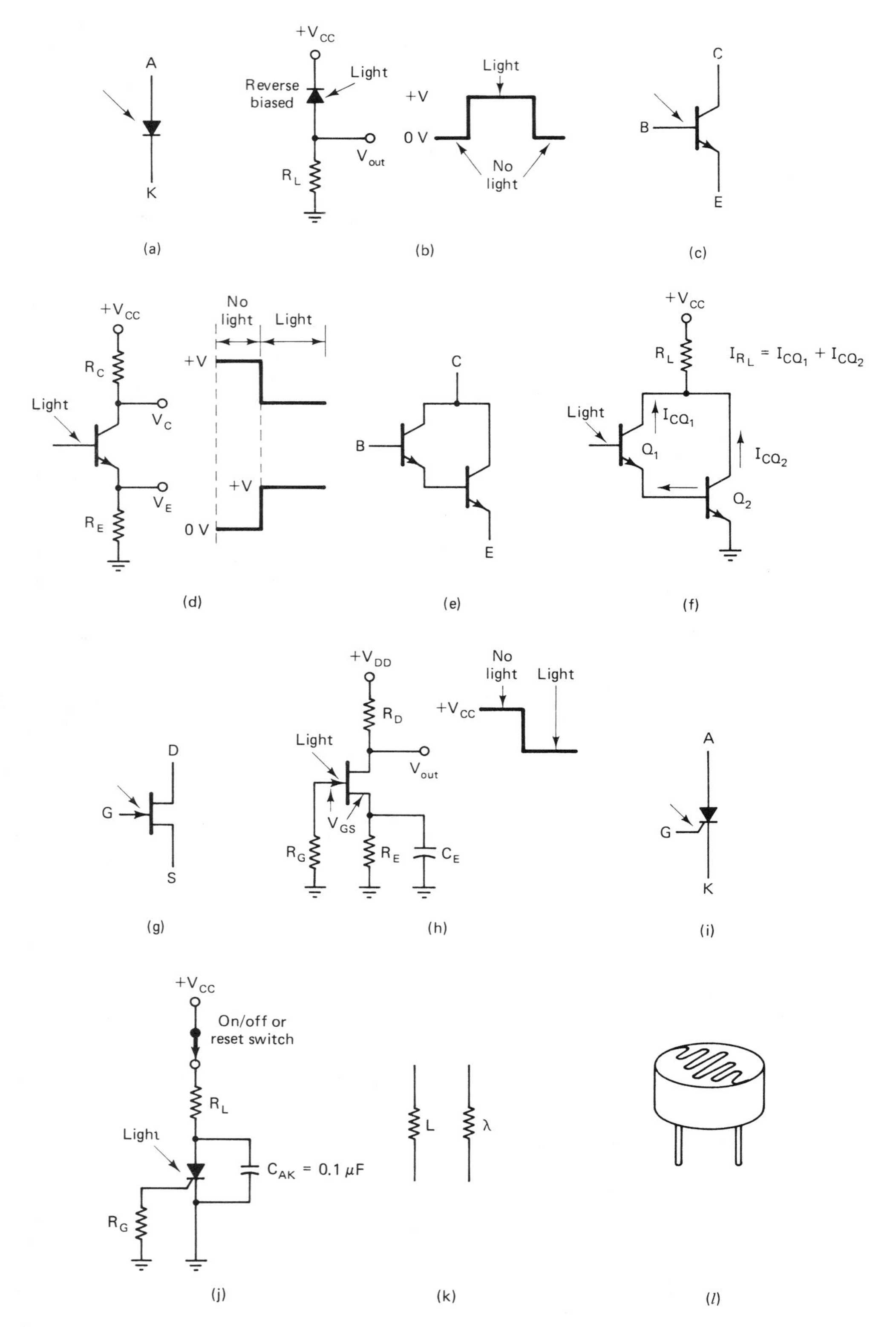

Figure 10–4 Photodetector symbols: (a) photodiode; (b) photodiode operation; (c) phototransistor; (d) phototransistor operation; (e) photo-Darlington transistor; (f) photo-Darlington operation; (g) photofet; (h) photofet operation; (i) LASCR; (j) LASCR operation; (d) photoresistor; (l) photoresistor package.

practical reasons, no current flows through R_L and the voltage output is 0 V. When light strikes the diode, its resistance decreases, allowing current to flow through R_L to V_{CC}. The voltage output now increases toward $+V_{CC}$ as a result of the IR drop across R_L.

10-1d PHOTOTRANSISTOR

A phototransistor, as shown in Figure 10-4c and d, is similar to a normal transistor except that it has a small glass window or a clear package, which allows light to strike the base junction region. It operates in a similar manner to a photodiode, but has current gain due to the effect of beta. A slight base current results when light strikes the phototransistor, which allows current to flow from emitter to collector. The amount of current flow is proportional to the light and is expressed as $I_C \approx I_E \approx \beta I_B$. Its schematic symbol also has a light-indicating arrow pointing toward it, as shown in Figure 10-4c.

In the circuit shown in Figure 10-4d, when light is blocked from the phototransistor, very little current flows and essentially the circuit is open. The voltage at the collector (V_C) is high, while the voltage at the emitter (V_E) is low. When light strikes the phototransistor, the current through it increases, which results in a decrease in V_C and an increase in V_E.

10-1e PHOTO-DARLINGTON TRANSISTOR

A photo-Darlington transistor, shown in Figure 10-4e and f, can provide more load current than a single phototransistor and serve as isolation between circuits. A phototransistor (Q_1) is combined with a regular transistor (Q_2) to produce a photo-Darlington transistor. The emitter of Q_1 is directly connected to the base of Q_2 and their collectors are connected. The physical package usually has the same appearance as a phototransistor or other photodetector device.

When light is blocked from the unit, the transistors are off and no current can flow to the load (R_L). When light strikes the phototransistor, both transistors turn on, allowing current to flow through R_L. The load current is the sum of the two collector currents, $I_{R_L} = I_{C_{Q1}} + I_{C_{Q2}}$. Expressed another way, the total collector current is the photoinduced base current of Q_1 multiplied by the beta of each transistor: $I_T = I_{B_1}\beta_1\beta_2$.

10-1f PHOTO FIELD-EFFECT TRANSISTOR

The photo field-effect transistor (photofet) shown in Figure 10-4g and h is similar to a conventional JFET with the exception of a lens for focusing light onto the gate junction. The photofet provides a photodetector with high input impedance and low noise in a single device. Usually, the photofet is negatively biased, as shown in Figure 10-4h. The source voltage is more positive than the gate (0 V). When light is blocked from the photofet, a small current flows through the circuit and V_{out} is high. When light strikes the gate area, valence electrons are released, which slightly increases nominal gate current I_G. Since I_G flows through R_G, a voltage drop is developed, making the gate voltage increase, which decreases the negative bias V_{GS}. This change in bias voltage causes the drain current through the photofet to increase, which causes V_{out} to decrease.

10-1g LIGHT-ACTIVATED SCR

The light-activated SCR (LASCR) shown in Figure 10-4i and j operates similarly to a conventional SCR, except for a window and lens that focuses light on the gate junction area. The LASCR operates like a latch. It can be triggered on by a light input to the gate area, but will not turn off after the light source is removed. It can be turned off only by reducing the current through it below its holding current. In an actual circuit, a capacitor from anode to cathode (C_{AK}) prevents false triggering when power is applied to the circuit, as shown in Figure 10-4j. On the basis of size, the LASCR is capable of handling larger amounts of current than can be handled by photodiodes and phototransistors.

10-1h PHOTORESISTOR

A photoresistor, shown in Figure 10-4k and l, is made of cadmium sulfide (CdS) or cadmium selenide (CdSe) semiconductor material. This material is deposited on a ceramic substrate with attached metallic leads and mounted in a case with a window. When light strikes the photoresistor, electron–hole carriers are released; the electrons travel toward the positive end and the holes travel toward the negative end. The current through the photoresistor has increased; therefore, the resistance of the photoresistor has decreased. The amount of decrease in resistance is proportional to the intensity of the light. Light resistance (R_{light}) is the minimum resistance for a given amount of light, and dark resistance (R_{dark}) is the maximum resistance when there is no light.

Photoresistors can be used in voltage-divider circuits to produce varying outputs with respect to light, as shown in Figure 10-5. If the output is taken across the photoresistor, the output voltage (V_{out}) decreases when light is present, as shown in Figure 10-5a. For example, assume the following values: $+V_{CC} = 12$ V, $R_S = 1$ MΩ with photoresistance, $R_{\text{dark}} = 1$ MΩ, and $R_{\text{light}} = 10$ kΩ.

When light is blocked, the total resistance of the circuit is

$$R_T = R_S + R_{\text{dark}} = 1\ \text{M}\Omega + 1\ \text{M}\Omega = 2\ \text{M}\Omega$$

The total current through the circuit is

$$I_{T_{\text{dark}}} = \frac{+V_{CC}}{R_T} = \frac{12\ \text{V}}{2\ \text{M}\Omega} = 6\ \mu\text{A}$$

The output voltage is then

$$\text{V}_{\text{out}} = I_{T_{\text{dark}}} R_{\text{dark}} = 6\ \mu\text{A} \times 1\ \text{M}\Omega = +6\ \text{V}$$

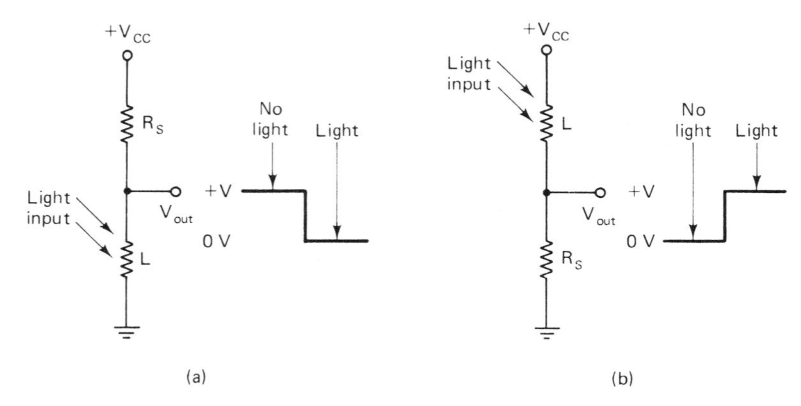

Figure 10–5 Photoresistor voltage dividers: (a) decreasing voltage output; (b) increasing voltage output.

When light strikes the photoresistor, the total resistance of the circuit is

$$R_T = R_S + R_{\text{light}} = 1\ \text{M}\Omega + 10\ \text{k}\Omega = 1.01\ \text{M}\Omega$$

The total current through the circuit is

$$I_{T_{\text{light}}} = \frac{+V_{CC}}{R_T} = \frac{12\ \text{V}}{1.01\ \text{M}\Omega} \approx 12\ \mu\text{A}$$

The output voltage is then

$$V_{\text{out}} = I_{T_{\text{light}}} R_{\text{light}} = 12\ \mu\text{A} \times 10\ \text{k}\Omega = +0.12\ \text{V}$$

If the resistors are switched and the output is taken across the fixed resistor (R_S), V_{out} increases when light is present, as shown in Figure 10-5b. With no light, $V_{\text{out}} = +6$ V as before, but with light

$$V_{\text{out}} = I_{T_{\text{light}}} R_S = 12\ \mu\text{A} \times 1\ \text{M}\Omega = +12\ \text{V}$$

10-1i SOLAR CELL

The solar cell or photocell is a photovoltaic photodetector that produces a small voltage and current when exposed to light. The silicon solar cell is a specially doped *PN* junction, shown in Figure 10-6a. When light strikes the cell, electron–hole pair carriers are released. Electrons are forced into the *N*-type region as holes are forced into the *P*-type region. The depletion region increases slightly, and a small voltage is stored between the positive and negative terminals as long as light is present. Up to a point, the stored voltage is directly proportional to the intensity of the light. A single solar cell may produce from 0.25 to 0.6 V across its terminals and be able to supply up to 50 mA of current.

The schematic symbol for a solar cell, shown in Figure 10-6b, resembles an ordinary voltage cell but includes the Greek symbol λ (lambda).

In a basic circuit, as shown in Figure 10-6c, when light is blocked, there is no voltage and current and $V_{\text{out}} = 0$ V. When light is present, a small voltage is created, which produces current flow through *R*, and V_{out} increases.

Similar to conventional voltaic cells, solar cells can be connected in series to produce a larger voltage (Figure 10-7a) and connected in parallel to provide more current capability (Figure 10-7b).

Figure 10-6 Silicon solar cell: (a) structure; (b) schematic symbol; (c) circuit operation.

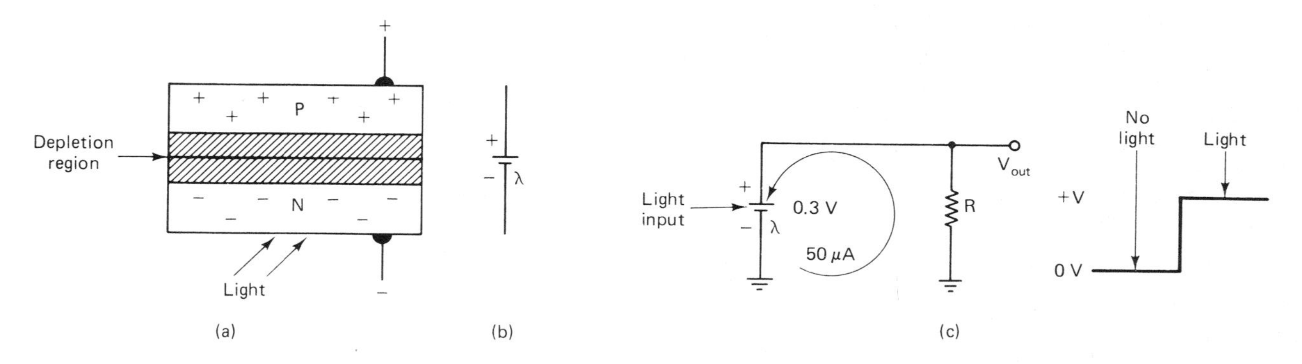

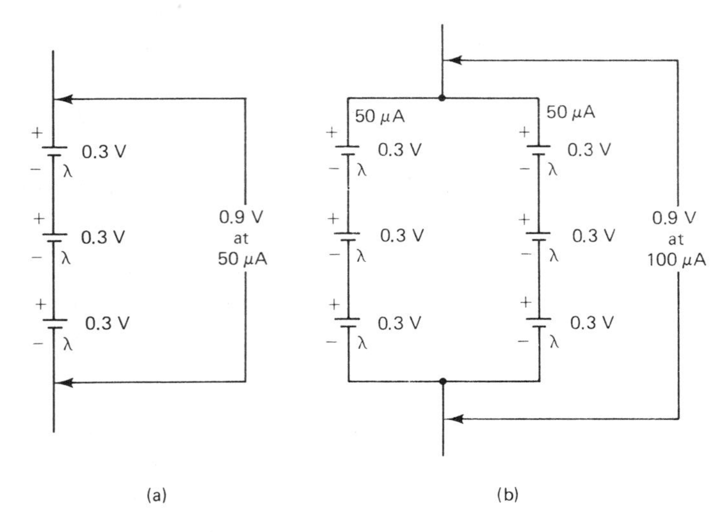

Figure 10–7 Solar cells connected to produce more voltage and current: (a) series connection increases voltage; (b) parallel connection increases current.

10-1j PHOTO-COUPLED ISOLATOR

A photo-coupled isolator, also called an optically coupled isolator, optoisolator, optocoupler, photon coupler, photocoupler, and other names, is used to electrically isolate one circuit from another, while allowing one circuit to control the other.

The photo-coupled isolator consists of a photosource, such as a LED, and a photodetector, such as a photoresistor, shown in Figure 10-8. Usually, both devices are fabricated into a light-tight container, such as a six-pin mini-dual-in-line IC package (DIP), which provides a light coupling only from the LED to the photodetector. A +5-V pulse turns on the LED, which causes light to strike the photoresistor connected to a +12-V source. The resistance of the photodetector decreases and the output voltage decreases.

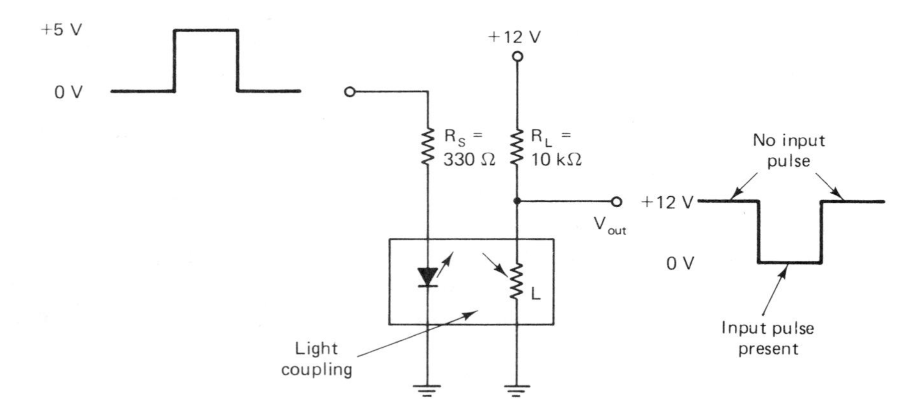

Figure 10–8 Photo-coupled isolator.

Figure 10–9 Photocoupler IC packages: (a) LED/phototransistor; (b) LED/photo-Darlington transistor; (c) LED/LASCR; (d) LED/photo-TRIAC.

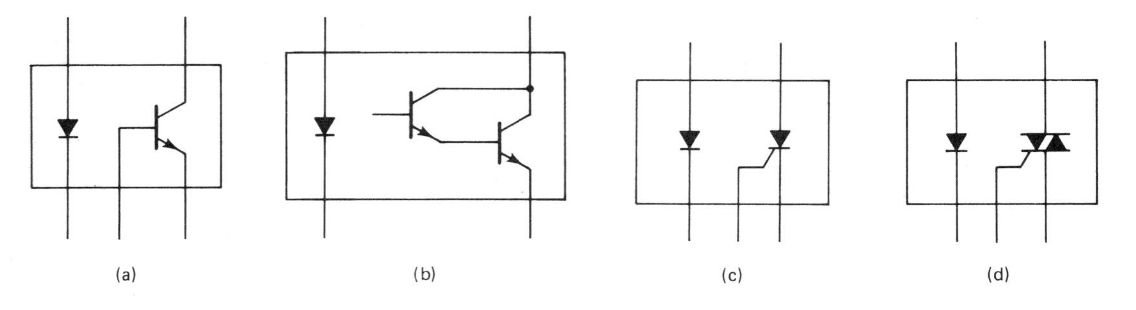

Some other combinations of photo-coupled isolators are shown in Figure 10-9: LED/phototransistor, LED/photo-Darlington transistor, LED/LASCR, and LED/photo-TRIAC. These devices not only provide interfacing between circuits, but are not plagued by the contact bounce, electrical noise, and physical wear associated with their electromechanical counterparts, such as relays.

10-1k OPTOELECTRONIC DEFINITIONS

The following definitions are associated with properties of light and optoelectronic devices.

Radiometric System: pertains to the measurement of the spectral emission characteristics of sources of electromagnetic radiation, including infrared, visible light, and ultraviolet wavelengths. The various terms belonging to this system are:

a. *Angstrom:* a unit of measurement of the wavelength of light or radiation equal to 10^{-8} cm.
b. *Radiance:* the radiation or emission of electromagnetic waves from a surface.
c. *Radiant energy (Q_e):* emitted energy in the form of electromagnetic radiation, measured in units such as kilowatt-hours, ergs, joules, or calories.
d. *Radiant flux (O_e):* the time rate of flow of radiant energy, expressed in ergs or watts per second.
e. *Radiant power:* same as radiant flux.
f. *Radiant intensity (I_e):* the radiant power that is traveling in a given direction per the unit of solid angle, measured in watts per steradian.
g. *Steradian:* a solid spherical angle, which includes a surface equal to the square of the radius of a sphere. A sphere contains a total of 4π steradians.
h. *Radiant exitance (M_e):* radiant energy that is emitted from a surface per unit area, measured in watts per square meter.
i. *Radiant incidence or irradiance (E_e):* radiant energy that strikes a surface per unit area, measured in watts per square meter.

Photometric System: pertains to the measurement of the spectral emission characteristics of sources of visible light. Although photometric terms also apply to light, they are not exactly equivalent to their corresponding terms, and it is often difficult to convert from one system to the other. The various terms associated with this system are the following:

a. *Luminous:* the emitting of light (photons) from a surface.
b. *Luminous energy (Q_v):* emitted energy in the form of photons, measured in units such as lumens or footcandles.
c. *Luminous flux (Φ_v):* the time rate of flow of light, expressed in lumens, candelas, or footcandles per second.
d. *Luminous power:* same as luminous flux.
e. *Luminous intensity (I_v):* the ratio of luminous power emitted by a source to the solid angle in which it is emitted, measured in candelas.
f. *Lumen:* a unit of luminous power produced by a uniform point source of one candle intensity on a unit area of surface.

g. *Footcandle:* a flux of one lumen uniformly distributed at a surface of one square foot or to a surface one foot from a uniform source of one candlepower.
h. *Lux:* the same as footcandle, but in the metric system equivalent to a flux of one lumen uniformly distributed at a surface of one square meter or to a surface one meter from a uniform source.
i. *Candela:* one lumen per steradian.
j. *Luminous exitance* (M_v)*:* luminous energy emitted from a surface per unit area, measured in lumens per square meter.
l. *Illuminance* (E_v)*:* similar to irradiance, the luminous energy that strikes a surface per unit area, measured in lux.

Photosensitive: materials that respond to radiant energy, especially light.

Photoconductive: materials whose electrical resistance varies inversely with the intensity of the light that strikes it.

Photoresistive: same as photoconductive.

Photoemissive: materials that emit electrons when struck by light.

Photovoltaic: materials that produce a voltage when struck by light.

Photometer: a device for measuring the intensity of light source.

SECTION 10-2 OPTOELECTRONIC DEFINITION EXERCISE

Refer to the previous sections and write a brief definition for each term.

1. Radiometric system ______________________________

____

____

2. Photometric system ______________________________

____

____

3. Photons ______________________________

____

____

4. Photosensitive ______________________________

____

____

5. Photoconductive

6. Photoresistive

7. Photoemissive

8. Photovoltaic

9. I_{dark}

10. I_{light}

11. R_{dark}

12. R_{light}

SECTION 10-3 EXERCISES AND PROBLEMS

Perform the following exercises before beginning Section 10-4.

1. Draw the schematic symbol of an LED.

2. Draw the schematic symbol of a photodiode.

3. Draw the schematic symbol of a phototransistor.

4. Draw the schematic symbol of a photofet.

5. Draw the schematic symbol of a LASCR.

6. Draw the schematic symbols of a photoresistor.

7. Draw the schematic symbol of a solar cell. (Indicate polarities.)

8. Draw the schematic diagram of a photocoupler containing a LASCR.

9. Draw the schematic diagram of a photocoupler containing a phototransistor.

10. Draw the circuit of Figure 10-3 and find R_S for the various values given. Use the formula $R_S = V_{CC} - V_F/I_F$.

a. $V_{CC} = +5$ V
$V_F = 1.8$ V
$I_F = 12$ mA

$R_S =$ ______ Ω

b. $V_{CC} = +9$ V
$V_F = 2.1$ V
$I_F = 10$ mA

$R_S =$ ______ Ω

c. $V_{CC} = +12$ V
$V_F = 1.7$ V
$I_F = 15$ mA

$R_S =$ ______ Ω

d. $V_{CC} = -15$ V
$V_F = 2$ V
$I_F = 20$ mA

$R_S =$ ______ Ω

11. Refer to Figure 10-4b and correctly draw a photodiode circuit showing the output voltage waveform with and without light.

12. Referring to Figure 10-4d, draw a phototransistor circuit using only an emitter resistor and indicate if the voltage across this resistor increases or decreases when light strikes the phototransistor.

13. Draw the circuit of Figure 10-4j and explain the purpose of C_{AK}.

14. What is the output voltage of the following circuits when light is present and then blocked from the photoresistors ($R_{dark} = 10$ MΩ, $R_{light} = 10$ kΩ).

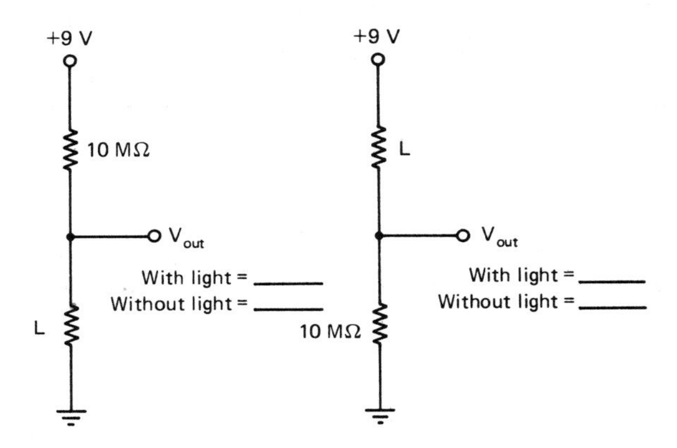

Figure 10–E.1

15. Draw a circuit using solar cells for each solar battery with the requirements given. Each cell produces 0.25 V and 70 μA when light strikes it.

a. $V_T = 0.75$ V
$I_T = 70\ \mu$A

b. $V_T = 0.25$ V
$I_T = 0.21$ mA

c. $V_T = 0.75$ V
$I_T = 0.14$ mA

SECTION 10-4 EXPERIMENTS

EXPERIMENT 1. TESTING AN LED WITH AN OHMMETER

Objective:

To show a practical method of testing LEDs with an ohmmeter.

Introduction:

An LED is tested in the same way as a regular diode. The ohmmeter will show low resistance in the forward-biased condition and high or infinite resistance in the reverse-biased condition. The LED may even glow slightly with forward bias if the ohmmeter can produce sufficient current. This is a go/no go test.

Materials Needed:

A standard or digital ohmmeter
One or several LEDs

Procedure:

1. Place the ohmmeter leads on the LED as shown in Figure 10-10a and record the reading in the ohmmeter circle indicated.
2. Place the ohmmeter leads on the LED as shown in Figure 10-10b and record the reading in the ohmmeter circle indicated.

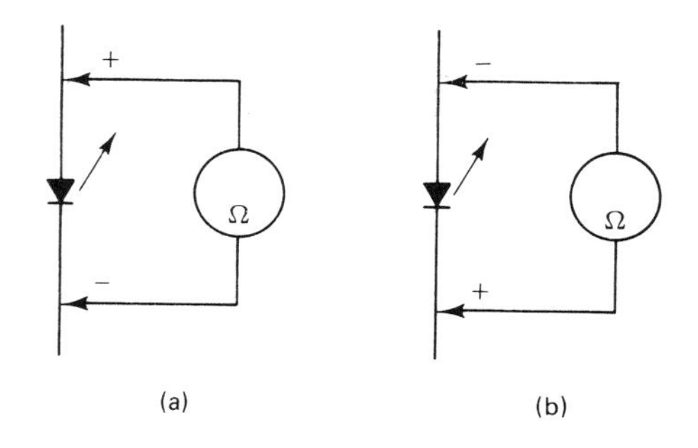

Figure 10–10 Testing an LED with an ohmmeter.

Fill-in Questions:

1. An LED is forward biased when the ohmmeter's ____________ lead is placed on the anode and the ____________ ____________ lead is placed on the cathode.

2. An LED is reverse biased when the ohmmeter's positive lead is placed on the ____________ and the negative lead is placed on the ____________ .

EXPERIMENT 2. LED OPERATION

Objective:

To demonstrate how an LED turns on with forward bias and turns off with reverse bias, and to show the procedure for calculating series resistor R_S.

Introduction:

An LED must be protected from too much current flowing through it. A current-limiting resistor in series with an LED will accomplish this, but enough current must be allowed to flow so that the LED is of sufficient brightness.

Materials Needed:

1 +5-V power supply
1 Standard or digital voltmeter
1 LED with a $V_F \approx 2$ V
1 220-Ω resistor at 0.5 W
1 470-Ω resistor at 0.5 W
1 1-kΩ resistor at 0.5 W
1 Breadboard for constructing circuit

Procedure:

1. Referring to Figure 10-11a, calculate the value of R_S using the formula

$$R_S = \frac{+V_{CC} - V_F}{I_F} = \_\_\_\_\_\_\_\_\_\_ = \_\_\_\_\_\_ \Omega$$

2. From the three resistors given in this experiment, select the one whose value

is nearest the calculation found in step 1. Write the value of this resistor in the spaces provided in Figure 10-11.

3. Construct the circuit of Figure 10-11a and record the condition of the LED.
4. Measure the voltage from anode to ground (V_F) and compare this value with the value of Figure 10-11a.
5. Turn off the power supply and turn the LED around as shown in Figure 10-11b.
6. Apply power to the circuit and record the condition of the LED.
7. Measure and record the voltage from the cathode to ground (V_R).
8. Determine the current (I_R) flowing in the circuit and record it in the appropriate space.
9. Turn off the power supply and turn the LED around as originally performed in Figure 10-11a.
10. Turn on the power supply and notice the brightness of the LED.
11. Replace R_S with the 470-Ω resistor. Is the LED brighter or dimmer? ______
12. Replace R_S with the 1-kΩ resistor. Is the LED brighter or dimmer? ______

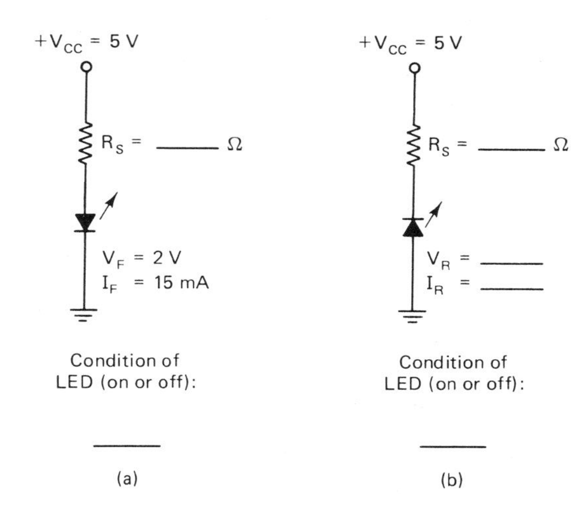

Figure 10–11 Calculating R_S and LED operation: (a) forward biased; (b) reverse biased.

Fill-in Questions:

1. A resistor is usually placed in ____________ to ____________ current flow through an LED.
2. The ____________ voltage drop of an LED is approximately 2 V.
3. A ____________-biased LED will not glow.

EXPERIMENT 3. TESTING PHOTODETECTORS

Objective:

To show how to test photodetectors with a multimeter.

Introduction:

Most photodetectors can be tested with an ohmmeter by reading the resistance of the device when light is present and then when light is blocked. When light strikes the photodetector, the resistance should be low, and when light is blocked, the resistance should go high or infinite. Using the same procedure, the voltage of a solar cell can be tested with a voltmeter. This is a go/no go test.

Materials Needed:

A standard or digital multimeter
Assorted types of photodetectors

Procedure:

1. Testing a photodiode:
   a. Connect the ohmmeter as shown in Figure 10-12a. Remember that the diode must be reverse biased.
   b. With light blocked from the diode, the ohmmeter reads ______ Ω.
   c. With light striking the diode, the ohmmeter reads ______ Ω.
2. Testing a phototransistor:
   a. Connect the ohmmeter as shown in Figure 10-12b.
   b. With light blocked from the phototransistor, the ohmmeter reads ______ Ω.

c. With light striking the phototransistor, the ohmmeter reads ______ Ω.

3. Testing a photo-Darlington transistor:
   a. Connect the ohmmeter as shown in Figure 10-12c.
   b. With light blocked from the transistor, the ohmmeter reads ______ Ω.
   c. With light striking the transistor, the ohmmeter reads ______ Ω.
4. Testing a photofet:
   a. Connect the ohmmeter as shown in Figure 10-12d, using a 1-MΩ resistor from gate to source.
   b. With light blocked from the photofet, the ohmmeter reads ______ Ω.
   c. With light striking the photofet, the ohmmeter reads ______ Ω.
5. Testing a LASCR:
   a. Set the ohmmeter on the low range and connect it as shown in Figure 10-12e, using a 100-kΩ resistor from gate to cathode. Light must be blocked from the LASCR before and after the ohmmeter is connected to prevent false triggering.
   b. With light blocked from the LASCR, the ohmmeter reads ______ Ω.
   c. With light striking the LASCR, the ohmmeter reads ______ Ω.
   d. Again, with light blocked from the LASCR, the ohmmeter reads ______ Ω. (Is the ohmmeter supplying sufficient holding current to keep the LASCR on?)
6. Testing a photoresistor:
   a. Connect the ohmmeter as shown in Figure 10-12f. The polarity of the leads is not critical.
   b. With light blocked from the photoresistor, the ohmmeter reads ______ Ω.
   c. With light striking the photoresistor, the ohmmeter reads ______ Ω.
7. Testing a solar cell:
   a. Set the voltmeter on the low range and connect it as shown in Figure 10-12g.
   b. With light blocked from the solar cell, the voltmeter reads ______ V.

Figure 10–12 Testing photodetectors: (a) photodiode; (b) phototransistor; (c) photo-Darlington transistor; (d) photofet; (e) LASCR; (f) photoresistor; (g) solar cell.

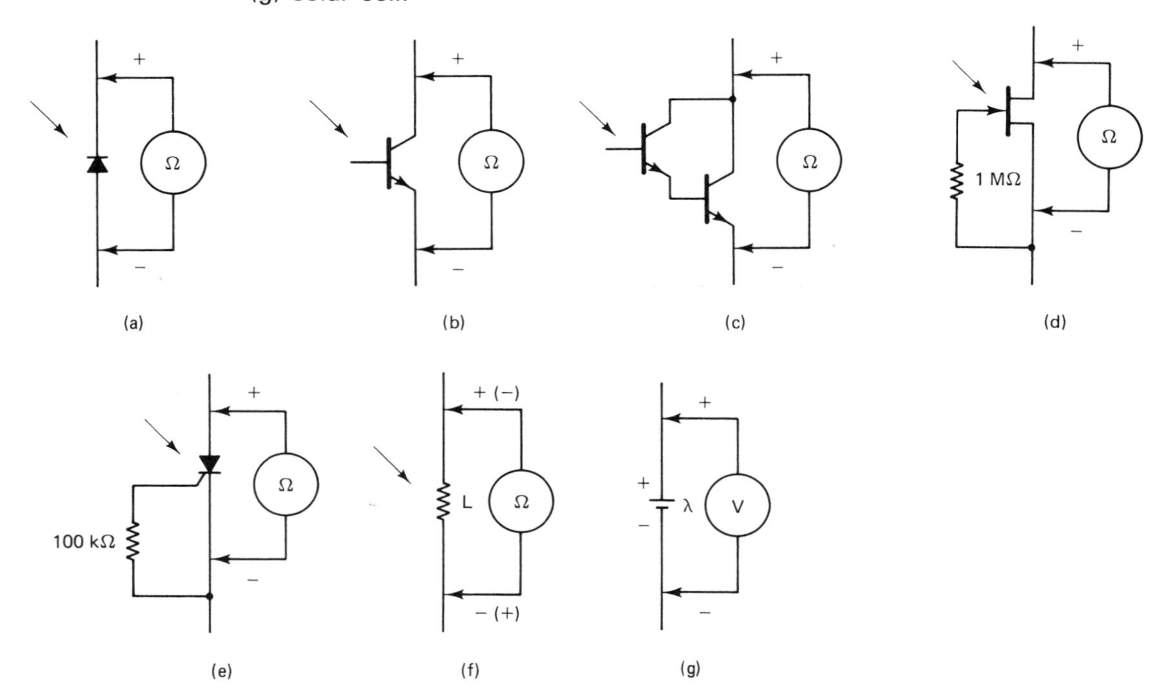

c. With light striking the solar cell, the voltmeter reads ______ V.

Fill-in Questions:

1. When light is blocked from a photodetector, its resistance is ______ .

2. When light strikes a photodetector, its resistance is ______ .

3. When light strikes a solar cell, it produces ____________ and ____________ .

EXPERIMENT 4. PHOTORESISTOR VOLTAGE DIVIDERS

Objective:

To demonstrate how light can vary the output voltage of a photoresistor voltage divider, causing it to increase or decrease depending on the position of the photoresistor.

Introduction:

Refer to Figure 10-13a and note that when light strikes the photoresistor, its resistance will decrease and V_{out} will also decrease. When light strikes the photoresistor in Figure 10-13c, its resistance will decrease, but V_{out} will increase.

Materials Needed:

1 Fixed +12-V power supply
1 Standard or digital voltmeter
1 100-kΩ resistor at 0.5 W (R_1)
1 R.S. 276-116 photoresistor or equivalent (R_2)
1 Breadboard for constructing circuit

Procedure:

1. Construct the circuit shown in Figure 10-13a.
2. With light blocked from the photoresistor, measure V_{out} and record the value in the proper place in the data table in Figure 10-13b.
3. Allowing light to strike the photoresistor, measure V_{out} and record the value in the proper place in the data table in Figure 10-13b.
4. Construct the circuit shown in Figure 10-13c.
5. With light blocked from the photoresistor, measure V_{out} and record the value in the proper place in the data table in Figure 10-13d.
6. Allowing light to strike the photoresistor, measure V_{out} and record the value in the proper place in the data table in Figure 10-13d.

Fill-in Questions:

1. When light is blocked from a photoresistor, its resistance ____________ .

2. When light strikes a photoresistor, its resistance ____________ .

Figure 10–13 Photoresistor voltage dividers: (a) decreasing output test circuit; (b) data table; (c) increasing output test circuit; (d) data table.

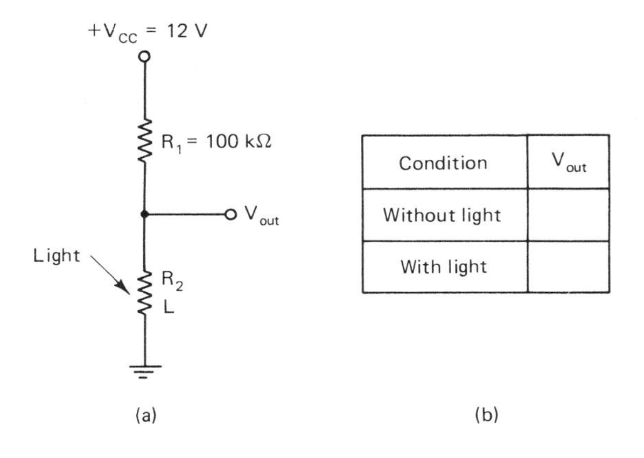

Condition	V_{out}
Without light	
With light	

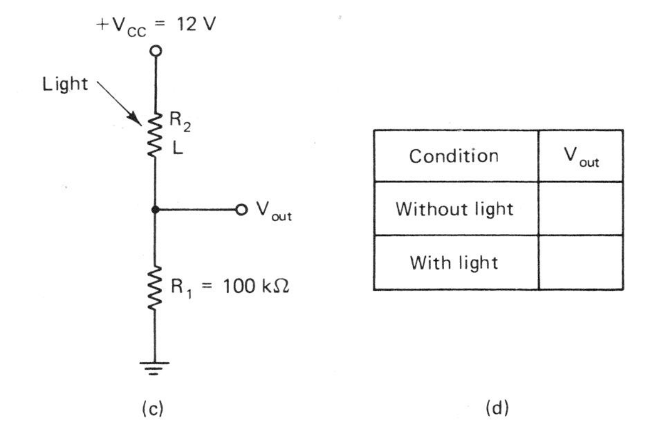

Condition	V_{out}
Without light	
With light	

3. In a voltage-divider circuit, when light strikes a photoresistor, the voltage across it ____________ .

4. In a voltage-divider circuit, when light is blocked from a photoresistor, the voltage across it ____________ .

EXPERIMENT 5. TRANSISTOR CONTROL WITH A PHOTODETECTOR

Objective:

To show how a photodetector can turn a transistor on or off depending on its position in the circuit and light conditions.

Introduction:

Refer to Figure 10-14a and note that, when light is blocked from the photodetector, its resistance is high, very little base current flows, Q_1 is cut off, and the LED is extinguished. When light strikes the photodetector, its resistance decreases, sufficient base current flows to turn on Q_1, and the LED glows.

In Figure 10-14c, when light is blocked from the photodetector, its resistance is high, the base of Q_1 is forward biased, Q_1 is on, and the LED glows. When light strikes the photodetector, its resistance decreases, causing reverse bias by pulling the base of Q_1, voltagewise, toward ground. Transistor Q_1 cuts off and the LED is extinguished.

Figure 10–14 Controlling a bipolar transistor with a photodetector: (a) turn on circuit; (b) data table; (c) turn off circuit; (d) data table.

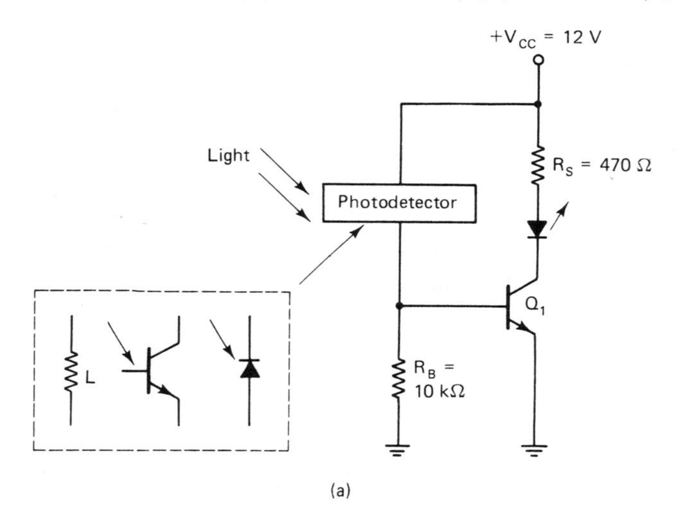

Light	Transistor (on or off)	LED (on or off)
Without		
With		

(b)

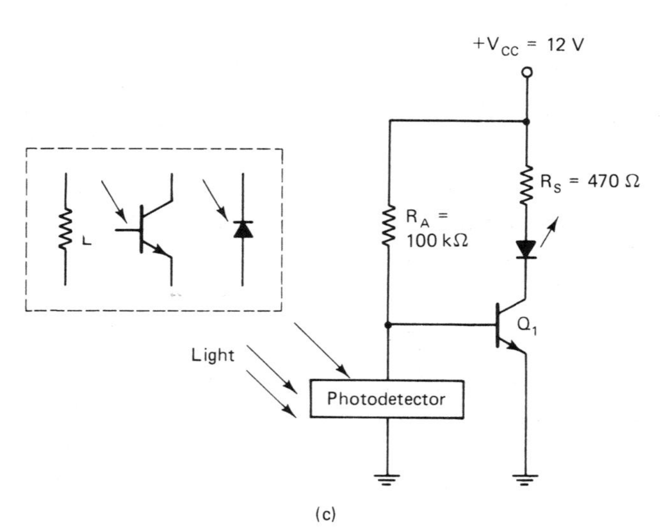

Light	Transistor (on or off)	LED (on or off)
Without		
With		

(d)

Materials Needed:

1 Fixed +12-V power supply
1 Standard or digital voltmeter
1 2N2222 transistor or equivalent (Q_1)
1 LED at $V_F \approx 2$ V
1 470-Ω resistor at 0.5 W (R_S)
1 10-kΩ resistor at 0.5 W (R_B)
1 100-kΩ resistor at 0.5 W (R_A)
1 Photodetector (photoresistor, phototransistor, or photodiode)
1 Breadboard for constructing circuit

Procedure:

1. Connect the circuit shown in Figure 10-14a.
2. With light blocked from the photodetector, indicate the condition of Q_1 and the LED in the data table of Figure 10-14b.
3. With light striking the photodetector, indicate the condition of Q_1 and the LED in the data table of Figure 10-14b.
4. Connect the circuit shown in Figure 10-14c.
5. With light blocked from the photodetector, indicate the condition of Q_1 and the LED in the data table of Figure 10-14d.
6. With light striking the photodetector, indicate the condition of Q_1 and the LED in the data table of Figure 10-14d.

Fill-in Questions:

1. Referring to Figure 10-14a, when light is blocked from the photodetector, Q_1 is ______ and the LED is ______ .

2. Referring to Figure 10-14a, when light strikes the photodetector, Q_1 is ______ and the LED ______ .

3. Referring to Figure 10-14c, when light is blocked from the photodetector, Q_1 is ______ and the LED ______ .

4. Referring to Figure 10-14c, when light strikes the photodetector, Q_1 is ______ and the LED is ______ .

EXPERIMENT 6. CIRCUIT ISOLATION USING A PHOTOCOUPLER

Objective:

To demonstrate how a photocoupler integrated circuit is used to isolate and control circuits electrically.

Introduction:

Referring to Figure 10-15a, when switch S_1 is closed, the LED in the photocoupler IC turns on. The photodetector in the IC senses the light from the LED and allows current to flow through LED 1 and R_2 to the +12-V power supply. When S_1 is opened, the LED in the IC turns off, light is extinguished, and the photodetector opens the +12-V circuit (unless it is a LASCR) and LED 1 turns off. In this case, a +5-V circuit is controlling a +12-V circuit without being electrically connected.

Materials Needed:

1 Fixed +5-V power supply
1 Fixed +12-V power supply
1 Standard or digital voltmeter
1 Photocoupler integrated circuit (LED/photoresistor, LED/phototransistor, LED/photo-Darlington transistor)
1 LED at $V_F \approx 2$ V (LED 1)
1 220-Ω resistor at 0.5 W (R_1)
1 470-Ω resistor at 0.5 W (R_2)
1 Single-pole single-throw (SPST) switch (S_1)
1 Breadboard for constructing circuit

Procedure:

1. Using the specification or data sheet given with your photocoupler IC, record the proper pin numbers on Figure 10-15a (refer to Figure 10-9 for the types of photocouplers).
2. Connect the circuit shown in Figure 10-15a.

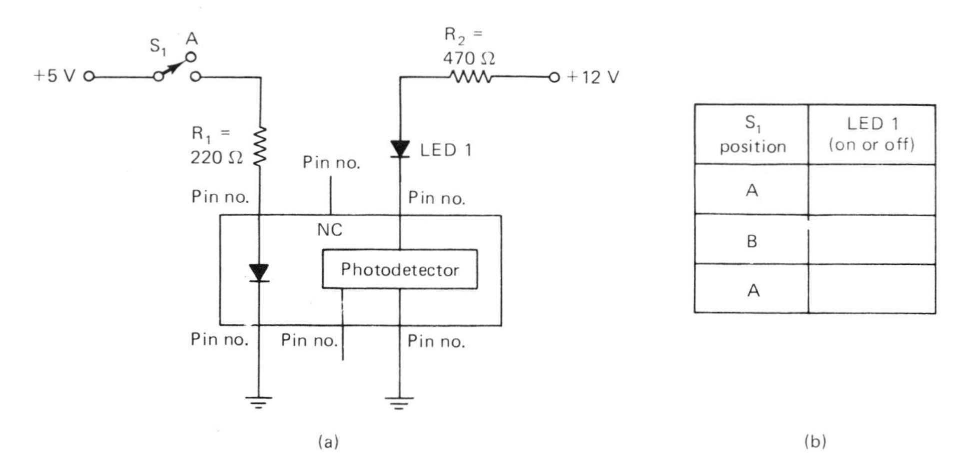

S_1 position	LED 1 (on or off)
A	
B	
A	

Figure 10–15 Photocoupler (optoisolator): (a) circuit; (b) data table.

3. Set switch S_1 to position A.
4. Indicate in the data table of Figure 10-15b if LED 1 is on or off.
5. Place S_1 to position B.
6. Indicate in the data table if LED 1 is on or off.
7. Place S_1 to position A.
8. Indicate in the data table if LED 1 is on or off.

Fill-in Questions:

1. A photocoupler electrically __________ __________ two circuits.

2. In a photocoupler IC, the circuit using the LED (photo source) __________ the circuit using the photodetector.

SECTION 10-5 PHOTODETECTOR BASIC TROUBLESHOOTING APPLICATION: PHOTODETECTOR CONTROL OF TRANSISTORS

Construct the circuit shown below. The photodetector can be a photoresistor (shown), photodiode (with cathode to base of Q_1 and anode to ground), or a phototransistor (with collector to base of Q_1 and emitter to ground). Open or short the components as listed and record the voltages in the proper place in the table. The abbreviations will help indicate the voltage conditions associated with each problem. All voltages are referenced to ground.

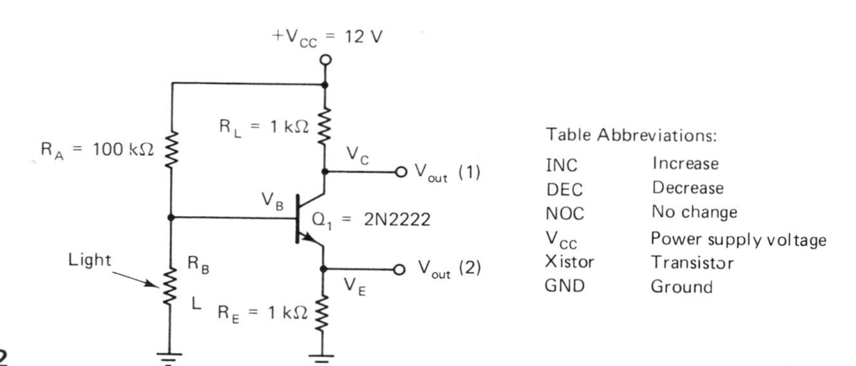

Table Abbreviations:

INC	Increase
DEC	Decrease
NOC	No change
V_{CC}	Power supply voltage
Xistor	Transistor
GND	Ground

Figure 10–E.2

Circuit condition	Without Light			With light			Comments
	V_C	V_B	V_E	V_C	V_B	V_E	
Normal							All voltages are proper
R_A open	V_{CC}	DEC	DEC	NOC	NOC	NOC	Open at V_{CC}: Xistor cut off
R_B open	NOC	INC	INC	DEC	INC	INC	Open at GND: Xistor on
R_B shorted	V_{CC}	DEC	DEC	NOC	NOC	NOC	Xistor cut off
R_L open	DEC	DEC	DEC	DEC	NOC	NOC	Open at V_{CC}: Xistor not conducting
R_E open	V_{CC}	INC	INC	NOC	NOC	INC	Open at GND: Xistor not conducting through collector
Xistor open	V_{CC}	INC	DEC	NOC	NOC	NOC	Remove Xistor: Xistor not conducting
Xistor shorted	INC	INC	INC	NOC	NOC	NOC	Short from C to E: R_L and R_E in series-parallel with R_A and R_B

Figure 10–E.2 (cont'd.)

SECTION 10-6 MORE EXERCISES

Perform the following exercises before beginning Section 10-7.

1. Draw the proper way to connect an ohmmeter for testing the following devices (refer to Figures 10-10 and 10-12).

a. LED

b. Photodiode

c. Phototransistor

d. LASCR

e. Photofet

f. Photoresistor

2. Draw a photoresistor voltage-divider circuit where:

a. when light strikes the photoresistor V_{out} increases.

b. when light strikes the photoresistor V_{out} decreases.

3. Draw a circuit where a LASCR controls the turning on of a LED. Use a LED, LASCR, series-limiting resistor of 220 Ω (for the LED) and a gate resistor of 68 kΩ (for the LASCR).

4. Indicate which input voltage waveform times (T_1, T_2, T_3, T_4, T_5) will cause V_{out} to be low for the following circuits (for example, V_{out} may be low for T_1, T_3, and T_5 or T_2 and T_4). V_{in} to all circuits = +5 V.

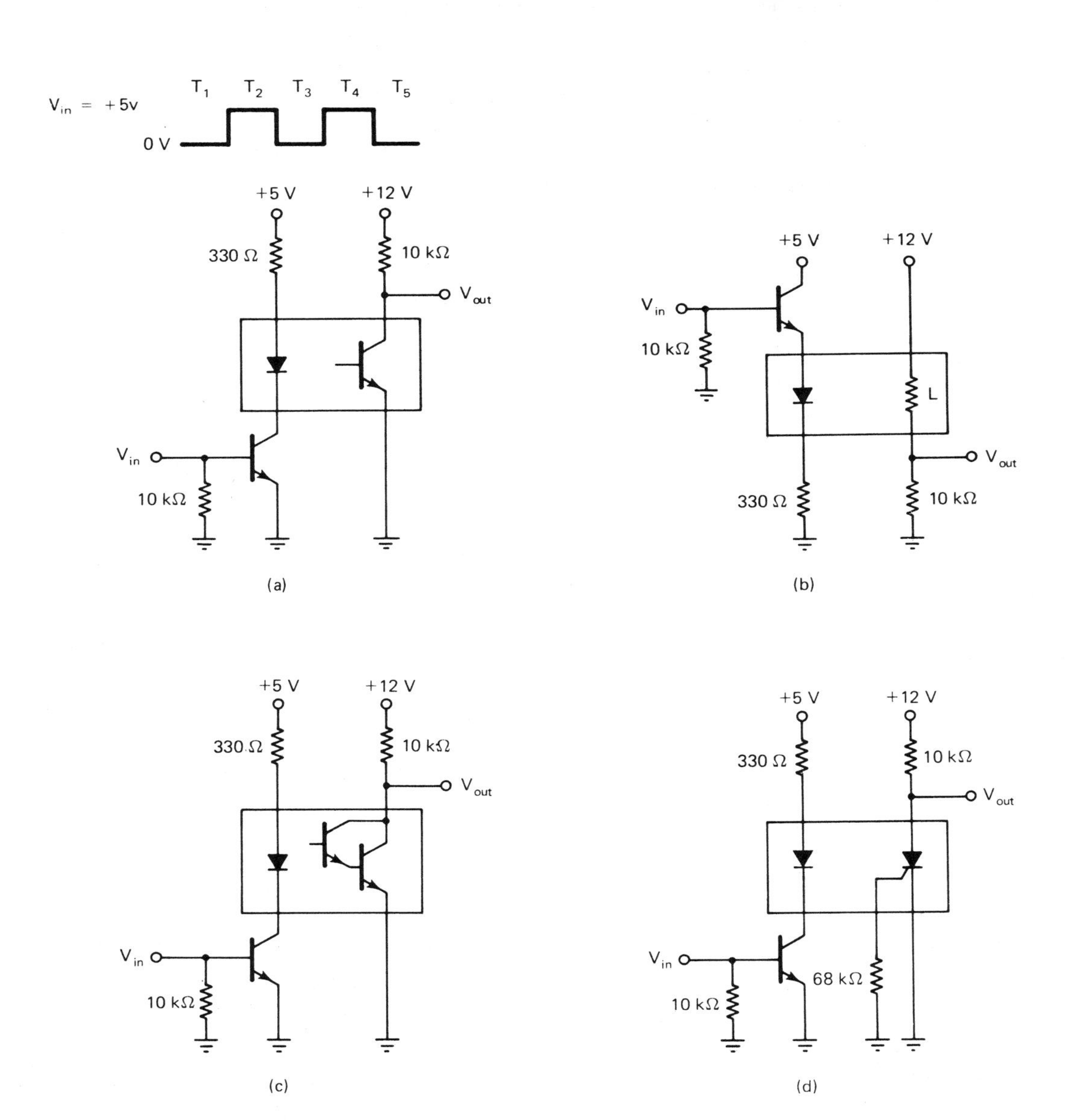

Figure 10–E.3

SECTION 10-7
OPTOELECTRONICS INSTANT REVIEW

Optoelectronics technology combines the fields of optics (light) and electronics. A photosource device emits light when electrical energy is applied, such as from an LED. An LED must be forward biased in order to conduct and emit light. A photosensor or photodetector device conducts electrical current when exposed to light, such as from a photodiode, phototransistor, photofet, LASCR, or photoresistor. The photodiode is reverse biased for proper operation, whereas the phototransistor, photofet, and LASCR are forward biased for proper operation. The photoresistor does not require biasing. The solar cell is a photodetector that produces voltage and current when exposed to light.

SECTION 10-8 SELF-CHECKING QUIZZES

10-8a OPTOELECTRONICS: TRUE–FALSE QUIZ

Place a T for true or an F for false to the left of each statement.

_____ **1.** Photons are released when electrons fall to a lower energy level, such as when they cross a *PN* junction and combine with holes.

_____ **2.** When light strikes a typical photodetector, its resistance increases.

_____ **3.** An LED has a nearly constant voltage drop across its *A–K* leads when forward biased.

_____ **4.** The anode lead of an LED is indicated by a notch or flat side on its case.

_____ **5.** A photodiode must be forward biased in an actual circuit to operate correctly.

_____ **6.** A phototransistor provides more current than a photodiode because of its beta effect.

_____ **7.** A photofet has a high impedance and low noise compared to a standard phototransistor.

_____ **8.** A LASCR operates the same as a standard SCR, but is triggered into conduction by light.

_____ **9.** Photoresistors have a dark resistance and a light resistance.

_____ **10.** Solar cells connected in series will produce more current than a single cell.

_____ **11.** To increase the voltage required by a circuit, solar cells are connected in parallel.

_____ **12.** A photo-coupled isolator requires an input voltage to cause it to operate correctly.

10-8b OPTOELECTRONICS: MULTIPLE-CHOICE QUIZ

Circle the correct answer for each question.

1. Particles of light are called:

a. ions **b.** neutrons

c. photons **d.** none of the above

2. When light is blocked from the diode, the output voltage of Figure 10-A is about:

a. 0 V **b.** 0.7 V

c. 8.3 V **d.** 9.0 V

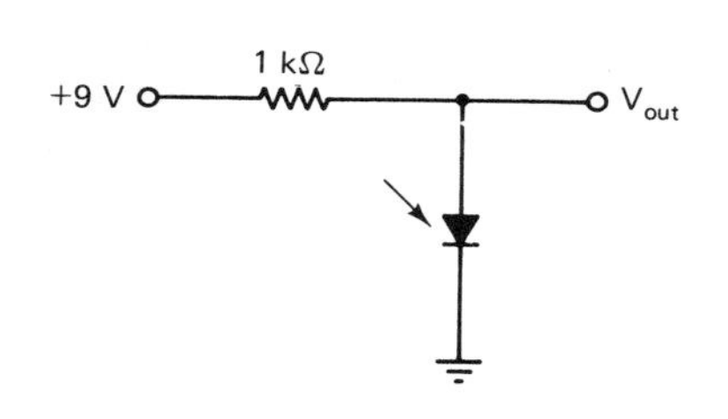

Figure 10–A

3. To obtain the most current from a photodetector, a circuit would use:
   - **a.** a photodiode
   - **b.** a phototransistor
   - **c.** an LED
   - **d.** a photo-Darlington transistor

4. The photodetector that can be turned off only by reducing the current through it below its holding level is the:
   - **a.** photodiode
   - **b.** LASCR
   - **c.** solar cell
   - **d.** LED

5. If each solar cell produces 0.4 V and 80 μA, the output voltage and current of Figure 10-B is:
   - **a.** 0.4 V and 160 μA
   - **b.** 0.8 V and 80μA
   - **c.** 0.8 V and 160 μA
   - **d.** 0.4 V and 80 μA

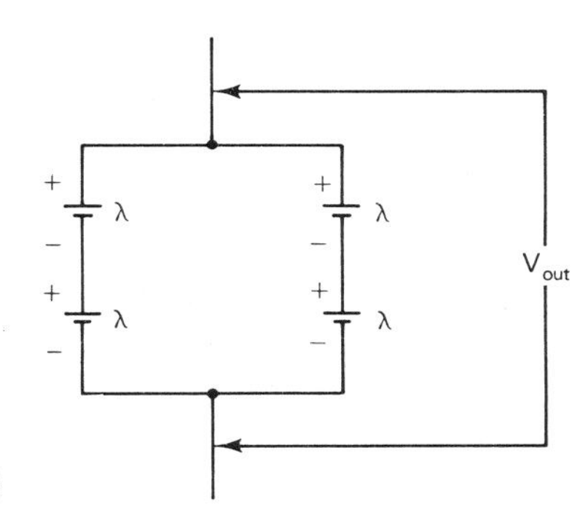

Figure 10–B

Questions 6 and 7 refer to Figure 10-C (R_{dark} = 500 kΩ, R_{light} = 100 kΩ).

6. When light strikes the photoresistor, V_{out} equals:
   - **a.** 0 V
   - **b.** +2.5 V
   - **c.** +12.5 V
   - **d.** +15 V

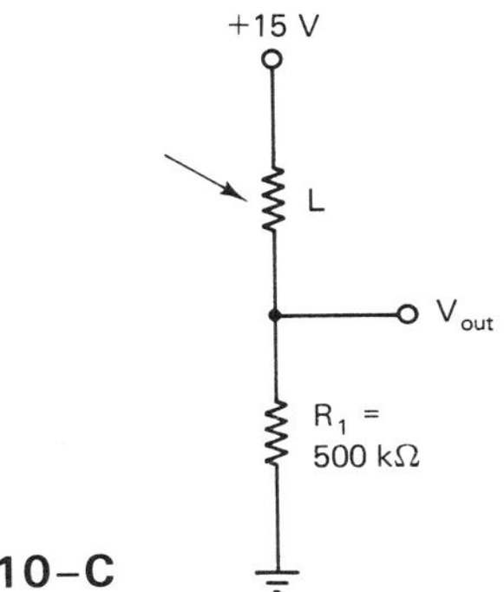

Figure 10–C

7. When light is blocked from the photoresistor, V_{out} equals:
   - **a.** 0 V
   - **b.** +1.5 V
   - **c.** +7.5 V
   - **d.** +15 V

Questions 8, 9, and 10 refer to Figure 10-D.

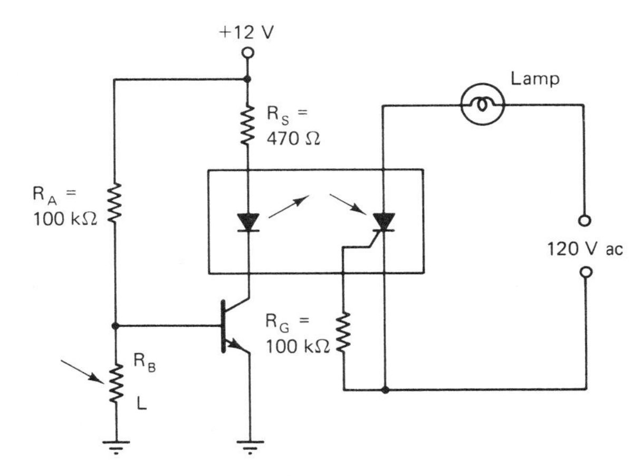

Figure 10–D

8. le circuit shown:
   - **a.** connects the +12-V dc supply to the 120-V ac supply
   - **b.** isolates the +12-V dc supply from the 120-V ac supply
   - **c.** steps up the voltage
   - **d.** none of the above

9. When light strikes the photoresistor:
   - **a.** Q_1 is on, the LED is off, and the LASCR is off
   - **b.** Q_1 is on, the LED is on, and the LASCR is off
   - **c.** Q_1 is off, the LED is off, and the LASCR is off
   - **d.** Q_1 is off, the LED is on, and LASCR is on

10. When light is blocked from the photoresistor:
    - **a.** Q_1 is off, the LED is off, and the LASCR is off
    - **b.** Q_1 is on, the LED is on, and the LASCR is on
    - **c.** Q_1 is on, the LED is off, and the LASCR is on
    - **d.** Q_1 is off, the LED is on, and the LASCR is off

11. If the ohmmeter readings across an LED indicate low ohms in the forward-biased condition and high ohms in the reverse-biased condition, the LED appears to be:

a. functioning (okay)

b. malfunctioning (not okay)

c. neither a nor b

12. If an LED having a V_F of 1.8 V is connected in series with a 1-kΩ resistor to a +15-V power source, the current through the circuit is:

a. 15 mA **b.** 13.2 mA

c. 8.33 mA **d.** 1.8 mA

ANSWERS TO EXPERIMENTS AND QUIZZES

Experiment 1.	**(1)** positive, negative **(2)** cathode, anode
Experiment 2.	**(1)** series, limit **(2)** forward **(3)** reverse
Experiment 3.	**(1)** high **(2)** low **(3)** voltage, current
Experiment 4.	**(1)** increases **(2)** decreases **(3)** decreases **(4)** increases
Experiment 5.	**(1)** off, extinguished (off) **(2)** on, glows (is on) **(3)** on, glows (is on) **(4)** off, extinguished (off)
Experiment 6.	**(1)** isolates **(2)** controls
True-False.	**(1)** T **(2)** F **(3)** T **(4)** F **(5)** F **(6)** T **(7)** T **(8)** T **(9)** T **(10)** F **(11)** F **(12)** T
Multiple Choice.	**(1)** c **(2)** b **(3)** d **(4)** b **(5)** c **(6)** c **(7)** c **(8)** b **(9)** c **(10)** b **(11)** a **(12)** b

Unit 11

Operational Amplifier

INTRODUCTION Operational amplifiers (op amps) are the heart of many types of electronic equipment, including medical, radio–TV–stereo, computer, and industrial devices.

UNIT OBJECTIVES Upon completion of this unit you will be able to:

1. Describe the basic elements of an op amp.
2. Draw the schematic symbol for an op amp.
3. Define the terms offset nulling, output offset voltage, input offset voltage and current, slew rate, common-mode rejection ratio, latch up, and other terms associated with op amps.
4. List the most important characteristics of an op amp.
5. Identify op amp circuits, including a voltage comparator, inverting amplifier, noninverting amplifier, voltage follower, summing amplifier, difference amplifier, voltage-level detector, square-wave generator, low-pass filter, high-pass filter, bandpass filter, and bandreject filter.
6. Calculate various values needed in op-amp circuits.
7. Test op amps.
8. Troubleshoot an op-amp inverting amplifier.

SECTION 11-1 THEORY OF STRUCTURE AND OPERATION

11-1a THE BASIC OP AMP

A basic operational amplifier (op amp) is a solid-state device with several circuits within a single package capable of sensing and amplifying dc and ac input signals. Op amps can be used for various electronic circuit functions with only a few external components. Figure 11-1 shows the block diagram of a standard op amp consisting of a high-input-impedance differential amplifier, a high-gain voltage amplifier, and a low-impedance output amplifier. Notice that there are two inputs, one output, and dual (±) power supply connections, which enables the output to swing both positive and negative.

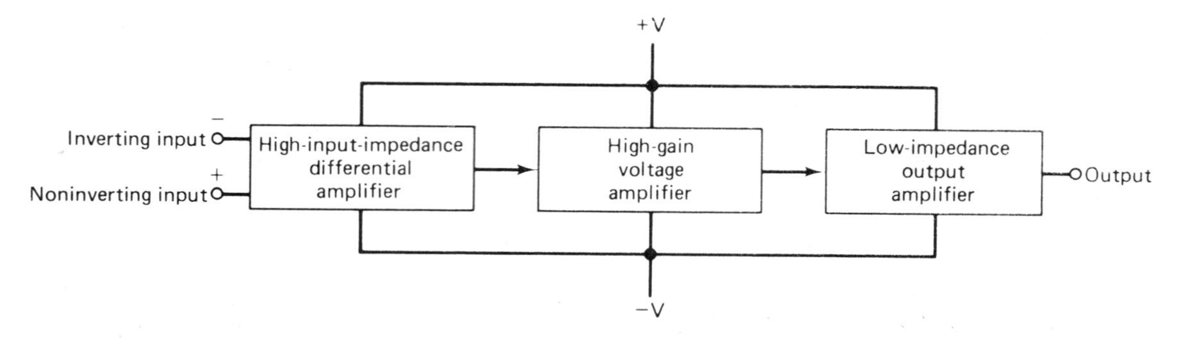

Figure 11–1 Block diagram of op amp. (From F. Hughes, *Op Amp Handbook*, Prentice-Hall, Englewood Cliffs, N.J., © 1981, Fig. 1–1, p. 2. Reprinted with permission.)

The most important characteristics of an op amp are:

1. *Very high input impedance,* which produces negligible currents at the inputs
2. *Very high open-loop gain,* which is useful for amplifying very small signals
3. *Very low output impedance,* so that it is affected very little by other circuit loads

Although it is possible to use the op amp without knowing exactly what goes on inside it, a better understanding of its operating characteristics can be gained with some knowledge of its internal circuitry. The popular 741 IC op amp is shown in Figure 11-2. In this IC design, resistors are held to an absolute minimum, using transistors whenever possible. No coupling capacitors are used, allowing the circuit to amplify dc as well as ac signals. The 30-pF capacitor shown provides internal frequency compensation and reduces the possibility of self-oscillations at higher frequencies. Transistors Q_1 and Q_2 form the differential amplifier inputs. Transistors Q_{16} and Q_{17} produce a high-gain Darlington amplifier. The output stage is complementary symmetry using transistors Q_{14} and Q_{20}. When Q_{14} is conducting heavily, the output is pulled up toward the positive supply voltage at pin 7 (Q_{20} is conducting less at this time). When Q_{20} is conducting heavily, the output is pulled down toward the negative supply voltage at pin 4 (Q_{14} is conducting less at this time). Output short-circuit protection is provided by current-limiting transistor Q_{15}. The remaining components provide bias and loads for the circuitry. An external potentiometer can have its ends connected to pins 1 and 5 (offset null) with its wiper connected to pin 4 ($-V_{cc}$) for nulling adjustments.

The schematic diagram for a standard op amp is represented as a triangle, as shown in Figure 11-3. The inverting input is represented by a minus sign. The voltage at this input causes the output voltage to be inverted 180°.

The noninverting input is represented by a plus sign. The voltage at this input causes the voltage at the output to be in phase. The output terminal is at the apex of the triangle. Power supply leads are shown above and below the triangle and must always be connected, even though they may not be indicated on a schematic diagram. Other leads coming out of the op amp may be used for frequency compensation or nulling components.

Integrated-circuit op amps are available in four commonly used packages, as shown in Figure 11-4a. The metal TO-type package is approximately 0.300 to 0.450 in. in diameter and 0.130 to 0.185 in. high, with eight to ten leads coming out of the bottom. The flat package has

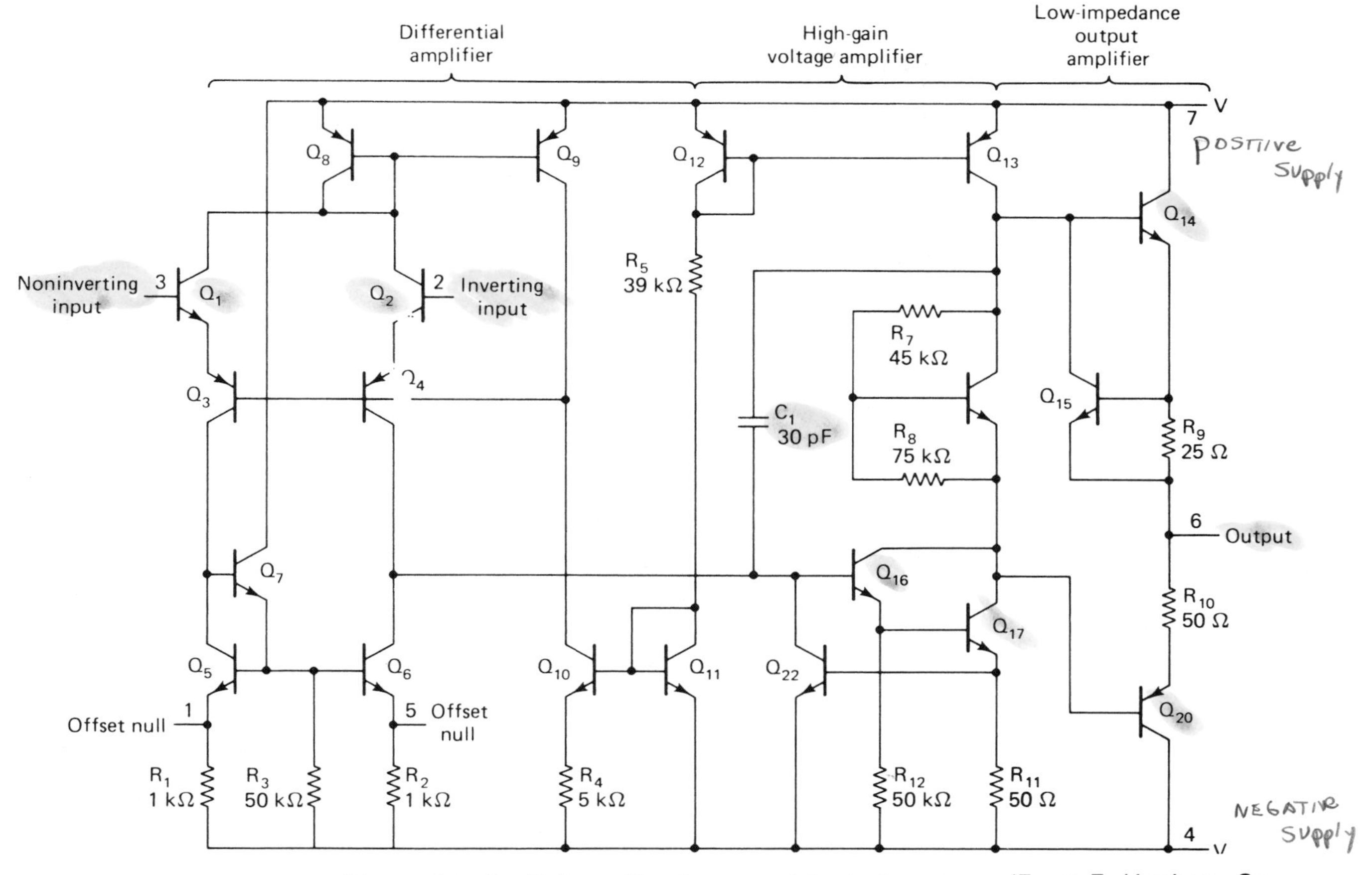

Figure 11–2 Schematic diagram of typical op amp. (From F. Hughes, *Op Amp Handbook*, Prentice-Hall, Englewood Cliffs, N.J., © 1981, Fig. 1–3, p. 4. Reprinted with permission.)

a body about 0.250 to 0.270 in. square and 0.50 to 0.70 in. thick, usually with five leads coming out of each side. This package may be of metal, glass, or plastic. The dual-in-line package (DIP) can be made of metal, glass, ceramic, or plastic. It measures about 0.750 in. long, 0.270 in. wide, and 0.190 in. thick, with seven leads protruding downward from each side. The mini-DIP is about half the size of a standard 14-pin DIP and has four leads protruding downward from each side.

The lead identification shown in Figure 11-4b is usually self-explanatory. The positive supply voltage is connected to the $+V$ terminal, and the negative supply voltage is connected to the $-V$ terminal. Input and output terminals are clearly indicated. The balance terminals (sometimes designated "Offset Null") are connected to a potentiometer

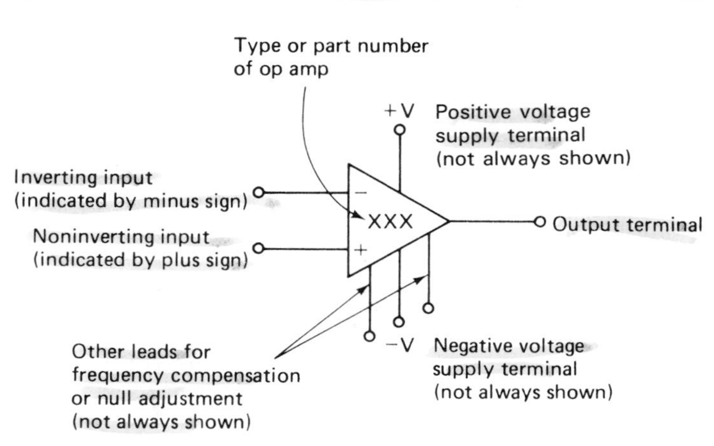

Figure 11–3 Schematic symbol of standard op amp. (From F. Hughes, *Op Amp Handbook*, Prentice-Hall, Englewood Cliffs, N.J., © 1981, Fig. 1–2, p. 3. Reprinted with permission.)

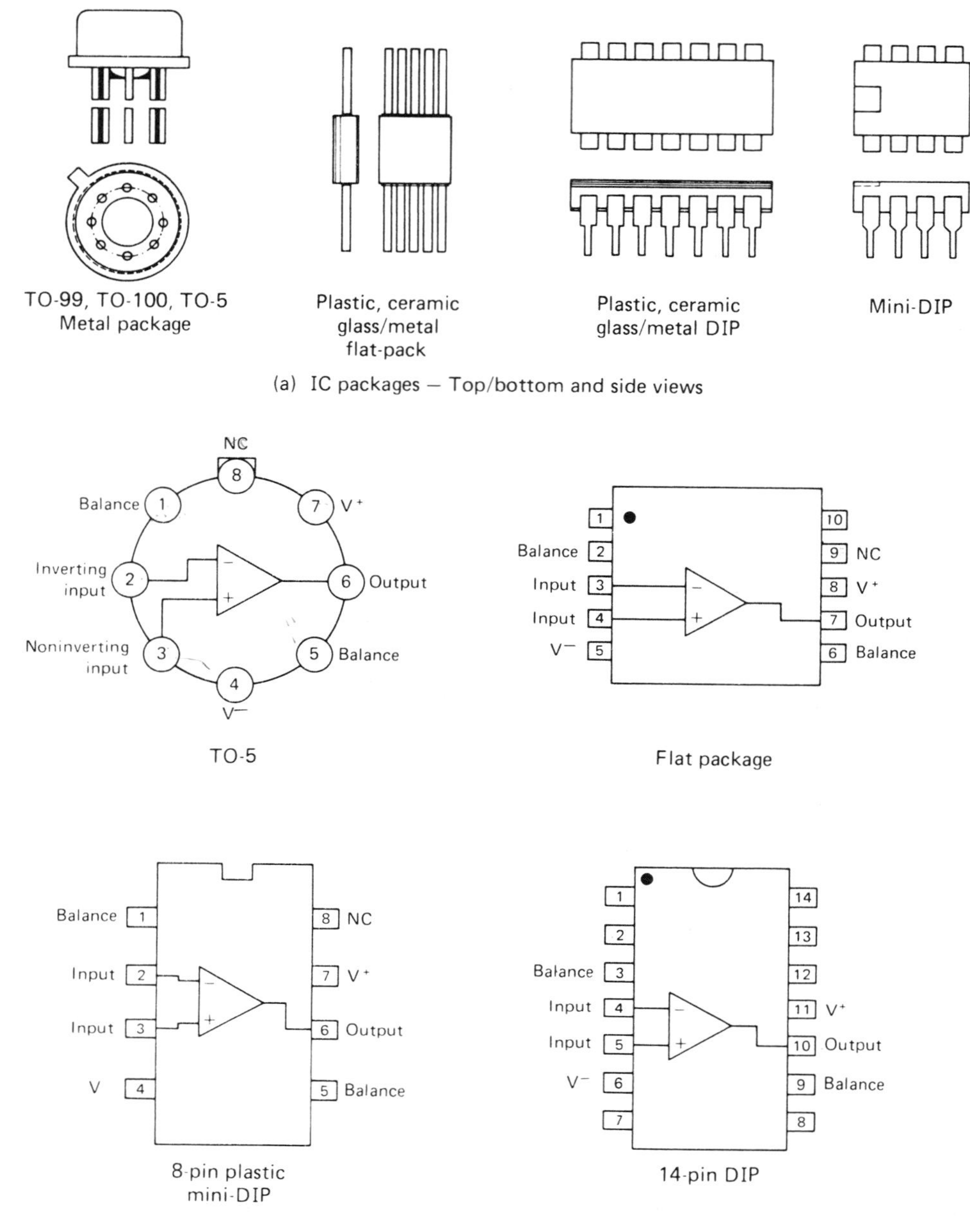

Figure 11–4 IC op-amp packages. (From F. Hughes, *Op Amp Handbook*, Prentice-Hall, Englewood Cliffs, N.J., © 1981, Fig. 1–14, pp. 21 and 22. Reprinted with permission.)

for null adjusting. Terminals marked "NC" (no connection) are included for physical ruggedness of the package.

11-1b INPUT/OUTPUT VOLTAGE POLARITY

An important function to remember about an op amp is the relationship of input voltage polarity to output voltage polarity. Figure 11-5 illustrates this relationship, where the noninverting input is at 0 V or ground. If the inverting input is more positive than the noninverting input, the output will be at a negative voltage potential. Similarly, if the inverting input is more negative than the noninverting input, the output voltage will be at a positive potential. This relationship remains the same even if both input voltages are positive or negative.

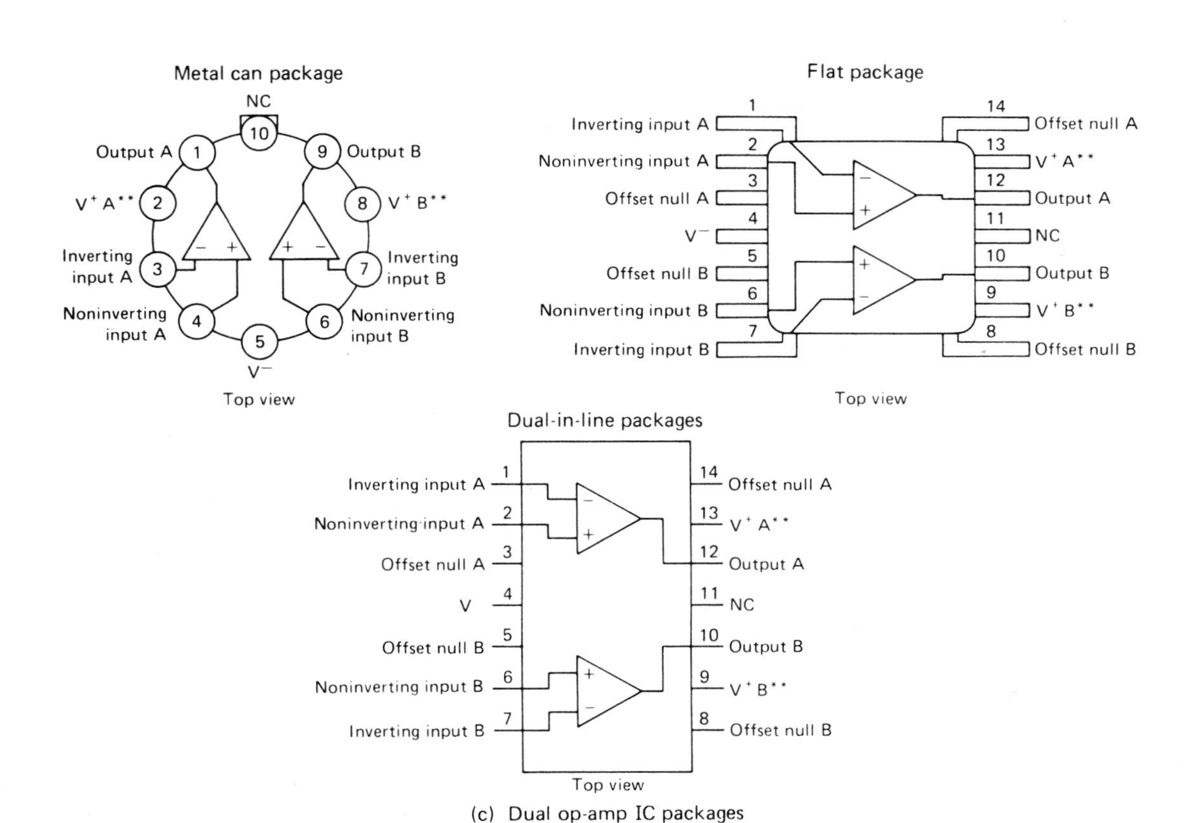

(c) Dual op-amp IC packages

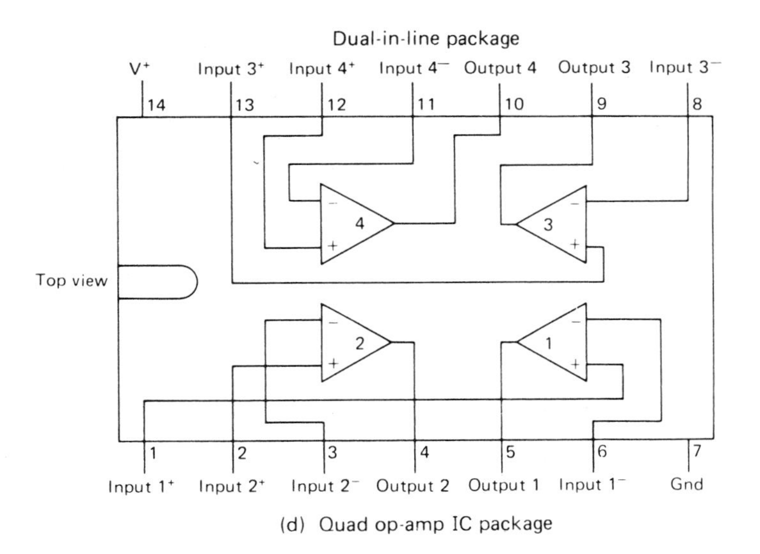

(d) Quad op-amp IC package

Figure 11–4 (cont'd.)

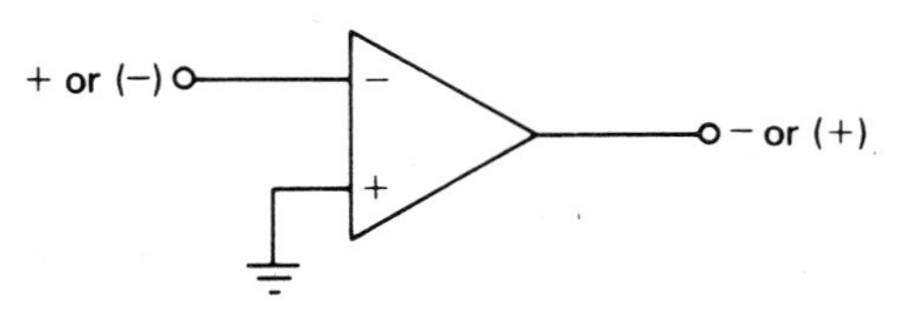

1. If − input is more positive than + input, output will be negative.
2. If − input is more negative than + input, output will be positive.

Figure 11–5 Input/output voltage polarity relationship. (From F. Hughes, *Op Amp Handbook*, Prentice-Hall, Englewood Cliffs, N.J., © 1981, Fig. 1–5, p. 6. Reprinted with permission.)

11-1c OP-AMP GAIN

Ideally, the gain of an op amp would be infinite; however, practically, the gain may exceed 200,000 in the open-loop mode. In the open-loop mode, there is no feedback from the output to the inputs and the voltage gain (A_v) is maximum, as shown in Figure 11-6a. In a practical circuit, the slightest voltage difference at the inputs causes the output voltage to attempt to swing to the maximum power-supply level. The maximum voltage at the output will be about 90% of the supply voltages, because of the internal voltage drops of the op amp. (Refer to Figure 11-2 and note components Q_{14}, R_9, R_{10}, and Q_{20}.) The output is said to be at saturation and can be represented (for either polarity) by $+V_{sat}$ and $-V_{sat}$. As an example, an op-amp circuit in the open-loop mode using a ±15-V supply would have its output swing from +13.5 V to −13.5 V. With this type of circuit the op amp is very unstable and the output is 0 V for a 0-V difference between the inputs, or the output voltage is at either extreme, with a slight voltage difference at the inputs. The open-loop mode is found primarily in voltage comparators and level-detector circuits.

The versatility of the op amp is demonstrated by the fact that it can be used in so many types of circuits in the closed-loop mode, as shown in Figure 11-6b. External components are used to feed back a portion of the output voltage to the inverting input. This feedback stabilizes most circuits and can reduce the noise level. The voltage gain (A_v) will be less (<) than maximum gain in the open-loop mode.

Closed-loop gain must be controlled to be of any value in a practical circuit. By adding a resistor R_{in} to the inverting input as shown in Figure 1-6c, the gain of the op amp can be controlled. The resistance ratio of R_F to R_{in} determines the voltage gain of the circuit and can be found by the formula

$$A_v = -\frac{R_F}{R_{in}}$$

Figure 11–6 Op-amp gain. (From F. Hughes, *Op Amp Handbook*, Prentice-Hall, Englewood Cliffs, N.J., © 1981, Fig. 1–4, p. 5. Reprinted with permission.)

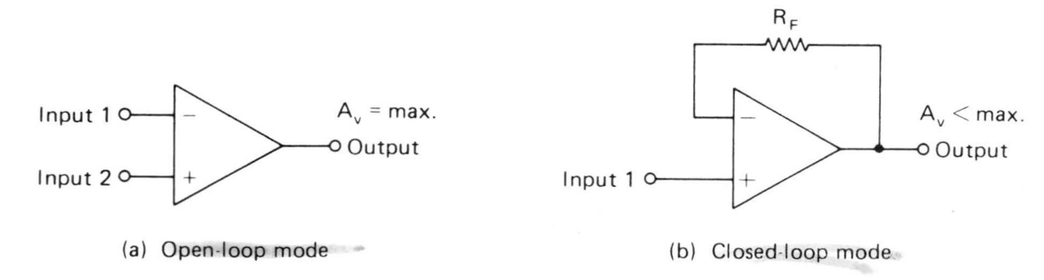

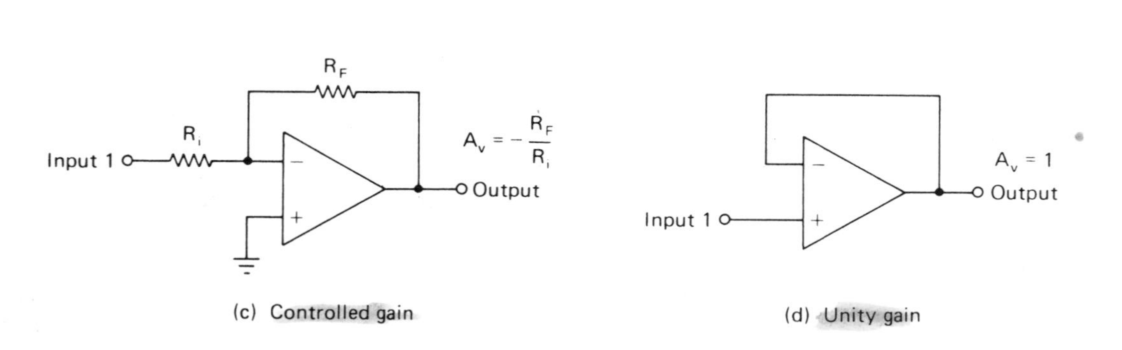

The minus sign indicates that the op-amp circuit is in the inverting configuration and is disregarded for calculations. For example, if R_{in} = 10 kΩ and R_F = 100 kΩ, then A_v would equal 10. An input voltage of 0.01 V would yield an output voltage of 0.1 V. If R_{in} were changed to 1 kΩ, the A_v would increase to 100. Now, an input voltage of 0.01 V would yield an output voltage of 1 V.

If both R_{in} and R_F are the same value, the A_v equals 1, or unity gain, as shown in Figure 1-6d. In this noninverting configuration, the voltage out equals the voltage in and A_v equals +1.

11-1d OFFSET NULLING

Ideally, the output voltage of an op amp should be zero when the voltages at both inputs are the same or zero. However, because of the high gain of the op amp, a slight circuit imbalance of the input transistors can cause a slight output offset voltage. In a critical circuit, this offset can cause error voltages at the output. One method of balancing the op-amp output, called *offset nulling,* is to have a potentiometer connected to one of the inputs so that it can be adjusted to bring back the output voltage to zero when the voltage difference at the inputs is zero. Many op amps have offset nulling pins, as shown in Figure 11-7. The ends of a potentiometer are connected to these pins with the wiper attached to the $-V$ supply. Often, null circuits are used with an op amp but are not shown on the schematic diagram.

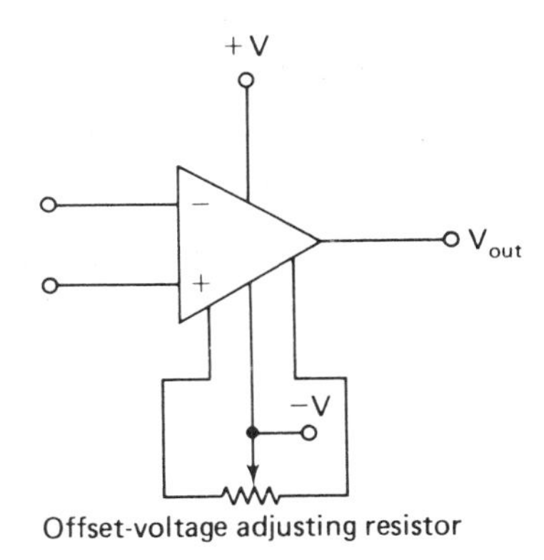

Figure 11–7 Offset nulling. (From F. Hughes, *Op Amp Handbook*, Prentice-Hall, Englewood Cliffs, N.J., © 1981, Fig. 1–6, p. 8. Reprinted with permission.)

11-1e OP-AMP FREQUENCY RESPONSE

The gain of an op amp decreases with an increase in frequency. The gain given by manufacturers is generally at zero hertz or dc. Figure 11-8 shows a voltage-gain versus frequency-response curve. In the open-loop mode, the gain falls off very rapidly as frequency increases. When the frequency increases tenfold, the gain decreases by 10. The breakover point occurs at 70.7% of the maximum gain. The frequency bandwidth is normally considered at the point where the gain falls to the breakover point. Therefore, the open-loop bandwidth is about 10 Hz for this example. Fortunately, op amps usually require degenerative feedback in amplifier circuits, and this feedback increases the bandwidth of the circuit. For a closed-loop gain of 100, the bandwidth has increased to about 10 kHz. Lowering the gain to 10 increases the bandwidth to about 100 kHz. The unit-gain point occurs at 1 MHz and is called the unity-gain frequency. The unity-gain frequency establishes the reference point at which many op amps are specified by manufacturers.

The gain–bandwidth product is equal to the unity-gain frequency. It not only tells us the upper useful frequency of a circuit, but allows us to determine the bandwidth for a given gain. For example (referring

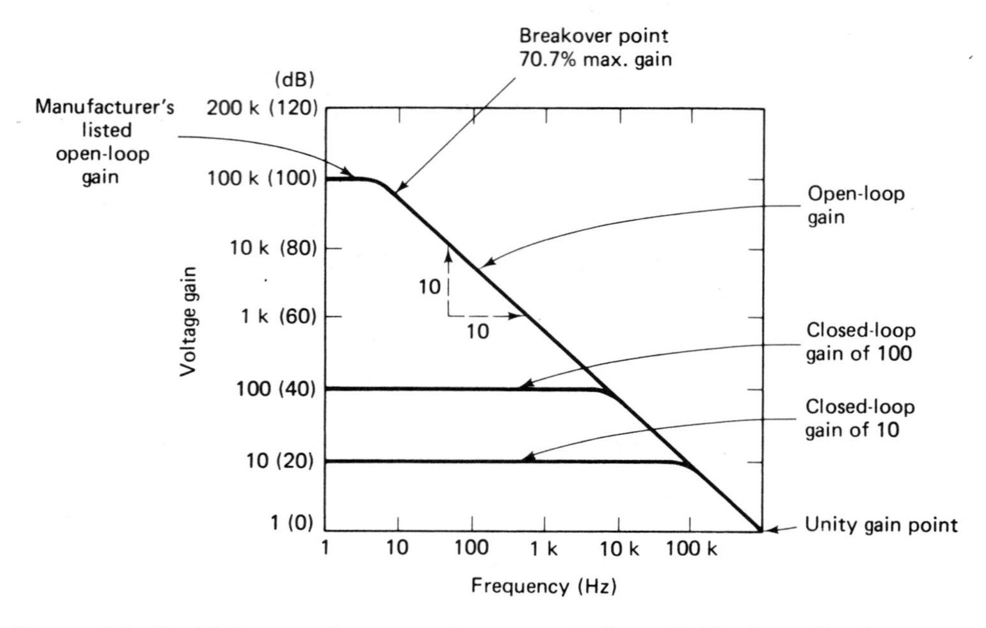

Figure 11–8 Voltage gain versus frequency. (From F. Hughes, *Op Amp Handbook,* Prentice-Hall, Englewood Cliffs, N.J., © 1981, Fig. 1–8, p. 10. Reprinted with permission.)

to Figure 11-8, which shows a frequency-response curve for a frequency-compensated op amp, such as the 741), if you multiply the gain and bandwidth of a specific circuit, the product will equal the unity-gain frequency:

$$\text{gain–bandwidth product} = \text{gain} \times \text{bandwidth} = \text{unity-gain frequency}$$
$$\text{GBP} = 100 \times 10\,\text{kHz} = 1{,}000{,}000\,\text{Hz}\,(1\,\text{MHz})$$
$$\text{GBP} = 10 \times 100\,\text{kHz} = 1{,}000{,}000\,\text{Hz}\,(1\,\text{MHz})$$

Therefore, if we wanted to know the upper frequency limit or bandwidth of a circuit with a gain of 100, we would divide the unity-gain frequency by gain:

$$\text{bandwidth} = \frac{\text{unity-gain frequency}}{\text{gain}}$$
$$\text{BW} = \frac{1{,}000{,}000}{100} = 10\,\text{kHz}$$

11-1f OP-AMP DEFINITIONS

The following definitions are useful to know when working with op amps.

Output Offset Voltage: a slight unwanted voltage at the output when the voltage between the inputs is zero. Ideally, V_{out} should be 0 V.

Input Offset Voltage: the voltage that is applied across the inputs to force V_{out} to 0 V.

Input Offset Current: the difference in the two dc input currents when the inputs are at the same potential.

Slew Rate: the maximum rate of change of the op amp's output voltage under large-signal conditions:

$$\text{slew rate} = \frac{\Delta V_{out}\,(\text{max})}{\Delta t}$$

Differential Input Voltage (V_d): the difference of voltage between the two inputs.

Differential gain (A_d): the ratio of the output voltage to the differential input voltage of a differential amplifier.

Common-Mode Voltage (V_{cm}): an unwanted, but unavoidable voltage on both inputs, such as 60-cycle hum.

Common-Mode Gain (A_{cm}): the ratio of the common-mode output voltage to the common-mode input voltage.

Common-Mode Rejection Ratio (CMRR): the ratio of the closed-loop gain (A_D) to the common-mode gain (A_{cm}) as given by the formula $\text{CMRR} = A_d/A_{cm}$.

Power Supply Voltage Rejection Ratio (PSRR): the ratio of the change in the power supply voltage to the resulting change in input offset voltage. Variation in power supply voltage will also affect the input offset voltage.

Power Supply Decoupling: capacitors in the range 0.1 to 1.0 μF connected from the power supply voltages to ground to bypass voltage variations to ground.

Input Voltage Range: the maximum voltage that can be applied between either input to ground without causing distortion at the output.

Input Protection: diodes, zener diodes, and/or resistors are used at the inputs to protect the op amp from excessively large input voltages.

Output-Voltage Maximum Swing: the maximum output-voltage swing without distortion.

Latch-Up: a condition where a large input signal causes the output to remain in $+V_{sat}$ or $-V_{sat}$. Diodes and resistors used in the output circuit can prevent this.

Output Protection: a low-value resistor connected in series with the output of an op amp to limit current during a short-circuit condition. Some op amps have this protection built in, such as transistor Q_{15} shown in Figure 11-2.

SECTION 11-2 OP-AMP DEFINITION EXERCISE

Refer to the previous sections and write a brief definition for each term.

1. $A_v = -R_F/R_{in}$ resistance ratio of the RF to Rin determines the voltage gain of the circuit called controlled gain

2. $A_v = R_F/R_{in} + 1$ unity gain;

3. Output offset voltage a slight unwanted voltage at the output when the voltage between the inputs is zero.

4. Offset nulling method to balance op-amp output *Have potentiometer connected to one of the inputs so ~~that~~ it can be adjusted to bring the output voltage to zero.

5. Input offset voltage voltage that is applied to inputs to force V_{OUT} to 0V

6. Input offset current Difference in the two dc input currents when the inputs are at the same potential

7. Slew rate maximum rate of change of the op-amp's output voltage under large-signal conditions.
slew rate = $\frac{\Delta V_{OUT}(max)}{\Delta t}$

8. V_d (Differential input voltage: difference of voltage between the inputs

9. A_d Differential gain = ratio of output voltage to the differential input voltage of a differential amplifier

10. V_{cm} Common mode voltage = unwanted, unavoidable voltage at both inputs.

11. A_{cm} common mode gain = Ratio of common mode output voltage to the common-mode input voltage

12. CMRR common-mode rejection ratio
Ratio of closed-loop gain to the common-mode gain
CMRR = A_d / A_{cm}

13. PSRR Power supply rejection ratio
ratio of change in power supply voltage to the resulting change in input offset voltage.

14. Power supply decoupling capacitors from 0.1 to 1.0 uF connected from power supply voltages to ground to bypass voltage variations to ground.

15. Input voltage range maximum voltage that can be applied between either input to ground without causing distortion

16. Input protection devices such as diodes, zener diodes, and/or resistors to protect op-amp from excessively large input voltages

17. Output voltage maximum swing maximum output voltage swing without distortion

18. Latch-up large input signal causes output to remain in +Vsat or -Vsat.

19. Output protection low-value resistor connected in series with the output of an op-amp to limit current during a short-circuit condition

SECTION 11-3 EXERCISES AND PROBLEMS

Perform the following exercises before beginning Section 11-4.

1. Draw a basic block diagram of an op amp.

2. Draw the schematic symbol for a standard op amp.

3. Draw an op-amp circuit showing controlled gain.

4. Draw an op-amp circuit showing unity gain.

5. Draw an example of offset nulling.

6. Draw a typical frequency-response curve for an op amp showing a gain of 10 and 100.

7. Without feedback the output of an op amp will saturate with the slightest difference of potential between its two inputs. For the input voltages given in the following circuits, indicate if the output goes to positive voltage saturation ($+V_{sat}$) or negative voltage saturation ($-V_{sat}$).

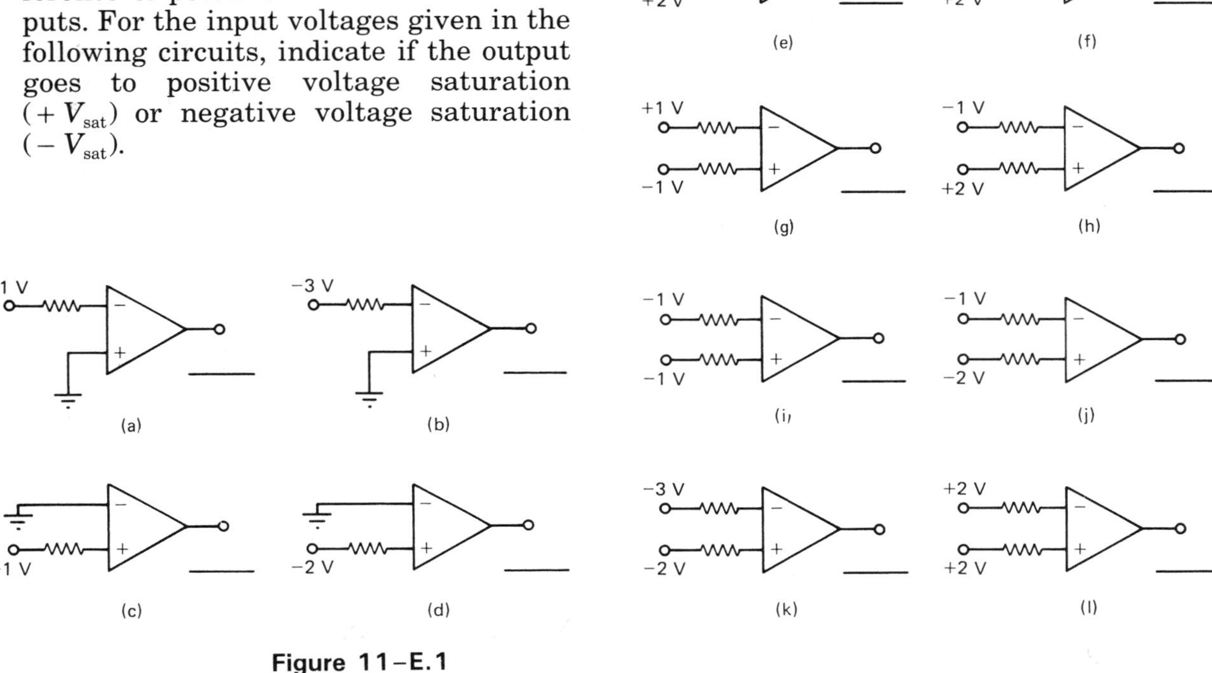

Figure 11–E.1

SECTION 11-4 EXPERIMENTS

EXPERIMENT 1. BASIC OP-AMP OPERATION

Objective:

To demonstrate the basic operation of an op amp and to understand a comparator circuit.

Introduction:

The polarity of the voltage at the output of an op amp depends on the relationship of the polarity between the voltages at the inputs. Remember that the inverting (−) input is referenced to the noninverting (+) input. When the − input is more positive than the + input, the output will be negative, and when the − input is more negative than the + input, the output will be positive. Without a feedback path, the output will either be at $+V_{sat}$ or $-V_{sat}$.

Materials Needed:

1 Dual ±15-V power supply
1 Standard or digital voltmeter
1 741 op amp or equivalent (see Figure 11-4 for pin identification)
2 10-kΩ linear potentiometers (R_1 and R_2)
3 10-kΩ resistors at 0.5 W (R_3, R_4, and R_L)
1 Breadboard for constructing circuit

Procedure:

1. Construct the circuit shown in Figure 11-9a.
2. Set input voltages V_1 and V_2 according to the data table shown in Figure 11-9b.
3. Record V_{out} in the data table, indicating polarity.
4. Repeat steps 2 and 3 for the various values of V_1 and V_2.

Fill-in Questions:

1. Without feedback, the output of an op amp will be at either +V SAT or − VSAT, depending on the polarity relationship of the input voltages.

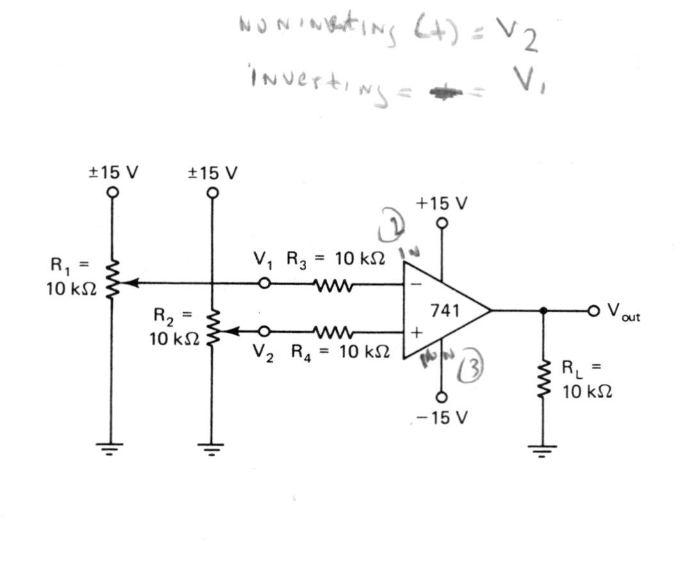

(a)

V_1 (V)	V_2 (V)	V_{out} (V)
+1	0	−12.78
−1	0	+14.16
0	+1	+14.16
0	−1	−12.78
+2	+1	−12.78
+1	+2	+14.16
+1	−1	−12.78
−1	+1	+14.16
−1	−2	−12.78
−2	−1	+14.16

(b)

Figure 11–9 Voltage comparator: (a) schematic diagram; (b) data table.

2. The output of a comparator will be at +V SAT when the inverting input is more negative than the noninverting input.

3. If the output of a comparator is $-V_{sat}$, the inverting input voltage must be more POSITIVE than the noninverting input voltage.

EXPERIMENT 2. OP-AMP INVERTING AMPLIFIER

Objective:

To show the operation of an op-amp inverting amplifier with dc and ac voltages and be able to calculate the gain of the circuit.

Introduction:

The inverting amplifier shown in Figure 11-10a consists mainly of resistors R_{in}, R_F, R_n, and R_L. Resistors R_1, R_2, and R_3 are used as a voltage divider to set the desired dc voltage at the inverting input. The gain of the circuit can be calculated by the formula $A_v = -R_F/R_{in}$ (where the minus sign indicates only that the polarity of the output voltage is opposite to the polarity of the input voltage) or can be found by $A_v = V_{out}/V_{in}$. Resistor R_n is used to reduce offset bias currents and is equal to the value of R_{in} and R_F in parallel ($R_n = R_{in}R_F/R_{in} + R_F$). The junction of R_{in} and R_F at the inverting input is about the same voltage as the noninverting input and is referred to as virtual ground. The input impedance is usually equal to R_{in}.

When the inverting amplifier is used for ac signals, as shown in Figure 11-10c, capacitors are used to block any dc voltage from the circuit that might cause distortion. The frequency response of an op-amp circuit depends on its gain. The lower the gain, the wider the frequency response is.

Materials Needed:

- *1* Dual ± 15-V power supply
- *1* Standard or digital voltmeter
- *1* Oscilloscope (dual trace preferred)
- *1* AC signal generator (up to 1 MHz)
- *1* 741 op amp or equivalent
- *1* 4.7-kΩ resistor at 0.5 W (R_{in})
- *1* 6.8-kΩ resistor at 0.5 W (R_n)
- *4* 10-kΩ resistors at 0.5 W (R_1, R_3, R_{in}, R_L, and R_n)
- *1* 22-kΩ resistor at 0.5 W (R_L)
- *1* 47-kΩ resistor at 0.5 W (R_L)
- *1* 100-kΩ resistor at 0.5 W (R_L)
- *1* 10-kΩ linear potentiometers (R_2)
- *2* 1-μF capacitors at 25 WV dc (C_{in}, C_{out})
- *1* Breadboard for constructing circuits

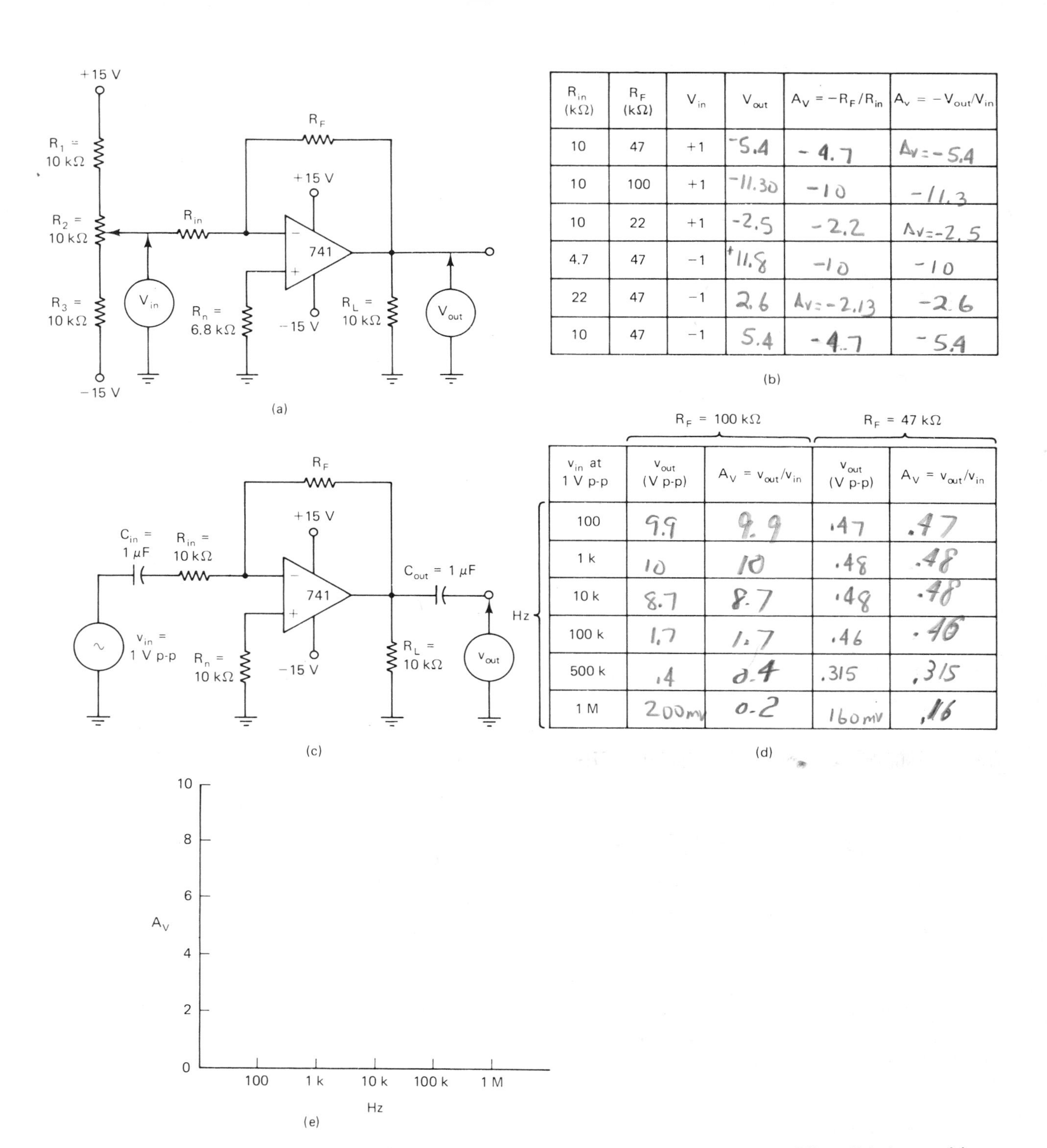

R_{in} (kΩ)	R_F (kΩ)	V_{in}	V_{out}	$A_V = -R_F/R_{in}$	$A_v = -V_{out}/V_{in}$
10	47	+1	−5.4	−4.7	Av=−5.4
10	100	+1	−11.30	−10	−11.3
10	22	+1	−2.5	−2.2	Av=−2.5
4.7	47	−1	+11.8	−10	−10
22	47	−1	2.6	Av=−2.13	−2.6
10	47	−1	5.4	−4.7	−5.4

v_{in} at 1 V p-p (Hz)	R_F = 100 kΩ: v_{out} (V p-p)	R_F = 100 kΩ: $A_V = v_{out}/v_{in}$	R_F = 47 kΩ: v_{out} (V p-p)	R_F = 47 kΩ: $A_V = v_{out}/v_{in}$
100	9.9	9.9	.47	.47
1 k	10	10	.48	.48
10 k	8.7	8.7	.48	.48
100 k	1.7	1.7	.46	.46
500 k	.4	0.4	.315	.315
1 M	200mV	0.2	160mV	.16

Figure 11–10 Op-amp inverting amplifier: (a) dc amplifier; (b) data table; (c) ac amplifier; (d) data table; (e) frequency response curve.

Procedure:

1. Construct the circuit shown in Figure 11-10a using the value of R_{in} and R_F of the first line from the data table of Figure 11-10b.
2. Using the voltmeter, set V_{in} according to the data table.
3. Using the voltmeter, measure and record V_{out} in the data table.
4. Calculate the gain of the circuit from the formula $A_v = -R_F/R_{in}$ and record in the data table.
5. Calculate the gain of the circuit from the formula $A_v = V_{out}/V_{in}$ and record in the data table.

6. Repeat steps 2 through 5 for all values of R_{in} and R_F given in the data table.
7. Construct the circuit shown in Figure 11-10c using R_F = 100 kΩ.
8. Using the oscilloscope, set the signal generator for V_{in} to the frequency given in the first line of the data table shown in Figure 11-10d (maintain 1 V p-p).
9. Using the oscilloscope, measure and record v_{out} in the data table.
10. Calculate the gain using the formula $A_v = v_{out}/v_{in}$ given in the data table.
11. Repeat steps 9 and 10 for the various frequencies of v_{in} given in the data table.
12. Change R_F to 47 kΩ and repeat steps 8 through 11.
13. Using the results of the data table shown in Figure 11-10d, draw the frequency-response curves for the two circuit gains in Figure 11-10e.

Fill-in Questions:

1. The gain of an op-amp inverting amplifier can be found by dividing ______ Rf ______ by ______ Rin ______.

2. With the inverting amplifier, the output voltage is ___180___ degrees out of phase with the input voltage.

3. V_{out} can be found by multiplying ______ Vin ______ with A_v.

4. Capacitors are used with an ac inverting amplifier to block ___dc___ voltage.

5. An inverting amplifier with a lower gain has a wider ______ frequency response ______.

EXPERIMENT 3. OP-AMP NONINVERTING AMPLIFIER

Objective:

To demonstrate the operation of an op-amp noninverting amplifier with dc and ac voltages, and to learn how to calculate the gain of the circuit.

Introduction:

The noninverting amplifier shown in Figure 11-11a consists primarily of resistors R_{in}, R_F, R_n, and R_L. Resistors R_1, R_2, and R_3 are used as a voltage divider to set the desired dc voltage at the noninverting input. The gain of the circuit is calculated by the formula $A_v = R_F/R_{in} + 1$ or $A_v = V_{out}/V_{in}$.

When the noninverting amplifier is used for ac signals as shown in Figure 11-11c, capacitors are used to block any dc voltage from the circuit that might cause distortion. Even though the input voltage changes, an amplifier's gain remains the same. A noninverting amplifier is used for high input impedance, where R_{in} cannot be made larger, because of affecting the gain of the circuit and creating more noise.

Materials Needed:

1 Dual ± 15-V power supply
1 Standard or digital voltmeter
1 Oscilloscope (dual trace preferred)
1 AC signal generator
1 741 op amp or equivalent
1 4.7-kΩ resistor at 0.5 W (R_{in})
1 6.8-kΩ resistor at 0.5 W (R_n)
4 10-kΩ resistors at 0.5 W (R_1, R_3, R_{in}, R_L, and R_n)
1 22-kΩ resistor at 0.5 W (R_{in}, R_F)
1 47-kΩ resistor at 0.5 W (R_F)
1 100-kΩ resistor at 0.5 W (R_F)
1 10-kΩ linear potentiometer (R_2)
2 1-μF capacitors at 25 WV dc (C_{in}, C_{out})
1 Breadboard for constructing circuit

Procedure:

1. Construct the circuit shown in Figure 11-11a, using the values of R_{in} and R_F of the first line from the data table of Figure 11-11b.

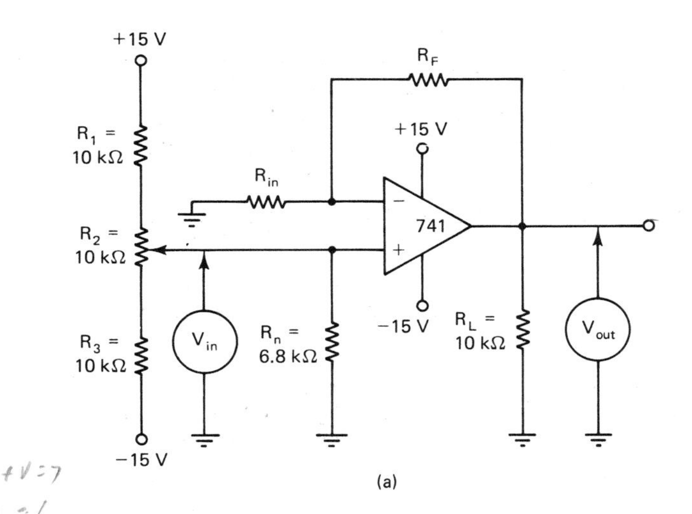

R_{in} (kΩ)	R_F (kΩ)	V_{in}	V_{out}	$A_V = R_F/R_{in} + 1$	$A_V = V_{out}/V_{in}$
10	47	+1	1.8	5.7	1.8
10	100	+1	13.3	11	13.3
10	22	+1	3.9	3.2	3.9
4.7	47	−1	-2.38	11	2.38
22	47	−1	-1.47	3.13	1.47
10	47	−1	-1.77	5.7	1.77

(b)

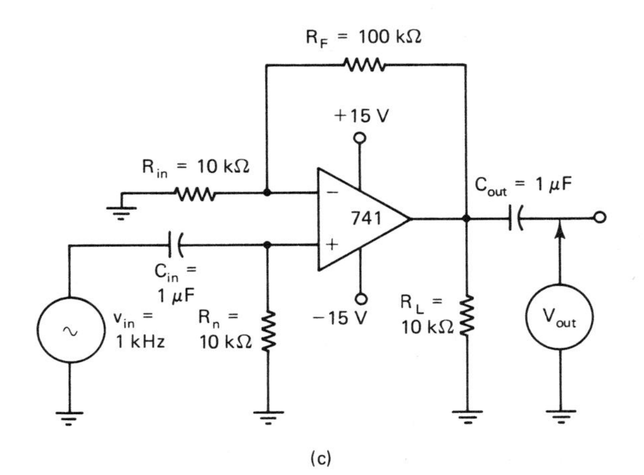

V_{in} (V p-p)	V_{out} (measured)	$V_{out} = A_V V_{in}$ (calculated)
0.1	1.38	(11)- 1.1 = 1.33
0.2	2.2	(11)→2.2 = 2.66
0.5	5.6	(11)→5.5 = 6.65
1.0	11	(11)→ 11 = 13.3
1.5	17	(11)→16.5 = 19.95

(d)

(use a 13.3 Av from above) use a (11) 1.1 2.2 5.5 11 16.5

Figure 11–11 Op-amp noninverting amplifier: (a) dc amplifier; (b) data table; (c) ac amplifier; (d) data table.

2. Using the voltmeter, set V_{in} according to the data table.
3. Using the voltmeter, measure and record V_{out} in the data table.
4. Calculate the gain of the circuit from the formula $A_v = R_F/R_{in} + 1$ and record in the data table.
5. Calculate the gain of the circuit from the formula $A_v = V_{out}/V_{in}$ and record in the data table.
6. Repeat steps 2 through 5 for all values of R_{in} and R_F given in the data table.
7. Construct the circuit shown in Figure 11-11c.
8. Calculate the gain of the circuit using the formula $A_v = R_F/R_{in} + 1$ and record here: 11.
9. Using the oscilloscope, set the signal generator for v_{in} (V p-p) given in the first line of the data table shown in Figure 11-11d.
10. Using the oscilloscope, measure and record v_{out} in the data table.
11. Calculate v_{out} using the formula $v_{out} = A_v v_{in}$ and record in the data table.
12. Repeat steps 10 and 11 for the various amplitudes of v_{in} given in the data table.
13. For the last measurement recorded in the data table of Figure 11-11d, explain why measured v_{out} was less than calculated v_{out} and why the signal was distorted.

Signal was too large.

Fill-in Questions:

1. The gain of an op-amp noninverting amplifier can be found by dividing ___Rf___ by ___RIN___ and adding ___1___.
2. With the noninverting amplifier, the output voltage is ___IN___ phase with the input voltage.
3. V_{out} can be found by multiplying v_{in} with ___Av___.
4. Capacitors are used with an ac noninverting amplifier to ___block___ dc voltage.
5. The output signal of an amplifier may distort if the input signal is too ___large___.

EXPERIMENT 4. OP-AMP VOLTAGE FOLLOWERS

Objective:

To demonstrate the operation of op-amp voltage followers, and to show differences between the inverting and noninverting types.

Introduction:

The noninverting voltage follower shown in Figure 11-12a has a gain of 1 because of the zero-resistance feedback loop. Its output voltage is in phase with the input voltage. The input impedance to this circuit can be made very high.

The inverting voltage follower shown in Figure 11-12b is similar to a standard inverting amplifier, except that its gain is 1 because R_{in} and R_F are the same value ($A_v = -R_F/R_{in}$). Its output voltage is 180° out of phase with the input voltage. The input impedance to this circuit is usually lower, being limited by the value of R_{in}.

Voltage followers are used to match circuit impedances and act as buffer amplifiers, isolating one circuit from another.

Material Needed:

1 Dual ± 15-V power supply
1 Standard or digital voltmeter
1 Oscilloscope (dual trace preferred)
1 AC signal generator
1 741 op amp or equivalent
1 4.7-kΩ resistor at 0.5 W (R_n)
3 10-kΩ resistors at 0.5 W (R_{in}, R_F, R_L)
1 100-kΩ resistor at 0.5 W (R_n)
2 1-μF capacitors at 25 WV dc (C_{in}, C_{out})
1 Breadboard for constructing circuit

Procedure:

1. Construct the circuit shown in Figure 11-12a.
2. Using the oscilloscope, set the signal generator for 1 kHz at 2 V p-p for v_{in}.
3. Using the oscilloscope, measure v_{out}.
4. Draw the output voltage waveform in the space provided, indicating the voltage peak to peak. (Assume that the input signal initially goes positive.)
5. Construct the circuit shown in Figure 11-12b.
6. Using the oscilloscope, set the signal generator for 1 kHz at 2 V p-p for v_{in}.
7. Using the oscilloscope, measure v_{out}.
8. Draw the output voltage waveform in the space provided, indicating the voltage peak to peak. (Assume that the input signal initially goes positive.)

Fill-in Questions:

1. Voltage followers are used for ________ ________ matching and ________.
2. An advantage of the noninverting voltage follower over the inverting type is that it has a higher ________ impedance.
3. The simplest voltage follower of the two types is the ________.

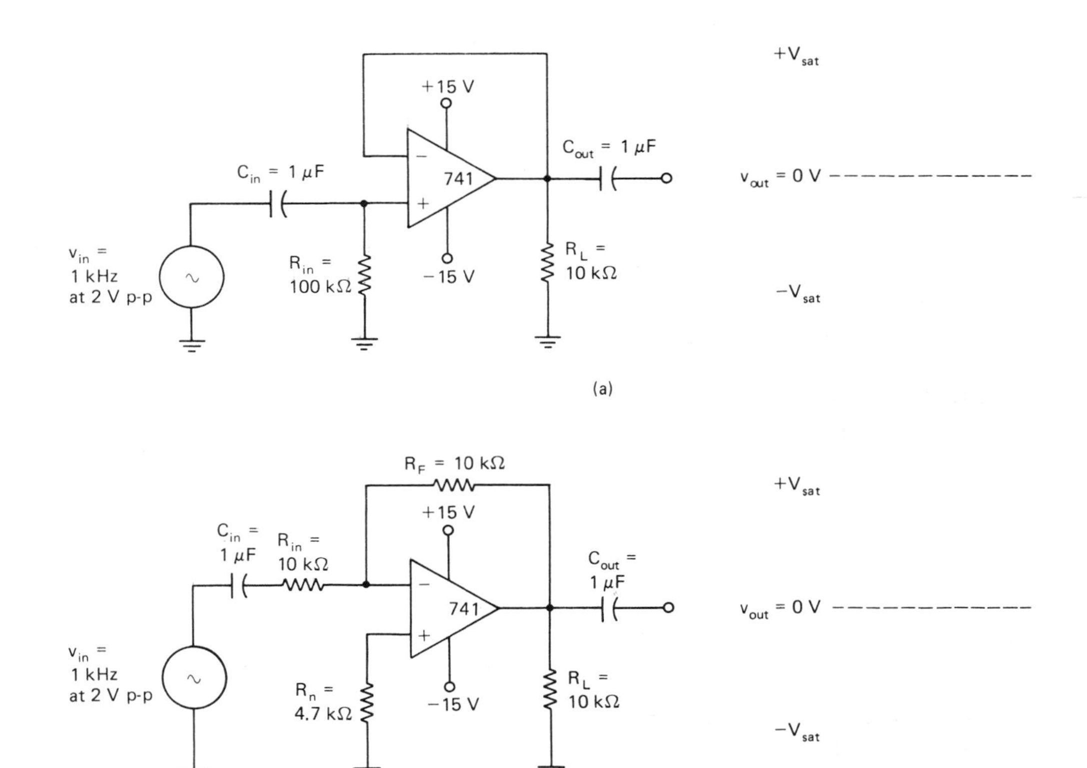

Figure 11–12 Op-amp voltage followers: (a) noninverting; (b) inverting.

EXPERIMENT 5. OP-AMP SUMMING AMPLIFIERS

Objective:

To demonstrate how an op amp can be used to sum algebraically various input voltages.

Introduction:

If more than one input is used on an inverting amplifier, it becomes a summing circuit or adder, as shown in Figure 11-13a and c. The output voltage is the algebraic sum of the inputs, but inverted, and can be found by the formula

$$V_{out} = -\left(\frac{R_F}{R_1}V_1 + \frac{R_F}{R_2}V_2 + \cdots + \frac{R_F}{R_n}V_n\right)$$

where R_n and V_n are the number of input resistors and input voltages. If all the resistors are of the same value, the formula simplifies to $V_{out} = -(V_1 + V_2 + \cdots + V_n)$. All input resistors may be of the same value, with R_F a larger value, resulting in a summing amplifier with gain. It may be required that some inputs influence the output voltages more than the others; therefore, different gains require various input resistor values. This type of circuit is called a *scaling adder.*

The input currents and current through R_F add up to zero at the inverting input, referred to as the *current summing point.* The summing amplifier can also be used as an audio signal mixer.

Resistors R_1 through R_6 are used to form two voltage dividers for setting the input voltages at V_1 and V_2.

Materials Needed:

1 Dual ± 15-V power supply
1 Standard or digital voltmeter
1 741 op amp or equivalent
1 4.7-kΩ resistor at 0.5 W (R_2)
7 10-kΩ resistors at 0.5 W (R_1, R_2, R_5, R_6, R_7, R_8, R_F)
1 22-kΩ resistor at 0.5 W (R_F)
2 10-kΩ linear potentiometers (R_3, R_4)
1 Breadboard for constructing circuit

Procedure:

1. Construct the circuit shown in Figure 11-13a using all 10-kΩ resistors.

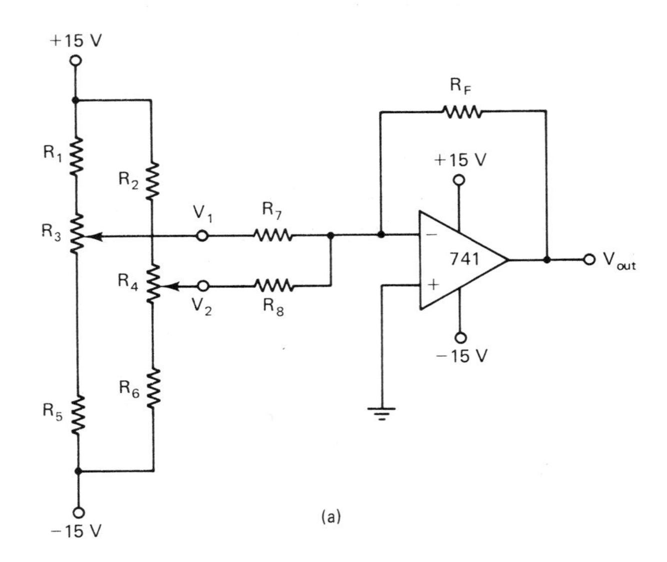

(a)

Input voltage		V_{out} algebraic sum (inverted), R_F = 10 kΩ		V_{out} algebraic sum (inverted), R_F = 22 kΩ	
V_1	V_2	Calculated	Measured	Calculated	Measured
+1	+2				
+1	−2				
+2	+1				
+2	−1				
−2	−2				

(b)

R1 = 10 kΩ, RF = 10 kΩ, R2 = 4.7 kΩ, +15 V, 741, −15 V, V1, V2

(c)

Input voltage		V_{out} algebraic sum (inverted)	
V_1	V_2	Calculated	Measured
+1	+2		
+1	−2		
+2	+1		
+2	−1		
−2	−2		

(d)

Figure 11–13 Op-amp summing amplifiers: (a) basic adder; (b) data table; (c) scaling adder; (d) data table.

2. Using the voltmeter, adjust R_3 and R_4 for the voltage values of V_1 and V_2, respectively, given in the first line of the data table of Figure 11-13b.
3. Calculate and record in the data table V_{out} using the formula

$$V_{out} = -(V_1 + V_2)$$

4. Using the voltmeter, measure V_{out} and record it in the data table.
5. Repeat steps 3 and 4 for the remaining values of V_1 and V_2 given in the data table.
6. Change R_F to 22 kΩ.
7. Calculate and record in the data table V_{out} using the formula

$$V_{out} = -\left(\frac{R_F}{R_1}V_1 + \frac{R_F}{R_2}V_2\right)$$

8. Using the voltmeter, set V_1 and V_2 to the values shown on the first line of the data table.
9. Using the voltmeter, measure V_{out} and record it in the data table.
10. Repeat steps 7 through 9 for the remaining values of V_1 and V_2 given in the data table.
11. Construct the circuit shown in Figure 11-13c using the voltage-divider circuits of the first part of this experiment to set V_1 and V_2.
12. Using the voltmeter, set V_1 and V_2 to the values given in the first line of the data table of Figure 11-13d.
13. Calculate and record in the data table V_{out} using the formula

$$V_{out} = -\left(\frac{R_F}{R_1}V_1 + \frac{R_F}{R_2}V_2\right)$$

14. Using the voltmeter, measure V_{out} and record it in the data table.
15. Repeat steps 13 and 14 for the remaining values of V_1 and V_2 given in the data table.

Fill-in Questions:

1. An inverting op-amp adder ___________ algebraically sums the voltages at the inputs, but the output voltage is inverted.
2. If a summing amplifier has gain, R_F will be higher than the other resistors in the circuit.
3. If all the resistors are the same value in an op-amp adder, the output-voltage formula is $V_{out} = -(V_1 + V_2 + V_3 \ldots + V_N)$.
4. A scaling adder has different-value input resistors.

EXPERIMENT 6. OP-AMP DIFFERENCE AMPLIFIER

Objective:

To show how an op amp can be used to find the algebraic differences between two input voltages.

Introduction:

Both inputs are used (or active) for a difference amplifier or subtractor, as shown in Figure 11-14a. The output voltage is found by the formula

$$V_{out} = -\frac{R_F}{R_A} V_1 + \left(\frac{R_n}{R_B + R_n}\right)\left(\frac{R_A + R_F}{R_A}\right) V_2$$

If all resistors are equal, the formula simplifies to $V_{out} = V_2 - V_1$; however, the polarity of the output voltage depends on the relationship of the inverting and noninverting inputs polarities, similar to a comparator circuit.

A difference amplifier may have gain or use a scaling input arrangement where one input has more influence on the output. Resistors R_1 through R_6 are used for voltage dividers to set the input voltages V_1 and V_2.

Materials Needed:

1 Dual ± 15-V power supply
1 Standard or digital voltmeter
1 741 op amp or equivalent
9 10-kΩ resistors at 0.5 W (R_1, R_2, R_5, R_6 R_A, R_B, R_n, R_F, R_L)

Figure 11–14 Op-amp voltage-difference amplifier: (a) schematic diagram; (b) data table.

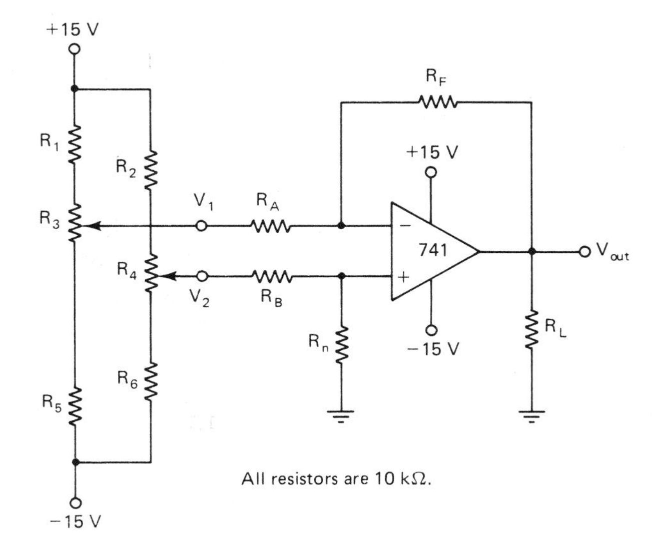

(a)

Input voltage		Output voltage algebraic difference (inverted)	
V_1	V_2	Calculated	Measured
+2	+4		
+4	+2		
+4	−2		
−2	+4		
−4	−2		

(b)

2 10-kΩ linear potentiometers (R_3, R_4)
1 Breadboard for constructing circuit

Procedure:

1. Construct the circuit shown in Figure 11-14a.
2. Using the voltmeter, set V_1 and V_2 to the values given in the first line of the data table shown in Figure 11-14b.
3. Calculate and record in the data table V_{out} using the formula

$$V_{out} = -(V_2 - V_1)$$

(Remember to indicate the proper polarity.)

4. Using the voltmeter, measure V_{out} and record it in the data table.
5. Repeat steps 3 and 4 for the remaining values of V_1 and V_2 given in the data table.

Fill-in Questions:

1. The output of an op-amp subtractor is the __________ difference at the inputs.

2. If the voltage at the inverting input is more negative than the voltage at the noninverting input for an op-amp subtractor, the output will be __________ .

EXPERIMENT 7. OP-AMP VOLTAGE-LEVEL DETECTORS

Objective:

To demonstrate how an op amp can sense a specific voltage level, and how to calculate the reference voltage (V_{ref}).

Introduction:

An op-amp comparator can be used to detect a positive voltage level and give a negative output-voltage indication, as shown in Figure 11-15a. The reference voltage (V_{ref}) placed on the noninverting input is found by the formula

$$V_{ref} = \frac{R_3}{R_2 + R_3}(+V)$$

When the voltage at the inverting amplifier is below V_{ref}, the output is at $+V_{sat}$. The instant the voltage at the inverting input increases above V_{ref}, the output swings to $-V_{sat}$.

When V_{ref} is at the inverting input as shown in Figure 11-15b, the output will swing from $-V_{sat}$ to $+V_{sat}$ the instant the voltage at the noninverting input is greater than V_{ref}.

If point A is moved to $-V$ power supply, the circuits will detect a negative voltage.

Materials Needed:

1 Dual ± 15-V power supply
1 Standard or digital voltmeter
1 741 op amp or equivalent
2 10-kΩ resistors at 0.5 W (R_3, R_L)
1 22-kΩ resistor at 0.5 W (R_2)
1 10-kΩ linear potentiometer
1 Breadboard for constructing circuit

Procedure:

1. Construct the circuit shown in Figure 11-15a.
2. Place the wiper of R_1 at ground (0 V).
3. Calculate V_{ref} and record here, using the formula

$$V_{ref} = \frac{R_3}{R_2 + R_3}(+V) = \_\_\_\_\_\_$$

4. Using the voltmeter, measure V_{ref} and record it in the place provided in Figure 11-15a.
5. Using the voltmeter, measure V_{out} and record it in the place provided.
6. Adjust R_1 until V_{out} changes and record the new reading in the place provided.
7. Remove the wire at point A from the $+V$ power supply and connect it to the $-V$ power supply.
8. Briefly repeat steps 2 through 6 to understand how a negative voltage is detected.
9. Construct the circuit shown in Figure 11-15b.
10. Place the wiper of R_1 at ground (0 V).

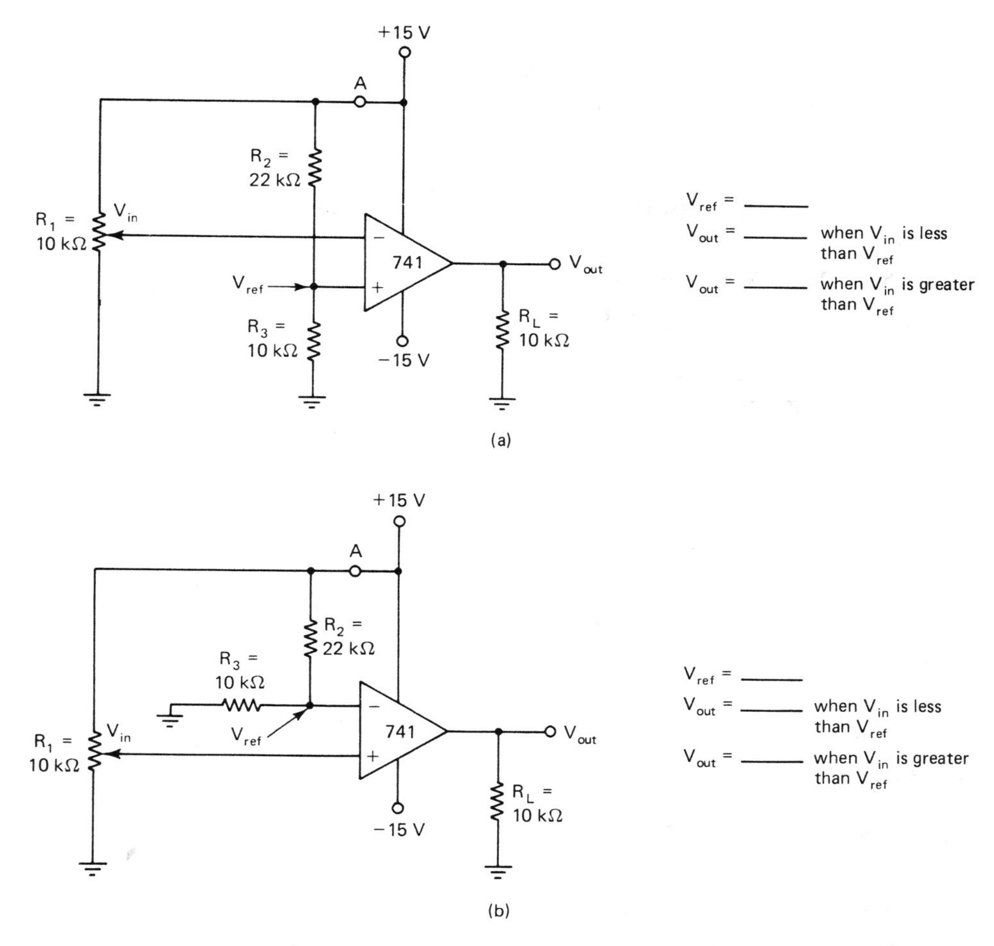

Figure 11–15 Op-amp voltage-level detectors: (a) inverting input sensor; (b) noninverting input sensor.

11. Using the voltmeter, measure V_{ref} and record it in the place provided in Figure 11-15b.
12. Using the voltmeter, measure V_{out} and record it in the place provided.
13. Adjust R_1 and V_{out} changes and record the new reading in the place provided.
14. Remove the wire at point A from the $+V$ power supply and connect it to the $-V$ power supply.
15. Briefly repeat steps 10 through 13 to understand how a negative voltage is detected.

Fill-in Questions:

1. For an op-amp voltage-level detector, when V_{in} becomes greater than V_{ref}, the ____________ voltage changes.
2. These circuits can detect ____________ and ____________ voltages.
3. Either input can be used for ____________ and ____________.

EXPERIMENT 8. OP-AMP SQUARE-WAVE GENERATOR

Objective:

To show how an op amp can be used as a square-wave generator, and how to calculate its output frequency.

Introduction:

An op amp can be constructed to produce a square-wave generator as shown in Figure 11-16. Resistors R_2 and R_3 form a voltage di-

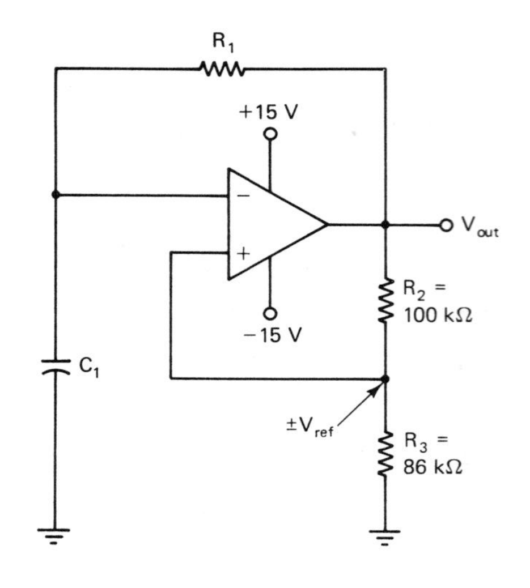

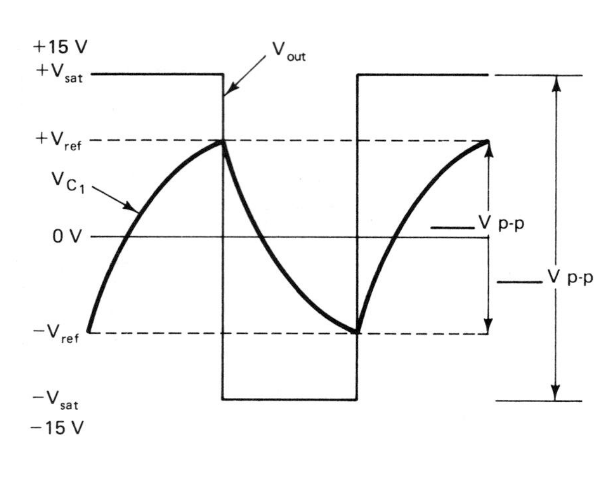

(a)

R_1 (kΩ)	C_1 (μF)	f_{out} (Hz)	
		Calculated	Measured
10	0.05		
22	0.05		
4.7	0.05		
10	0.02		
10	0.1		

(b)

Figure 11–16 Op-amp square-wave generator: (a) schematic diagram; (b) data table.

vider from the output of the op amp to ground and determine the $\pm V_{ref}$. Assume, initially, that V_{out} is at $+V_{sat}$. Capacitor C_1 begins to charge up through R_1 to $+V_{sat}$. The instant the voltage on the capacitor is greater than $+V_{ref}$ at the noninverting input, the output switches to $-V_{sat}$. The capacitor now charges toward $-V_{sat}$, and the instant V_{C1} is greater than $-V_{ref}$, the output switches back to $+V_{sat}$ and the process begins again. The square-wave output at V_{out} is $\pm V_{sat}$ in amplitude. The amplitude of V_{C1} is $\pm V_{ref}$ and can be found by the formula

$$+V_{ref} = \frac{R_3}{R_2 + R_3}(+V_{sat}) \quad \text{and}$$

$$-V_{ref} = \frac{R_3}{R_2 + R_3}(-V_{sat})$$

If R_3 is 86% of R_2, the approximate output frequency can be found by the formula

$$f_{out} = \frac{1}{2R_1C_1}$$

Materials Needed:

1 Dual ±15-V power supply
1 Standard or digital voltmeter
1 Oscilloscope (dual trace preferred)
1 4.7-Ω resistor at 0.5 W (R_1)
1 10-kΩ resistor at 0.5 W (R_1)
1 22-kΩ resistor at 0.5 W (R_1)
1 86-kΩ resistor at 0.5 W (R_3)
1 100-kΩ resistor at 0.5 W (R_2)
1 0.02-μF capacitor at 25 WV dc (C_1)
1 0.05-μF capacitor at 25 WV dc (C_1)
1 0.1-μF capacitor at 25 WV dc (C_1)
1 Breadboard for constructing circuit

Procedure:

1. Construct the circuit shown in Figure 11-16a using the values given in the first line of the data table of Figure 11-16b for R_1 and C_1.
2. Calculate $\pm V_{ref}$ using the formulas

$$+V_{ref} = \frac{R_3}{R_2 + R_3}(+V_{sat}) = \_\_\_\_\_\_$$

and

$$-V_{ref} = \frac{R_3}{R_2 + R_3}(-V_{sat}) = \_\_\_\_\_\_$$

3. Using the oscilloscope, measure $+V_{sat}$, $-V_{sat}$, $+V_{ref}$, and $-V_{ref}$ and record on the figure.
4. Calculate the frequency of the generator and record in the data table using the formula

$$f_{out} = \frac{1}{2R_1C_1}$$

5. Measure the f_{out} with the oscilloscope and record it in the data table.
6. Repeat steps 4 and 5 for the remaining values of R_1 and C_1 given in the data table.

Fill-in Questions:

1. The amplitude of the square-wave output of the op-amp generator is ____________.
2. The voltage waveform across the capacitor is a ____________.
3. The formula for calculating f_{out} = ____________.
4. When R_1 or C_1 increases, the f_{out} ____________.
5. When R_1 or C_1 decreases, the f_{out} ____________.

EXPERIMENT 9. OP-AMP LOW-PASS FILTER

Objective:

To show how an op-amp low-pass filter passes frequencies below the cutoff frequency (f_c) and attenuates the frequencies above this point.

Introduction:

A low-pass filter has a constant output voltage from dc up to a specific cutoff frequency (f_c), which may also be called the *corner* or *breakpoint frequency*. The f_c occurs at 0.707 V_{max} of the output voltage, sometimes referred to as the *half-power* or *−3-dB point*. Frequencies above the f_c are attenuated. Frequencies below the f_c point are called the *passband*, and frequencies above this point are called the *stopband*, as shown in Figure 11-17.

A low-pass filter is shown in Figure 11–18a. The resistors are not affected by frequency; however, the reactance of the capacitors is based on frequency. For low frequencies, X_C is high and C_2 represents a high impedance to ground. Capacitor C_1 also offers a high impedance to the feedback current; therefore, V_{out} remains high. As the frequency increases, X_C decreases and C_2 begins to shunt more of the input signal to ground, while C_1 allows more feedback current and the gain of the circuit decreases.

Materials Needed:

1 Dual ±15-V power supply
1 Standard or digital voltmeter
1 Dual-trace oscilloscope
1 Sine-wave generator or function generator
2 10-kΩ resistors at 0.5 W
1 22-kΩ resistor at 0.5 W
2 0.01-μF capacitors at 25 WV dc
1 741 op-amp IC or equivalent
1 Breadboard for constructing circuit

Procedure:

1. Construct the circuit shown in Figure 11-18a.
2. Apply power to the circuit.
3. Adjust the sine-wave generator for the first frequency given in the data log shown in Figure 11-18b.
4. Set the sine-wave generator output amplitude for a 1-V p-p signal for each frequency given in the data log. Make

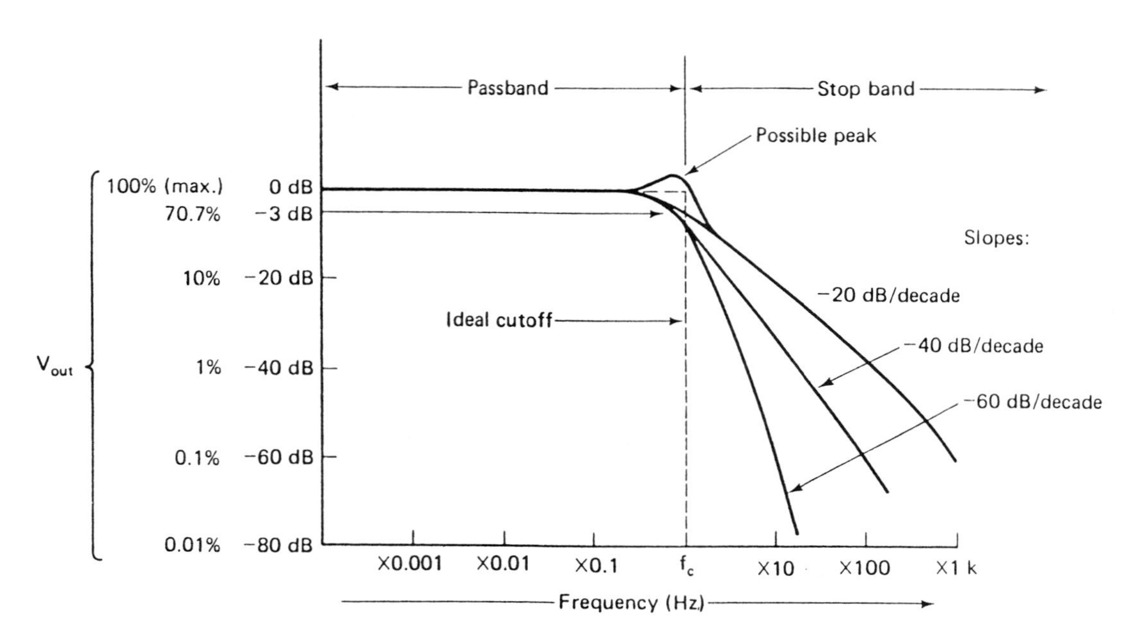

Figure 11–17 Low-pass filter frequency response curve.

sure this voltage level is maintained for each frequency setting.

5. Measure the V p-p of v_{out} and record its value in the data log.
6. Repeat steps 3, 4, and 5 for all the frequencies in the data log.
7. Turn off the power supply.
8. Calculate f_c from the formula

$$f_c = \frac{1}{2\pi\sqrt{R_1 R_2 C_1 C_2}}$$

9. Compare your readings with the graph shown in Figure 11-17.

Fill-in Questions:

1. The f_c for this low-pass filter is about _____________ .
2. Frequencies below f_c are _____________ by this circuit.
3. Frequencies above f_c are _____________ by this circuit.

Figure 11–18 Low-pass filter.

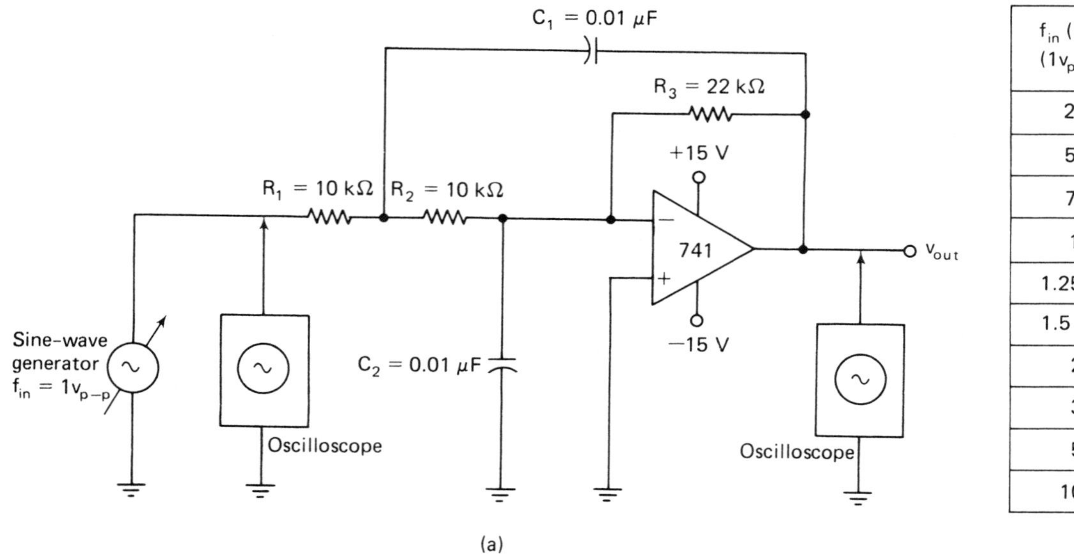

(a)

f_{in} (Hz) ($1v_{p-p}$)	V_{out} (v_{p-p})
250	
500	
750	
1 k	
1.25 k	
1.5 k	
2 k	
3 k	
5 k	
10 k	

(b)

EXPERIMENT 10. OP-AMP HIGH-PASS FILTER

Objective:

To demonstrate how an op-amp high-pass filter passes frequencies above the cutoff frequency and attenuates frequencies below this point.

Introduction:

A high-pass filter performs the opposite function from that of a low-pass filter. The high-pass filter attenuates all frequencies below f_c and passes all the frequencies above f_c. Figure 11-19 shows the frequency response curve for a high-pass filter.

An op-amp high-pass filter circuit is shown in Figure 11-20a. At lower frequencies, the X_C of C_1 and C_2 is large and drops a significant amount of input signal voltage. This means a smaller input signal reaches the op amp, and the resulting output is very small. As the input signal increases in frequency, the X_C decreases, which allows more signal to reach the op amp, therefore producing a larger output.

Materials Needed:

1 Dual ±15-V power supply
1 Standard or digital voltmeter
1 Dual-trace oscilloscope
1 Sine-wave generator or function generator
1 10-kΩ resistor at 0.5 W
2 22-kΩ resistors at 0.5 W
2 0.01-μF capacitors at 25 WV dc
1 741 op-amp IC or equivalent
1 Breadboard for constructing circuit

Procedures:

1. Construct the circuit shown in Figure 11-20a.
2. Apply power to the circuit.
3. Adjust the sine-wave generator for the first frequency given in the data log shown in Figure 11-20b.
4. Set the sine-wave generator output amplitude for a 1-V p-p signal for each frequency given in the data log. Make sure this voltage level is maintained for each frequency setting.
5. Measure the V p-p of v_{out} and record its value in the data log.
6. Repeat steps 3, 4, and 5 for all the frequencies in the data log.
7. Turn off the power supply.
8. Calculate f_c from the formula

$$f_c = \frac{1}{2\pi\sqrt{R_1 R_2 C_1 C_2}}$$

9. Compare your readings with the graph shown in Figure 11-19.

Figure 11-19 High-pass filter frequency response curve.

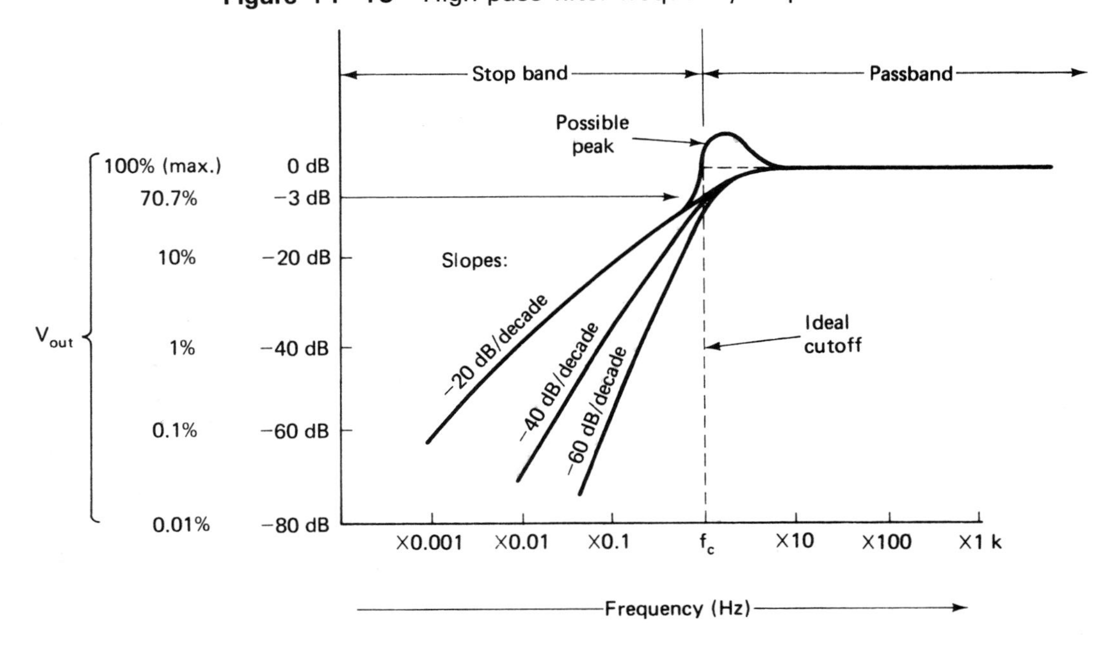

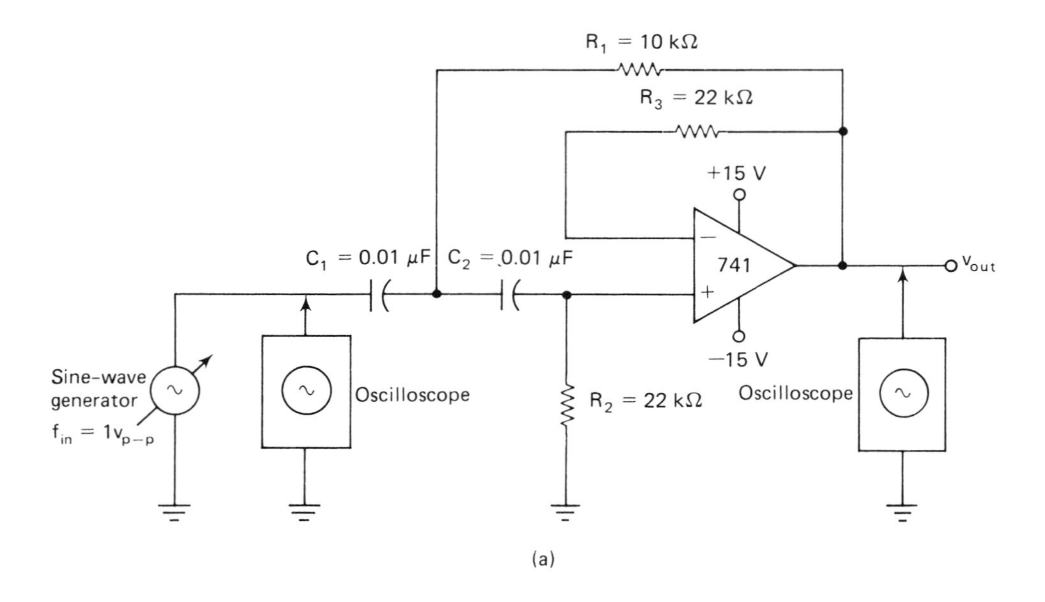

f_{in} (Hz) (1v_{p-p})	V_{out} (v_{p-p})
250	
500	
750	
1 k	
1.25 k	
1.5 k	
2 k	
3 k	
5 k	
10 k	

(b)

Figure 11–20 High-pass filter.

Fill-in Questions:

1. The f_c for this high-pass filter is about ______________ .

2. Frequencies below f_c are ______________ by this circuit.

3. Frequencies above f_c are ______________ by this circuit.

EXPERIMENT 11. OP-AMP BANDPASS FILTER

Objective:

To show how an op-amp bandpass filter passes a certain group of frequencies and rejects those above and below these limits.

Introduction:

The bandpass filter passes the group of frequencies that exist between the half-power points on each side of the resonant frequency (f_r). Other frequencies beyond these limits are attenuated. The greatest output voltage occurs at f_r, and it is referred to as *peaking* at this point. The width of the band of frequencies, referred to as *bandwidth*, depends on the Q of the circuit. The Q of a circuit is inversely proportional to the resistance of the circuit. A circuit with very little resistance has a high Q and produces a very narrow band of frequencies. More resistance in a circuit produces less amplitude output voltage and increases the bandwidth. Figure 11-21 shows a frequency response curve for a bandpass filter. The bandwidth occurs between f_L and f_H on the graph.

Figure 11-22a shows a bandpass filter circuit. The circuit combines the features of a low-pass and high-pass filter with one capacitor in the feedback path and one capacitor in series with the input, respectively. The action of these capacitors in this arrangement produces a bandpass filter.

Materials Needed:

1 Dual ±15-V power supply
1 Standard or digital voltmeter
1 Dual-trace oscilloscope
1 Sine-wave generator or function generator
2 10-kΩ resistors at 0.5 W
1 100-kΩ resistor at 0.5 W
2 0.01-μF capacitors at 25 WV dc
1 741 op-amp IC or equivalent
1 Breadboard for constructing circuit

Procedure:

1. Construct the circuit shown in Figure 11-22a.
2. Apply power to the circuit.
3. Adjust the sine-wave generator for the first frequency given in the data log shown in Figure 11-22b.
4. Set the sine-wave generator output amplitude for a 1-V p-p signal for each frequency given in the data log. Make sure this voltage level is maintained for each frequency setting.
5. Measure the V p-p of v_{out} and record its value in the data log.
6. Repeat steps 3, 4, and 5 for all the frequencies in the data log.
7. Adjust the frequency generator until the output peaks (has the highest output). Record this frequency. This is the actual f_r of the circuit.
8. Turn off the power supply.
9. Calculate f_r from the formula

$$f_r = \frac{1}{2\pi\sqrt{R_p R_3 C_1 C_2}}$$

where

$$R_p = \frac{R_1 R_2}{R_1 + R_2}$$

10. Compare your readings with the graph shown in Figure 11-21.

Fill-in Questions:

1. The f_r of this circuit is about **712** Hz.

Figure 11-21 Bandpass filter frequency response curve.

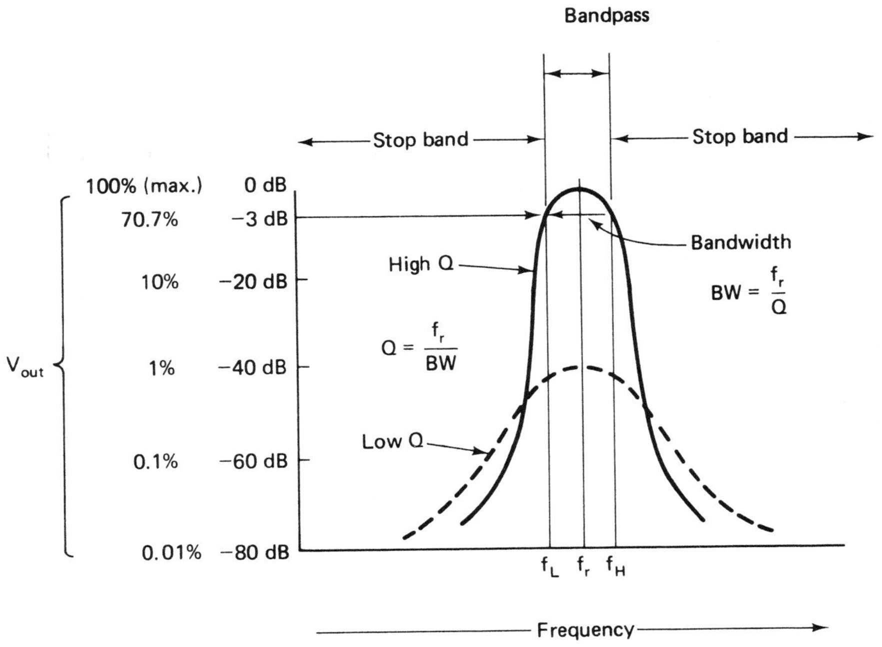

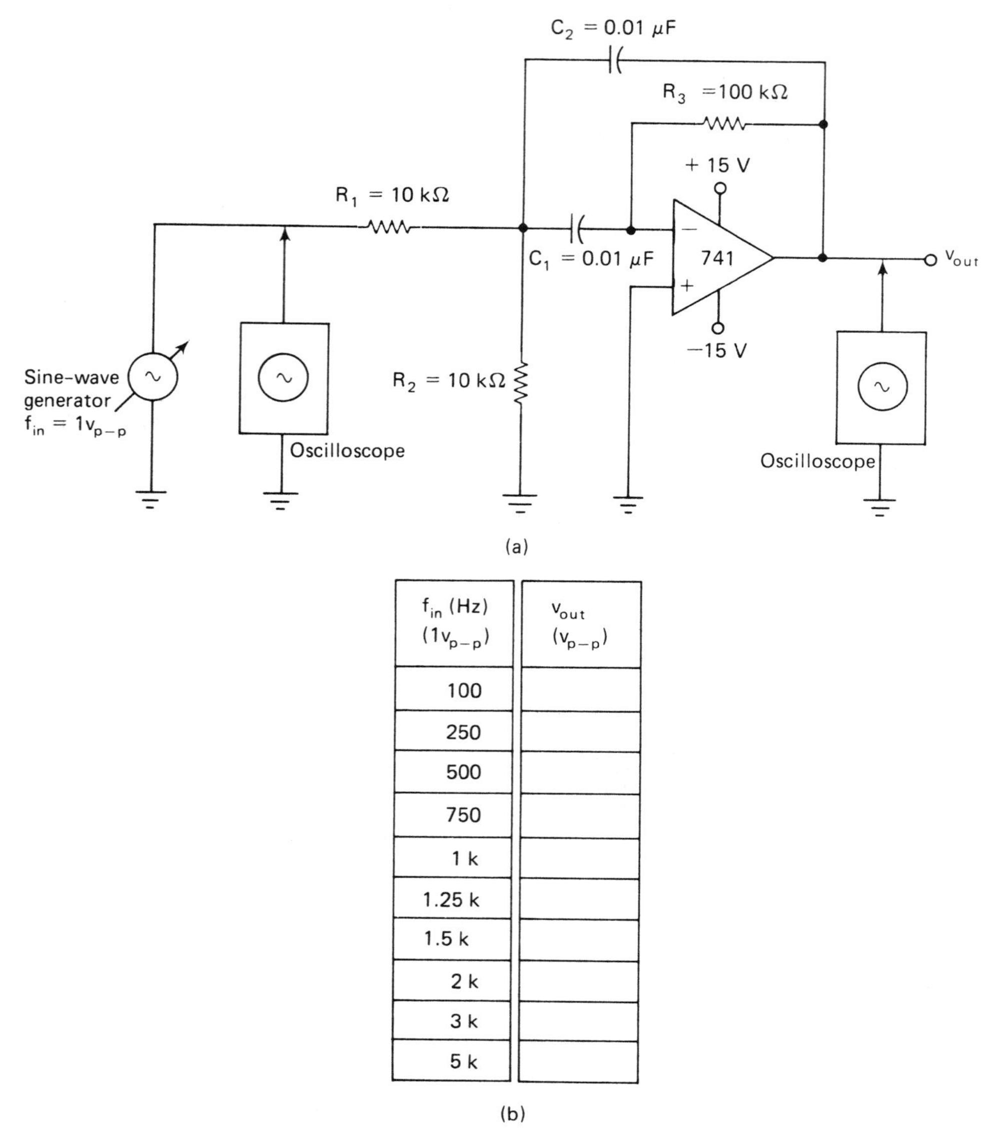

f_{in} (Hz) ($1v_{p-p}$)	V_{out} (v_{p-p})
100	
250	
500	
750	
1 k	
1.25 k	
1.5 k	
2 k	
3 k	
5 k	

(b)

Figure 11–22 Bandpass filter.

2. The approximate bandwidth of the circuit is 318 Hz.

3. Frequencies below the bandwidth are blocked.

4. Frequencies above the bandwidth are blocked.

EXPERIMENT 12. OP-AMP BANDREJECT (NOTCH) FILTER

Objective:

To demonstrate how an op-amp bandreject filter passes all frequencies above and below a specific bandwidth while rejecting the frequencies within the bandwidth.

Introduction:

A bandreject filter's function is opposite to that of a bandpass filter. This type of filter passes all frequencies except a specific group within the bandwidth. Figure 11-23

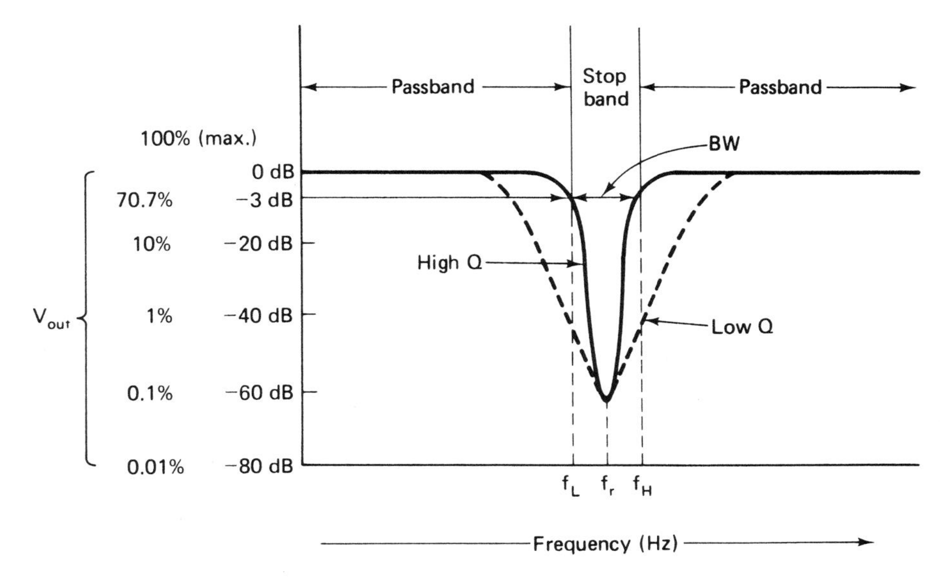

Figure 11–23 Notch filter frequency response curve.

shows a bandreject filter response curve. The rejected bandwidth may appear as a *notch* in the frequencies that are passed, and the circuit is often referred to as a notch filter. The Q of the circuit determines how narrow the notch will be.

Figure 11-24a shows a notch filter circuit. It is very similar to a bandpass filter, except a portion of the input signal is also applied to the noninverting input. The action of the capacitors and both inputs being active produce the notch filter characteristics.

Materials Needed:

1 Dual ±15-V power supply
1 Standard or digital voltmeter
1 Dual-trace oscilloscope
1 Sine-wave generator or function generator
1 10-kΩ resistor at 0.5 W
1 47-kΩ resistor at 0.5 W
1 1-MΩ resistor at 0.5 W
2 0.01-μF capacitors at 25 WV dc
1 741 op-amp IC or equivalent
1 Breadboard for constructing circuit

Procedure:

1. Construct the circuit shown in Figure 11-24a.
2. Apply power to the circuit.
3. Adjust the sine-wave generator for the first frequency given in the data log shown in Figure 11-24b.
4. Set the sine-wave generator output amplitude for a 1-V p-p signal for each frequency given in the data log. Make sure this voltage level is maintained for each frequency setting.
5. Measure the V p-p of v_{out} and record its value in the data log.
6. Repeat steps 3, 4, and 5 for all the frequencies in the data log.
7. Adjust the frequency generator until the output is at its lowest amplitude. This is the actual f_r of the circuit.
8. Turn off the power supply.
9. Calculate f_r from the formula

$$f_r = \frac{1}{2\pi\sqrt{R_1 R_4 C_1 C_2}}$$

10. Compare your readings with the graph shown in Figure 11-23.

Fill-in Questions

1. The f_r of this circuit is about ______ Hz.

2. The approximate bandwidth of the circuit is ______ Hz.

3. Frequencies below the notch are ______ ______.

4. Frequencies above the notch are ______ ______.

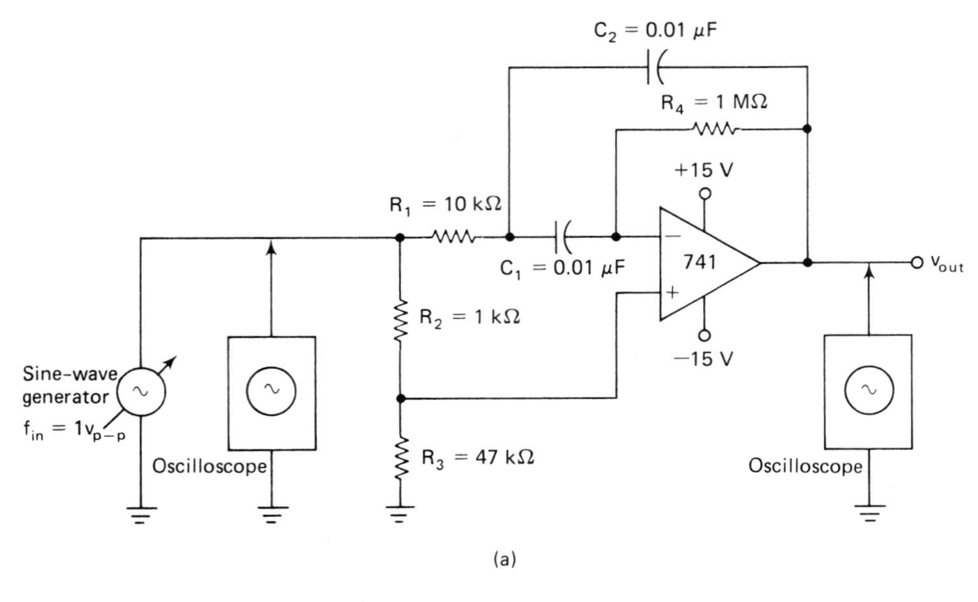

(a)

f_{in} (Hz) ($1v_{p-p}$)	v_{out} (v_{p-p})
50	
75	
100	
125	
150	
175	
200	
225	
250	

(b)

Figure 11–24 Notch filter.

SECTION 11-5 OP-AMP BASIC TROUBLESHOOTING APPLICATION: OP-AMP AC INVERTING AMPLIFIER

Construct the circuit shown below. Place the indicated signal at the input. Measure V_{out} (dc) with a voltmeter and v_{out} (signal) with an oscilloscope for each of the problem conditions. All voltages are referenced to ground.

Figure 11–E.2

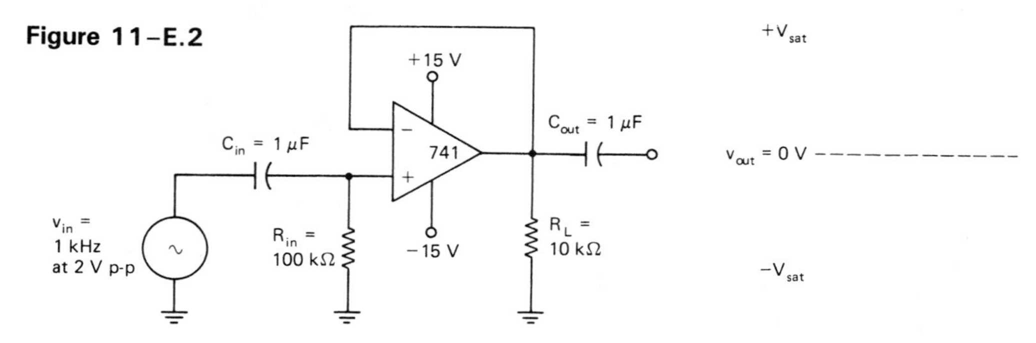

Condition	v_{out}	V_{out}	Comments
Normal			All voltages are proper
+15-V supply open			$-V_{sat}$ at output of op amp
−15-V supply open			$+V_{sat}$ at output of op amp
R_{in} open			No signal applied to op amp
R_F open			C_{in} charges up and op amp saturates
R_n open			$-V_{sat}$ at output of op amp
R_L open			No apparent trouble
C_{in} open			No signal applied to op amp
C_{in} shorted			V_{out} changes, maybe signal distorts
C_{out} open			No v_{out}
C_{out} shorted			Perhaps no problem, but maybe signal loading

Figure 11–E.2 (cont'd.)

SECTION 11-6 MORE EXERCISES

Perform the following exercises before beginning Section 11-7.

1. Draw a simple op-amp circuit of the following circuits:

a. comparator **b.** inverting amplifier **c.** noninverting amplifier **d.** summing amplifier

e. difference amplifier

f. voltage follower

g. oscillator

h. voltage-level detector

2. Draw a basic op-amp inverting amplifier and calculate the gain (A_V) and V_{out} for the various values given. Use the formulas $A_V = -R_F/R_{in}$ and $V_{out} = A_V V_{in}$. (Indicate polarity.) The power supply = ±15 V.

a. R_{in} = 10 kΩ
R_F = 47 kΩ
V_{in} = +1.75 V

A_V = ______

V_{out} = ______

b. R_{in} = 22 kΩ
R_F = 100 kΩ
V_{in} = −0.5 V

A_V = ______

V_{out} = ______

c. R_{in} = 100 kΩ
R_F = 100 kΩ
V_{in} = +2.5 V

A_V = ______

V_{out} = ______

3. Draw a basic op-amp noninverting amplifier and calculate the gain (A_V) and V_{out} for the various values given. Use the formulas $A_V = R_F/R_{in} + 1$ and $V_{out} = A_V V_{in}$. (Indicate polarity.) The power supply = ±15 V.

a. R_{in} = 4.7 kΩ
R_f = 47 kΩ
V_{in} = 1 V

A_V = ______

V_{out} = ______

b. R_{in} = 22 kΩ
R_F = 68 kΩ
V_{in} = +1.5 V

A_V = ______

V_{out} = ______

c. R_{in} = 100 kΩ
R_F = 100 kΩ
V_{in} = −1.5 V

A_V = ______

V_{out} = ______

4. Draw the circuit in Figure 11-13a and calculate V_{out} for the various values of input voltages (V_1 and V_2) given. Use the formulas $V_{out} = -R_F/R_7(V_1) + R_F/R_8(V_2)$ and/or $V_{out} = V_1 + V_2$. (Indicate polarity.) The power supply = ±15 V.

a. R_7 = 10 kΩ
R_8 = 10 kΩ
R_F = 10 kΩ
V_1 = +2 V
V_2 = +3 V

V_{out} = ______

b. R_7 = 10 kΩ
R_8 = 10 kΩ
R_F = 22 kΩ
V_1 = +4 V
V_2 = −1 V

V_{out} = ______

c. R_7 = 10 kΩ
R_8 = 4.7 kΩ
R_F = 100 kΩ
V_1 = +0.5 V
V_2 = +0.02 V

V_{out} = ______

5. Calculate V_{out} for the following circuit with various input voltages using the formula $V_{out} = V_2 - V_1$. (Indicate polarity.) The power supply voltage = ±15 V.

a. V_1 = +4 V
V_2 = +3 V

b. V_1 = −2 V
V_2 = +5 V

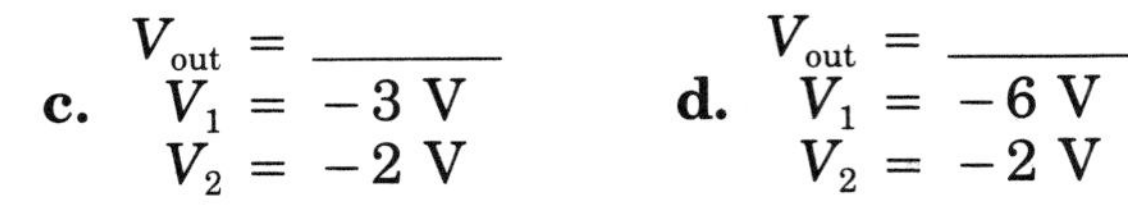

V_{out} = ______

c. $V_1 = -3$ V
$V_2 = -2$ V

V_{out} = ______

d. $V_1 = -6$ V
$V_2 = -2$ V

V_{out} = ______

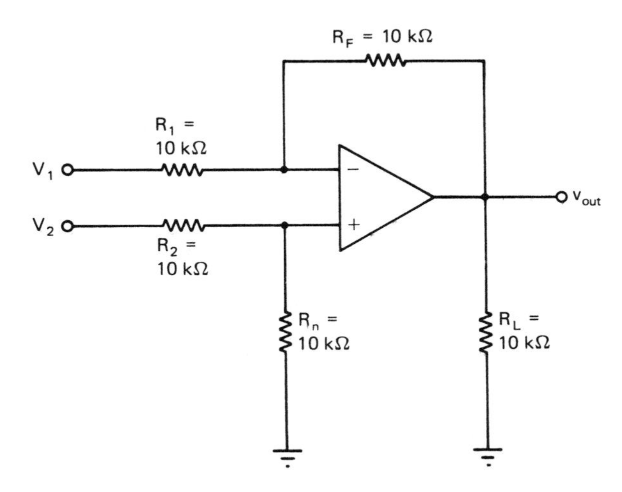

Figure 11–E.3

6. Calculate V_{ref} for the following circuit for various values of R_2, R_3, and $\pm V$ using the formula $V_{ref} = R_3/R_2 + R_3(+V)$.

a. $\pm V = 9$ V
$R_2 = 10$ kΩ
$R_3 = 22$ kΩ

V_{ref} = ______

b. $\pm V = 12$ V
$R_2 = 47$ kΩ
$R_3 = 22$ kΩ

V_{ref} = ______

c. $\pm V = 15$ V
$R_2 = 15$ kΩ
$R_3 = 10$ kΩ

V_{ref} = ______

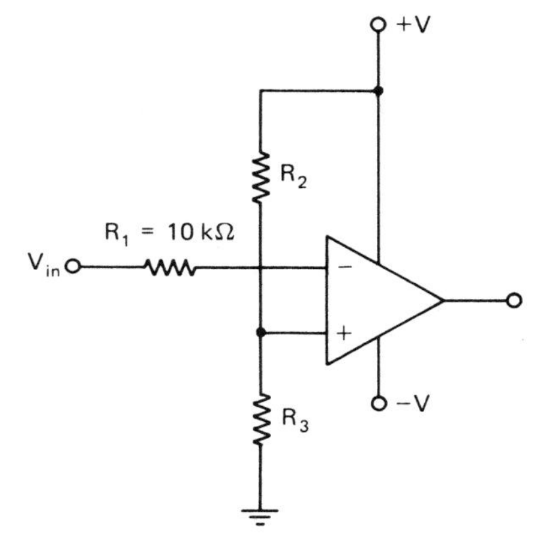

Figure 11–E.4

7. Calculate f_{out} for the following circuit for various values of R_1 and C_1 using the formula $f_{out} = 1/2R_1C_1$.

a. $R_1 = 100$ kΩ
$C_1 = 0.02\ \mu$F

f_{out} = ______

b. $R_1 = 1$ kΩ
$C_1 = 0.05\ \mu$F

f_{out} = ______

c. $R_1 = 3.3$ kΩ
$C_1 = 0.01\ \mu$F

f_{out} = ______

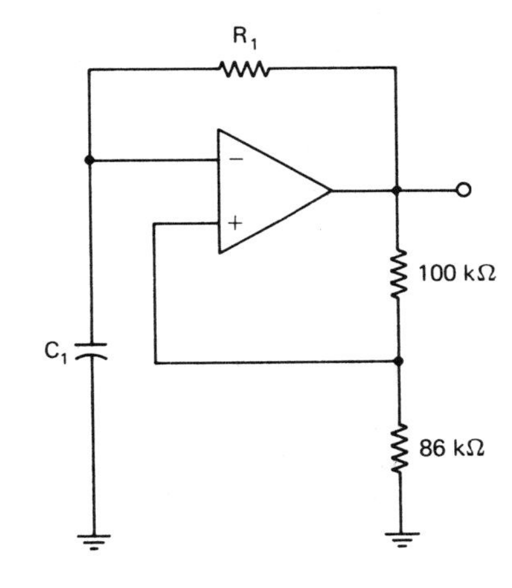

Figure 11–E.5

SECTION 11-7 OP AMP INSTANT REVIEW

An op amp is a solid-state device with several circuits basically consisting of a high-input-impedance differential amplifier, a high-gain voltage amplifier, and a low-impedance output amplifier. A standard op amp has an inverting input, a noninverting input, and a single output. Very often, an op amp is used with a dual ± voltage power supply.

The open loop gain of an op amp is extremely high. A slight change in voltage at the input will cause the output to saturate in one of two extremes: $+V_{sat}$ or $-V_{sat}$.

In the closed loop mode, a feedback resistor from the output to the inverting input determines the gain of the circuit. The gain for an inverting amplifier can be calculated by the formula: $A_V = -R_F/R_{in}$. The gain for a noninverting amplifier is calculated by the formula: $A_V = 1 + R_F/R_{in}$.

This device is very versatile and can perform many electronic functions with a minimum of external components. Some basic op amp circuits are the following: voltage comparator, inverting amplifier, noninverting amplifier, voltage follower, summing amplifier, difference amplifier, square-wave generator, low-pass filter, high-pass filter, bandpass filter, and bandreject filter.

SECTION 11-8 SELF-CHECKING QUIZZES

11-8a OP AMP: TRUE–FALSE QUIZ

Place a T for true or an F for false to the left of each statement.

____ **1.** If the noninverting input is grounded and a negative voltage is placed on the inverting input, the output of an op amp will swing positive.

____ **2.** An op amp has low input impedance.

____ **3.** An op amp has very high voltage gain.

____ **4.** The higher the CMRR is, the better the quality of an op amp.

____ **5.** Self-oscillations of an op amp can be canceled by offset nulling.

____ **6.** An op amp has high output impedance.

____ **7.** The closed-loop gain formula for an inverting op-amp amplifier is $A_V = -R_F/R_{in}$.

____ **8.** Errors in the output voltage of an op amp are caused by input-offset voltages and currents.

____ **9.** The output of an op amp can swing positive and negative when used with a dual (±) power supply.

____ **10.** The op-amp noninverting amplifier gain can be found by the formula $A_V = R_F/R_{in} + 1$.

11-8b OP AMP: MULTIPLE-CHOICE QUIZ

Circle the correct answer for each question.

1. An op amp may:

a. operate on a ± dual power supply

b. have two inputs

c. have one output

d. all of the above

2. The maximum rate of change of an op amp's output voltage is called:

a. common-mode rejection ratio

b. slew rate

c. bandwidth

d. latch-up

3. If a signal of the same amplitude and phase angle is applied to both inputs of an op amp simultaneously and the output is nearly zero, the op amp is said to have a:
   - **a.** low gain
   - **b.** high slew rate
   - **c.** high common-mode rejection ratio
   - **d.** none of the above

4. A standard op amp has:
   - **a.** low input impedance, high gain, and low output impedance
   - **b.** low input impedance, high gain, and high output impedance
   - **c.** high input impedance, high gain, and high output impedance
   - **d.** high input impedance, high gain, and low output impedance

5. The output voltage of the circuit shown in Figure 11-A is about:
   - **a.** +8.1 V
   - **b.** +5.25 V
   - **c.** −5.25 V
   - **d.** +0.25 V

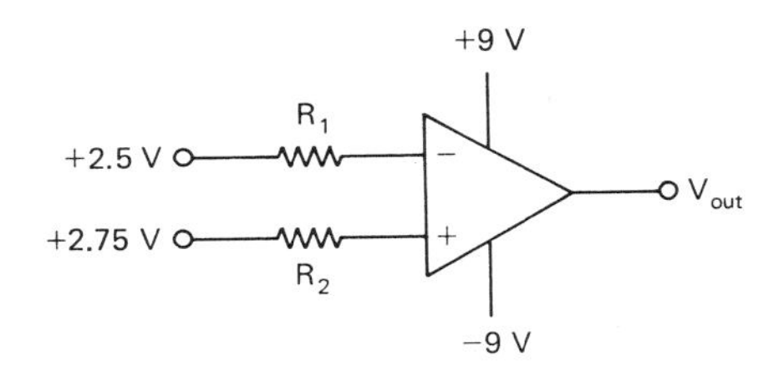

Figure 11–A

6. Error voltages at the output of an op amp can be caused by:
   - **a.** insufficient voltage at the inputs
   - **b.** input offset voltages and currents
   - **c.** both of the above
   - **d.** none of the above

7. Offset nulling can be accomplished by:
   - **a.** adjusting special potentiometers at the inputs
   - **b.** using a potentiometer connected between special pins on the op amp
   - **c.** both of the above
   - **d.** none of the above

8. Referring to Figure 11-B, when the input voltage reaches +5.0 V, the output voltage will be:
   - **a.** +0.1 V
   - **b.** −0.1 V
   - **c.** −9.9 V
   - **d.** −10.8 V

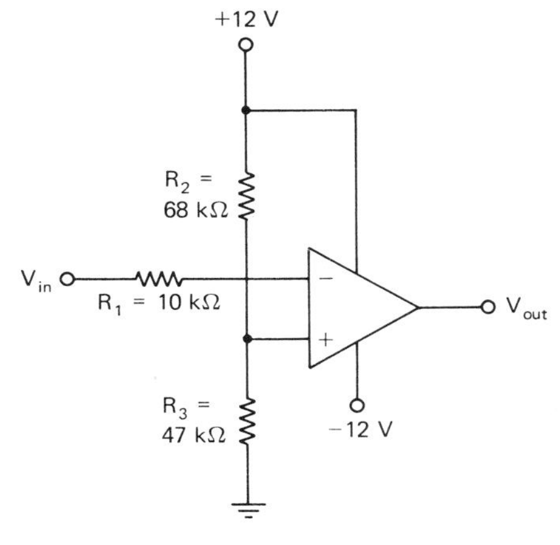

Figure 11–B

9. The type of feedback generally applied to an op amp to control gain and stability is:
   - **a.** negative feedback
   - **b.** positive feedback
   - **c.** regenerative feedback
   - **d.** none of the above

10. The final output voltage for the circuit shown in Figure 11-C is:
    - **a.** +4.4 V
    - **b.** −4.4 V
    - **c.** +6.4 V
    - **d.** −6.4 V

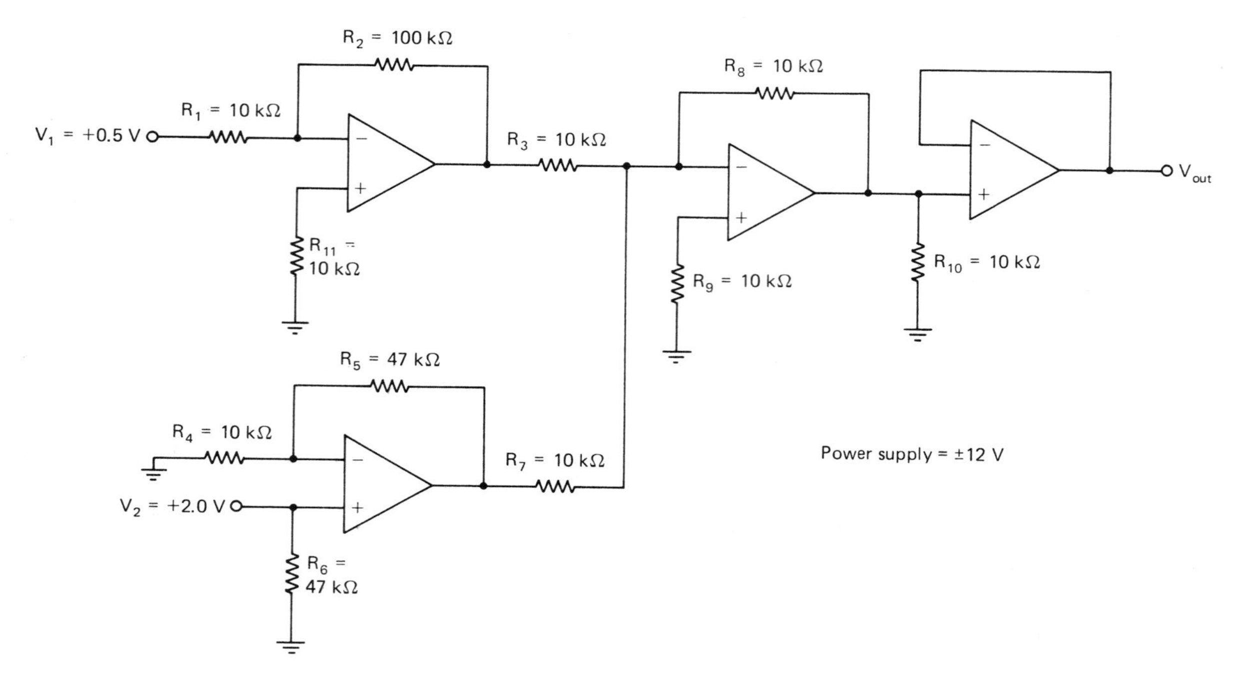

Figure 11-C

ANSWERS TO EXPERIMENTS AND QUIZZES

Experiment 1. (1) $+V_{sat}$, $-V_{sat}$ (or vice versa) (2) $+V_{sat}$ (3) positive

Experiment 2. (1) R_F, R_{in} (2) 180 (3) V_{in} (4) dc (5) frequency response or bandwidth

Experiment 3. (1) R_F, R_{in}, 1 (2) in (3) A_v (4) block (5) large

Experiment 4. (1) impedance, buffering (2) input (3) noninverting

Experiment 5. (1) algebraically, inverted (2) higher (or larger) (3) $V_{out} = -(V_1 + V_2 + \cdots + V_n)$ (4) input

Experiment 6. (1) algebraic (2) positive

Experiment 7. (1) output (2) positive, negative (3) V_{ref}, V_{in}

Experiment 8. (1) $\pm V_{sat}$ (2) sawtooth (3) $(1/2R_1C_1)$ (4) decreases (5) increases

Experiment 9. (1) 1,592 Hz (2) passed (3) blocked

Experiment 10. (1) 6,742 Hz (2) blocked (3) passed

Experiment 11. (1) 712 Hz (2) 318 Hz (3) blocked (4) blocked

Experiment 12. (1) 159 Hz (2) 5 Hz (3) passed (4) passed

True–False. (1) T (2) F (3) T (4) T (5) F (6) F (7) T (8) T (9) T (10) T

Multiple Choice. (1) d (2) b (3) c (4) d (5) a (6) b (7) c (8) d (9) a (10) d

Unit 12

555 Precision Timer IC

INTRODUCTION The 555 precision timer IC is a device found in many of today's modern electronic circuits. It is widely used and discussed in many trade journals and magazines.

UNIT OBJECTIVES Upon completion of this unit you will be able to:

1. Identify a 555 timer IC.
2. List the basic circuits within the 555 timer IC.
3. Describe the function and operation of the basic circuits within the 555 timer IC.
4. Define the terms totem-pole output, duty cycle, and monostable and astable multivibrator.
5. Construct circuits using the 555 timer IC.
6. Understand how to connect output indicators to a 555 timer IC depending on the required output.
7. Construct a circuit to test 555 timer ICs.
8. Troubleshoot an astable multivibrator using a 555 timer IC.

SECTION 12-1 THEORY OF STRUCTURE AND OPERATION

12-1a INSIDE THE 555 TIMER IC

The 555 precision timer IC contains a group of circuits used for timing and producing square-edge pulses with the minimum amount of external resistive and capacitive components. As shown in Figure 12-1, it is comprised of five distinct circuits, two voltage comparators, a three-resistor reference voltage divider, a bistable flip-flop, a discharge transistor, and an output stage arranged in a *totem-pole* configuration (one transistor above another). The schematic diagram of Figure 12-2 shows the actual discrete components.

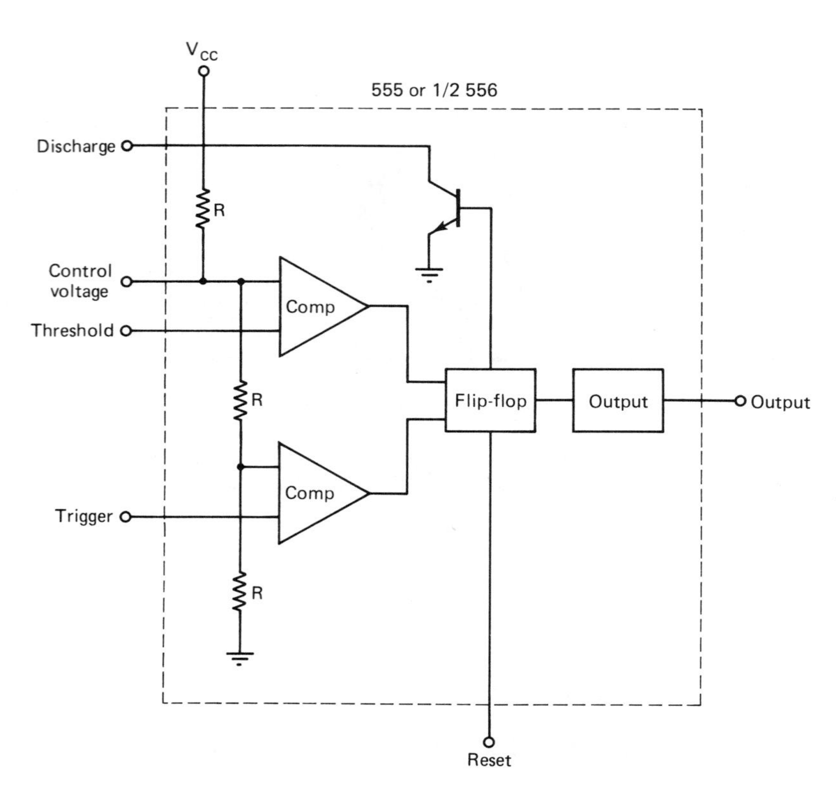

Figure 12–1 Block diagram of 555 precision timer. (Courtesy of Signetics Corporation, a subsidiary of U.S. Philips Corp., 811 E. Arque Ave., Sunnyvale, Calif. 94086, *Applications Manual*, 1979, p. 149.)

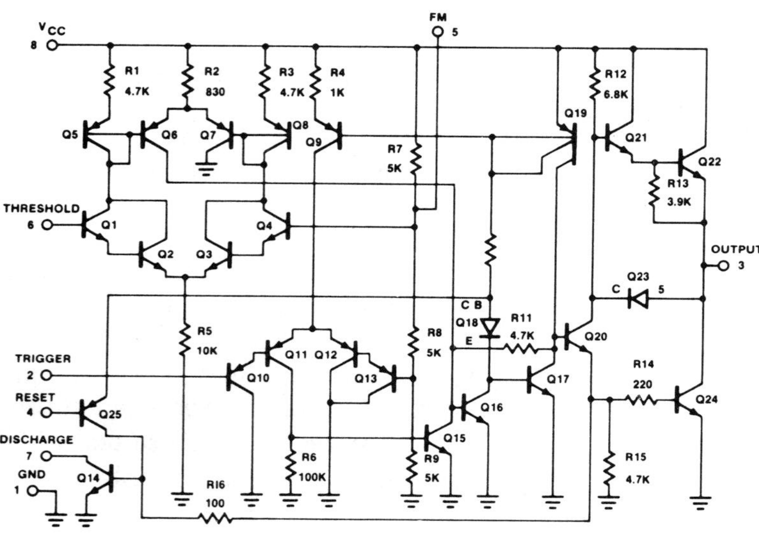

Figure 12–2 Schematic diagram of 555 timer. (Courtesy of Signetics Corporation, a subsidiary of U.S. Philips Corp., 811 E. Arque Ave., Sunnyvale, Calif. 94086, *Analog Data Manual*, 1981, p. 177.)

12-1b TERMINAL DESCRIPTIONS

Ground: Pin 1 is one of two power supply terminals that is connected to common or ground.

$+V_{cc}$: Pin 8 is the positive voltage supply terminal that is connected to $+V_{cc}$. The range of $+V_{cc}$ can be from +5 to +15 V for guaranteed circuit operation; however, some circuits can operate with as little as +3 V. The maximum $+V_{cc}$ should never exceed +18 V. Maximum power dissipation for the IC package is 600 mW.

Output: Pin 3 is the output terminal connected to other circuits. It has two states or conditions that are controlled by the flip-flop. In the low state, the output voltage is about 0 V or ground. In the high state, the output voltage is approximately $+V_{cc}$.

Trigger: Pin 2 applies a trigger voltage to the lower comparator. When the voltage at this terminal is greater than two-thirds of $+V_{cc}$, the output remains low. When a negative-going pulse below one-third of $+V_{cc}$ appears on pin 2, the lower-voltage comparator changes states, which causes the flip-flop to change states and the output goes to the high condition. If the trigger terminal is kept low, the output will remain in the high state (this condition is usually avoided).

Threshold: Pin 6 is the threshold voltage to the upper comparator. An external capacitor is usually connected from this terminal to ground. When the 555 timer is triggered into its high state, the external capacitor begins to charge toward $+V_{cc}$. When voltage across the capacitor increases to the threshold voltage, about two-thirds of $+V_{cc}$, the comparator changes states, which resets the flip-flop and the output goes to the low condition.

Discharge: Pin 7 is used to discharge the external capacitor. This pin is usually connected directly or through a resistor to pin 6 (threshold). When the output is high, the internal transistor (Q_{14}) connected to pin 7 is off, which allows the external capacitor to charge toward $+V_{cc}$. When the output goes low, transistor Q_{14} turns on and the capacitor has a discharge path to ground.

Reset: Pin 4 is a reset input that directly controls the flip-flop. This pin overrides any command pulses on the trigger input pin 2. When the reset terminal is brought to 0 V, both the output terminal, pin 3, and the discharge terminal, pin 7, are forced low or to 0 V. When the reset terminal is not used it should be connected to $+V_{cc}$.

Control voltage: Pin 5 is a control voltage (FM, frequency modulator) input. The internal connection of this terminal is to the three-resistor (R_7, R_8, and R_9) reference voltage divider that is used for the reference voltages of the comparators. Applying an external voltage to this terminal will change both the threshold and trigger voltage levels and can be used to modulate the output waveform. When pin 5 is not used, a 0.01-μF capacitor is usually connected from it to ground. This bypasses noise and/or power supply voltage ripple to ground, thereby minimizing their effects on the threshold voltage. Figure 12-3 shows the 555 timer pin locations for various ICs.

Figure 12–3 555 Timer IC packages: (a) 8-in mini-DIP; (b) 14-pin DIP; (c) 8-pin metal can. (Courtesy of Signetics Corporation, a subsidiary of U.S. Philips Corp., 811 E. Arque Ave., Sunnyvale, Calif. 94086, *Analog Data Manual*, 1981, p. 177.)

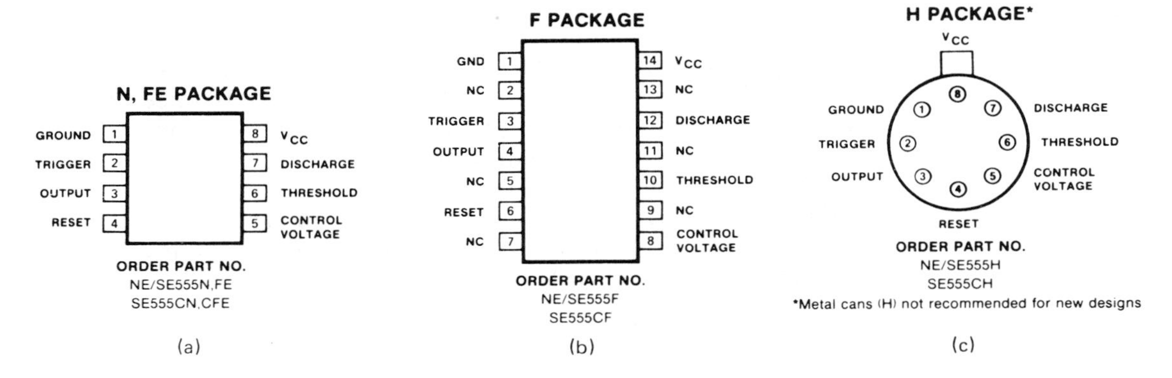

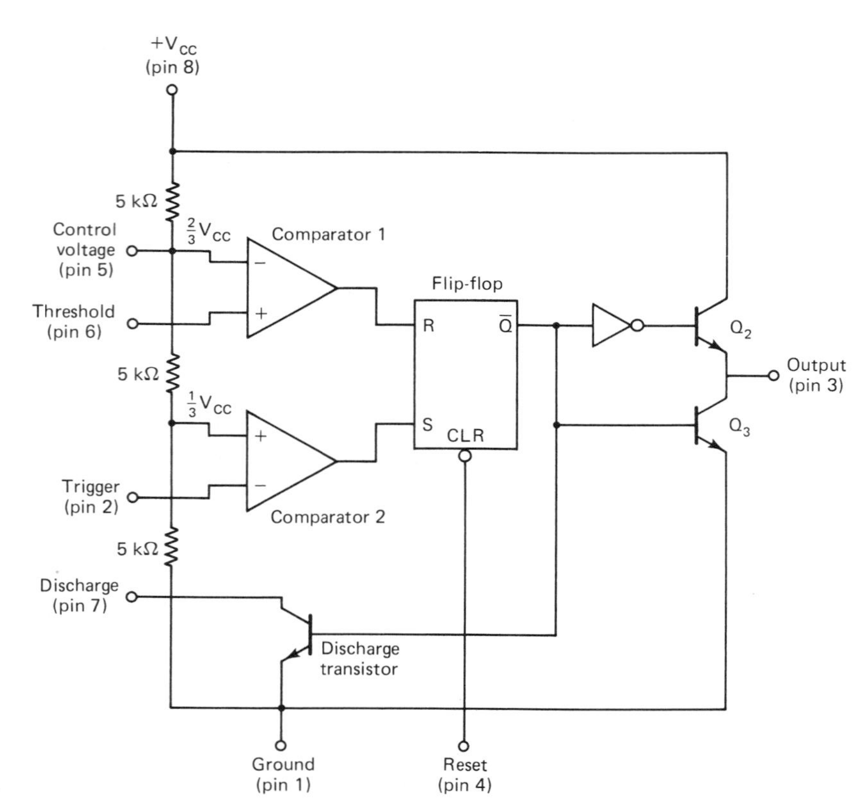

Figure 12–4 Functional diagram for the 555 precision timer IC (8 pins).

12-1c BASIC OPERATION OF THE 555 TIMER IC

The 555 precision timer IC is a special integrated circuit that can be used for timing, triggering, delaying, pulse reshaping, and clock circuits. A functional diagram is shown in Figure 12-4. Three 5-kΩ resistors form a voltage divider from $+V_{cc}$ to ground inside the IC. Two comparators use this voltage divider to establish voltage reference points at each of their inputs. The outputs of the comparators control the operation of an *R-S* flip-flop. The $\overline{Q}$ output of the flip-flop goes to an inverter stage, which controls the turning on and off of transistor Q_2. This transistor acts as a *source* transistor, connecting the output at pin 3 to the voltage source ($+V_{cc}$). Another output line from the *R-S* flip-flop controls the operation of transistor Q_3. This transistor acts as a *sink*, connecting the output pin to ground at pin 1. Only one of these transistors can be on at a single time and it is referred to as a *totem-pole output*. For example, if Q_2 is on, Q_3 is off, and vice versa. The output from the *R-S* flip-flop also controls the operation of Q_1, which is used to discharge an external capacitor used for timing sequences at pin 7. Pin 6 (threshold) and pin 2 (trigger) are used for external *RC* timing components. Pin 4 is used to reset the flip-flop externally and when not used is connected to $+V_{cc}$. Pin 5 can be used for externally controlling the reference voltage on comparator 1. When pin 5 is not used, a 0.01-μF capacitor is usually connected from it to ground to bypass noise and/or power supply voltage variations to ground, thereby reducing any effects on the reference voltage of comparator 1.

1. The reference voltage of comparator 1 is set at ⅔ V_{cc}. When the other input at pin 6 (threshold) is below the value of the reference voltage, the output of the comparator is low or 0, and the flip-flop

will not reset. Assuming that the flip-flop is on, there is a low or 0 condition at its output, which keeps Q_1 and Q_3 off, but because of the inverter, places a high or 1 condition at the base of Q_2. This transistor turns on and connects the output (pin 3) to $+V_{cc}$ or a high condition.

2. When the voltage at pin 6 reaches ⅔ V_{cc}, a high or 1 condition appears at the output of comparator 1, which resets the flip-flop (turning it off) and the $\overline{Q}$ output goes to a high or 1 condition. At this time, Q_1 and Q_3 turns on and Q_2 turns off. The output is now connected through Q_3 to ground or a low condition.
3. The reference voltage of comparator 2 is set at ⅓ V_{cc}. If the other input at pin 2 (trigger) is above the value of the reference voltage, the output of the comparator is low or 0 and the flip-flop will not set (turn on).
4. When the voltage at pin 2 falls to ⅓ V_{cc}, a high or 1 condition appears at the output of comparator 2, which sets (turns on) the flip-flop and its $\overline{Q}$ output goes to a low or 0 condition. Now Q_1 and Q_3 turn off and Q_2 turns on. The output of the 555 is again connected to $+V_{cc}$ or a high and Q_1 is off, allowing an external capacitor to charge toward $+V_{cc}$ and repeat the action.

12-1d MONOSTABLE OPERATION

A monostable (*one-shot*) multivibrator has only one stable state. It can be triggered to the other state for a predetermined time depending on an *RC* time constant, after which it returns to the original state. Once triggered, the one-shot multivibrator will not respond to any other input trigger pulses until it has completed its initial timing cycle. Figure 12-5a shows its logic or block diagram concept of operation.

The 555 timer makes a fine one-shot multivibrator, with the use of only two external components, a resistor and a capacitor, as shown in Figure 12-5b. The resistor is connected from $+V_{cc}$ to discharge (pin 7) and effectively to threshold (pin 6) since these two pins are connected. The capacitor is connected from threshold (pin 6) to ground.

Initially, the output is low, the discharge transistor is on, and there is no voltage charge on the capacitor. When a negative-going pulse appears on the trigger input (pin 2), the output goes high, the discharge transistor turns off, and the capacitor begins to charge up toward $+V_{cc}$. When the voltage on the capacitor increases to about two-thirds of $+V_{cc}$, the output goes low, which turns on the discharge transistor, allowing the voltage on the capacitor to discharge to ground. The timer has completed a cycle and will now await another trigger pulse. Ignoring capacitor leakage, the capacitor will reach two-thirds of the $+V_{cc}$ level in 1.1 time constants; therefore, the width of the output pulse can be found by the formula

$$T = 1.1RC$$

where T is in seconds, R is in ohms, and C is in farads.

Monostable multivibrators are used for time delay, reshaping a "ragged" input pulse, pulse stretcher (since the output pulse can be wider than the input pulse), and bounce-less switch (since it only responds to the first negative-going pulse in the on condition).

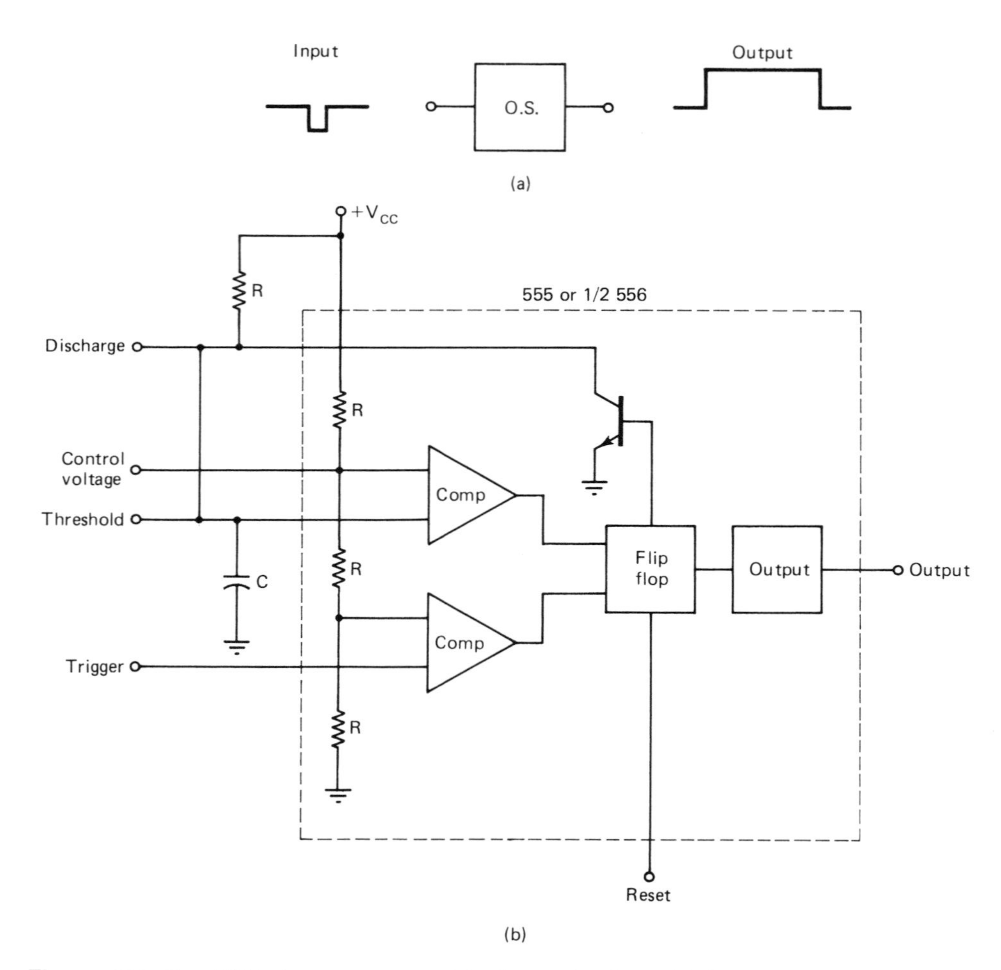

Figure 12–5 555 timer monostable (one-shot) multivibrator: (a) logic or block diagram; (b) circuit operation. (Courtesy of Signetics Corporation, a subsidiary of U.S. Philips Corp., 811 E. Arque Ave., Sunnyvale, Calif. 94086, *Applications Manual*, 1979, p. 151.)

12-1e ASTABLE OPERATION

An astable multivibrator, often referred to as the *clock* in a digital system, has no stable state, but switches back and forth between the two states, providing a square-edge output signal. Figure 12-6a shows a logic or block diagram concept of operation.

The 555 timer can be used as an astable multivibrator to provide a very stable output frequency, as shown in Figure 12-6b. In this case, three external components are used, two resistors (R_A and R_B) and one capacitor (C). Resistor R_A is connected from $+V_{cc}$ to discharge (pin 7), and R_B is connected from discharge (pin 7) to threshold (pin 6). The capacitor is connected from ground to trigger (pin 2) and effectively to threshold (pin 6), since these two pins are connected.

When power is applied to the circuit, the capacitor is discharged and holds the trigger input low, which causes the output to go high. The discharge transistor turns off, allowing the capacitor to charge up toward $+V_{cc}$ through R_A and R_B. When the voltage on the capacitor reaches two-thirds of $+V_{cc}$, the output goes low and turns on the discharge transistor. The timing capacitor now discharges through R_B to ground. When the voltage on the capacitor drops to one-third of $+V_{cc}$, the output again goes high, completing one cycle.

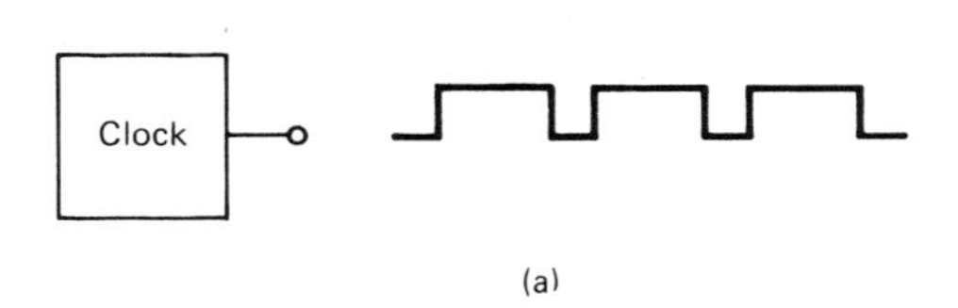

(a)

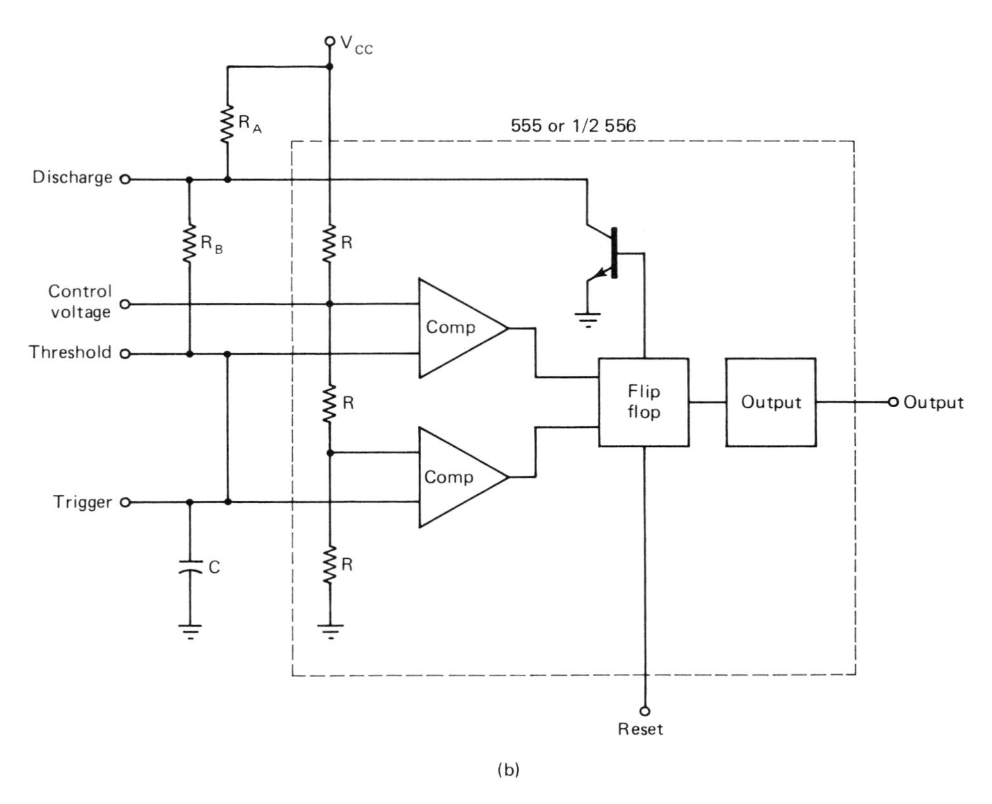

(b)

Figure 12–6 555 timer astable (clock) multivibrator: (a) logic or block diagram: (b) circuit operation. (Courtesy of Signetics Corporation, a subsidiary of U.S. Philips Corp., 811 E. Arque Ave., Sunnyvale, Calif. 94086, *Applications Manual*, 1979, p. 151.)

This action repeats itself and creates an oscillator whose frequency can be found by the formula

$$f = \frac{1.49}{(R_A + 2R_B)C}$$

12-1f DUTY CYCLE

The duty cycle of a device is the amount of time it operates, as opposed to its idle time. In the case of the 555 timer astable multivibrator, it is the ratio of time the output is low (t_{low}) to the total period of one cycle (T) and can be determined by the formulas

$$t_{\text{low}} = 0.693(R_B)C \qquad \text{(the time the output is low)}$$
$$t_{\text{high}} = 0.693(R_A + R_B)C \qquad \text{(the time the output is high)}$$

Thus the total period of time (T) for one cycle is

$$T = t_{\text{low}} + t_{\text{high}} = 0.693(R_A + 2R_B)C$$

The duty cycle is then

$$DC = \frac{t_{low}}{T} \quad \text{or simply} \quad DC = \frac{R_B}{R_A + 2R_B} \quad \text{(expressed as a percentage)}$$

Since the capacitor charges up through R_A and R_B, but discharges only through R_B, the duty cycle will never be greater than 50%. However, by placing a diode across R_B with the anode connected to discharge (pin 7) and the cathode connected to threshold (pin 6), the duty cycle can be extended from about 5% to greater than 95%.

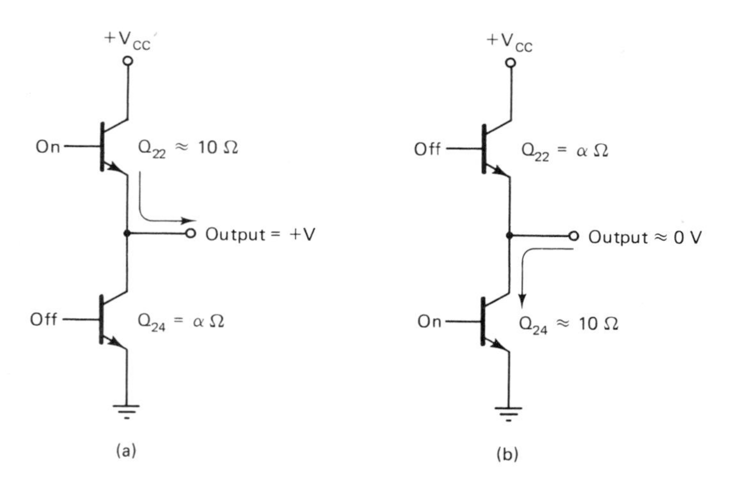

Figure 12–7 Totem-pole output-stage operation: (a) current source; (b) current sink.

12-1g 555 TIMER OUTPUT STAGE

Figure 12-7 shows the output stage of the 555 timer, consisting of transistors Q_{22} and Q_{24}. Only one of these transistors can be on at a time. When Q_{22} is on, Q_{24} is off and the output is connected to $+V_{cc}$ through about 10 Ω of the conducting transistor's resistance (Figure 12-7a). When Q_{24} is on, Q_{22} is off and the output is connected to ground through about 10 Ω of the conducting transistor's resistance (Figure 12-7b). If the output is high, Q_{22} acts like a source for current flow, and if the output is low, Q_{24} acts like a sink for current flow.

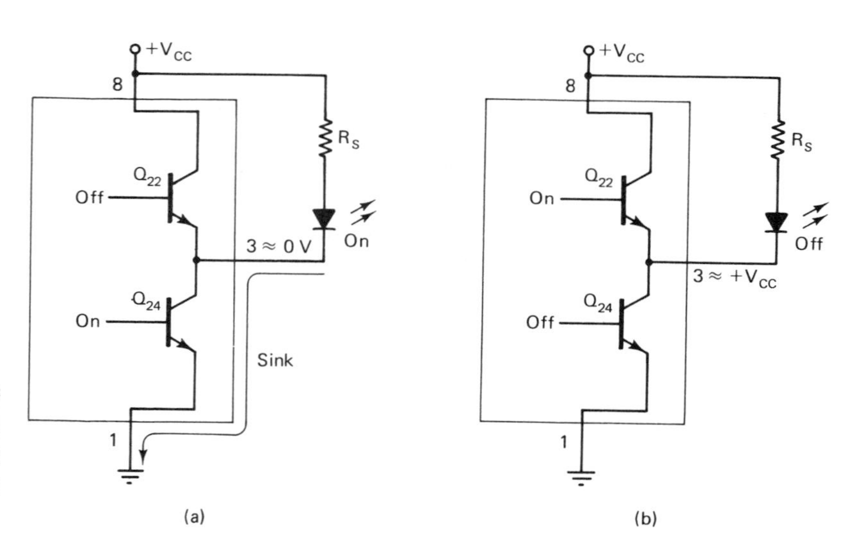

Figure 12–8 Methods of connecting output loads: (a) and (b) from $+V_{CC}$ to output pin; (c) and (d) from output pin to ground.

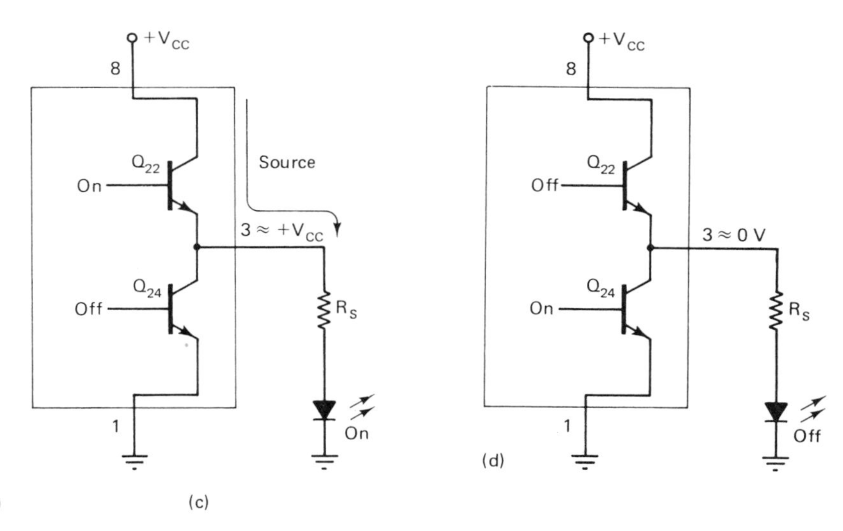

Figure 12–8 (cont'd.)

12-1h METHODS OF CONNECTING OUTPUT LOADS

There are two ways of connecting loads to the output of the 555 timer as shown in Figure 12-8. Figure 12-8a and b shows the load connected from $+V_{cc}$ to the output (pin 3) of the timer. When the output is low, current flows through the load, but when the output is high, no current flows through the load. Figure 12-8c and d shows the load connected from ground to the output (pin 3) of the timer. In this case, when the output is high, current flows through the load, but when the output is low, no current flows through the load.

SECTION 12-2
555 TIMER DEFINITION EXERCISE

Refer to the previous sections and write a brief definition for each term.

1. Astable multivibrator ______________________________

____

____

2. Monostable multivibrator ______________________________

____

____

3. One-shot multivibrator ______________________________

____

____

4. Totem-pole output ______________________________

____

____

5. Duty cycle ______

6. Low state or condition ______

7. High state or condition ______

8. DIP ______

9. Trigger pulse ______

10. Bypass capacitor ______

SECTION 12-3 EXERCISES AND PROBLEMS

Perform the following exercises before beginning Section 12-4.

1. Draw the functional block diagram of a 555 timer IC, identifying the pin terminals.

2. Draw and correctly number and label the pins on the eight-pin 555 timer mini-DIP IC.

3. Using a block diagram, draw the input pulse and the output waveform of a 555 monostable (one-shot) multivibrator. (Show pulse-width relationships.)

4. Using a block diagram, draw the output of a 555 astable multivibrator.

5. Draw the transistor totem-pole output stage of a 555 timer IC.

6. Draw two methods of connecting a load to the output of a 555 timer IC.

SECTION 12-4 EXPERIMENTS

EXPERIMENT 1. 555 TIMER BASIC OPERATION

Objective:

To understand the input terminal functions and basic operation of the 555 timer IC.

Introduction:

Trigger terminal (pin 2) initiates the timing sequence and overall operation of the 555 timer. When a negative-going pulse appears on this pin, the output goes high. The circuit shown in Figure 12-9a is arranged with a capacitor to the threshold (pin 6) and discharge (pin 7) terminals, which keeps the output high when the 555 is triggered. The reset terminal (pin 4) is normally high, but when it is brought low (0 V), the output goes low. This circuit serves as a latch and will "remember" that a negative-going pulse has occurred on pin 2 until it is reset.

Materials Needed:

1 +9-V power supply
1 Standard or digital voltmeter
1 555 timer IC
1 10-kΩ resistor at 0.5 W (R_1)
1 0.01-μF capacitor at 25 WV dc (C_2)
1 0.1-μF capacitor at 25 WV dc (C_1)
2 SPDT switches
1 Breadboard for constructing circuit

Procedure:

1. Construct the circuit shown in Figure 12-9a.
2. Close and open S_2 to make sure the output (pin 3) is low.
3. Measure the output voltage and record the value in the first line of the data table given in Figure 12-9b.
4. Momentarily close S_1 to position B and then back to position A.
5. Measure the output voltage and record the value in the second line of the data table.
6. Momentarily close S_2 to position B and then back to position A.
7. Measure the output voltage and record the value in the third line of the data table.

Fill-in Questions:

1. The 555 timer is triggered (the output goes high) when a __________ pulse appears on pin ______ .
2. The 555 timer is reset (the output goes low) when pin 4 is brought __________ .

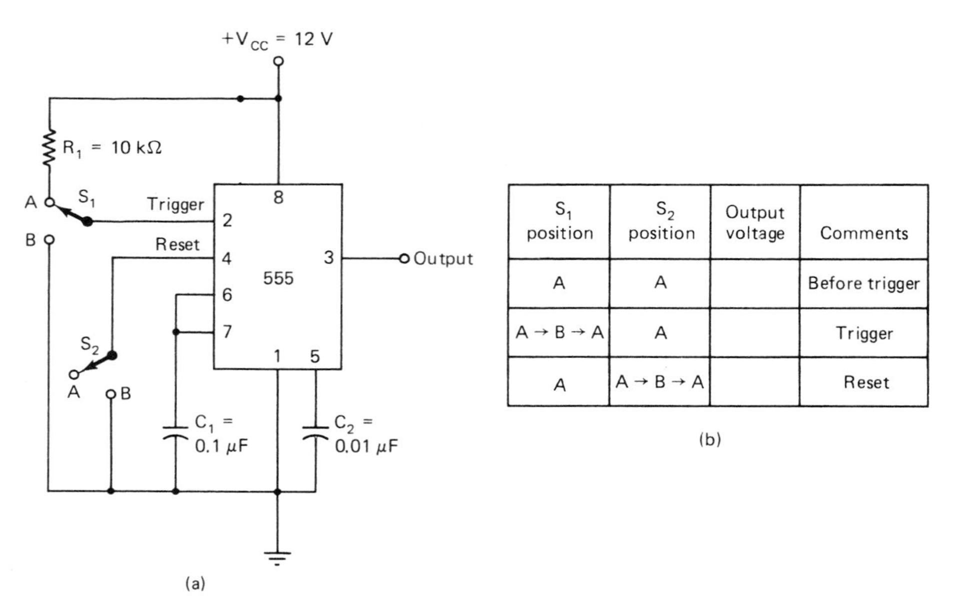

S_1 position	S_2 position	Output voltage	Comments
A	A		Before trigger
A → B → A	A		Trigger
A	A → B → A		Reset

Figure 12–9 555 timer basic operation: (a) schematic diagram; (b) data table.

EXPERIMENT 2. 555 MONOSTABLE (ONE-SHOT) MULTIVIBRATOR

Objective:

To demonstrate the operation of a 555 timer monostable multivibrator, and how to calculate the duration of the output voltage pulse.

Introduction:

Figure 12-10a shows the basic 555 timing circuit. Resistors R_1 and R_2 serve as only a method for triggering the circuit into operation. Components R_A and C_1 determine the width of the output voltage pulse.

Materials Needed:

1 +9-V power supply
1 Standard or digital voltmeter
1 Oscilloscope (dual trace preferred)
1 555 timer IC
1 1-kΩ resistor at 0.5 W (R_2)
2 10-kΩ resistors at 0.5 W (R_1, R_A)
1 100-kΩ resistor at 0.5 W (R_A)
1 470-kΩ resistor at 0.5 W (R_A)
1 1-MΩ resistor at 0.5 W (R_A)

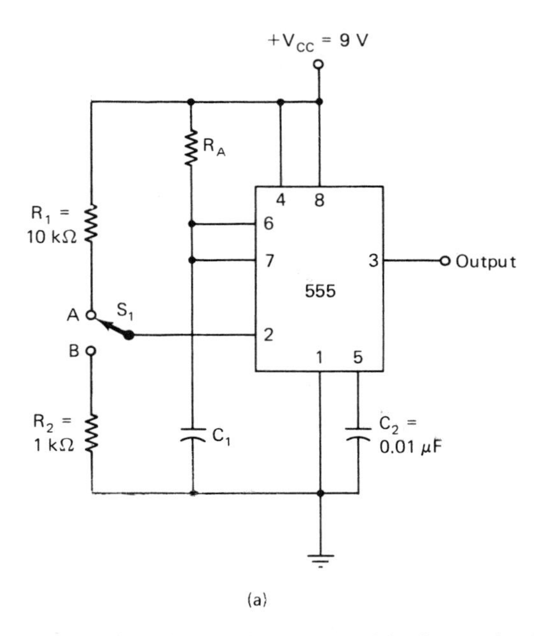

R_A (Ω)	C_1 (μF)	$T = 1.1\, R_A C_1$ calculated	T measured
1 M	10		
470 k	10		
100 k	10		
10 k	10		
470 k	1		
470 k	0.1		

(b)

Figure 12–10 555 monostable (one-shot) multivibrator: (a) circuit diagram; (b) data table.

1 0.01-μF capacitor at 25 WV dc (C_2)
1 0.1-μF capacitor at 25 WV dc (C_1)
1 1-μF capacitor at 25 WV dc (C_2)
1 10-μF capacitor at 25 WV dc (C_1)
1 SPDT switch
1 Breadboard for constructing circuit

Procedure:

1. Construct the circuit shown in Figure 12-10a, using the values given in the first line of the data table of Figure 12-10b for R_A and C_1, respectively.
2. Place the oscilloscope at pin 2 and pin 3 to make voltage measurements.
3. Calculate the output voltage pulse width and record in the data table using the formula $T = 1.1 R_A C_1$.
4. Momentarily move S_1 to position B and attempt to measure with a watch the width of the output pulse. Approximate the measured output pulse width and record it in the data table.
5. Initiate another timing cycle using S_1, and while the output is high, operate S_1 a few times to see that it will not interfere with the timing cycle set by R_A and C_1. (This timing cycle will last about 11 seconds.)
6. Repeat steps 3 and 4 for the values of R_A and C_1 given in the data table.

Fill-in Questions:

1. For a 555 monostable multivibrator, the time the output is high is determined by ______ and ______.
2. Once the monostable multivibrator is triggered, other trigger pulses will not ______ with the timing cycle.
3. The one-shot multivibrator can be used to ______, ______, and as a ______ (use statements).

EXPERIMENT 3. 555 ASTABLE MULTIVIBRATOR

Objective:

To show the operation of a 555 timer astable multivibrator, and how to calculate its output frequency and duty cycle.

Introduction:

With only two resistors (R_A and R_B) and a capacitor (C_1), the 555 timer IC can produce a highly accurate astable multivibrator as

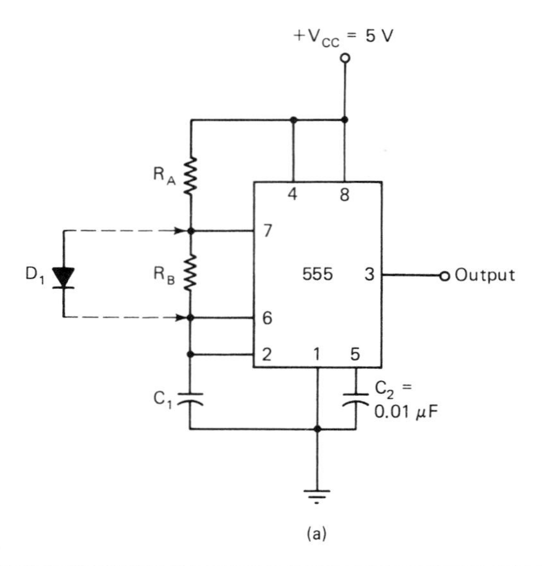

R_A (Ω)	R_B (Ω)	C_1 (μF)	$f_{out} = \frac{1.49}{(R_A + 2R_B)C}$ calculated (Hz)	f_{out} measured (Hz)	Output waveform	$dc = \frac{R_B}{R_A + 2R_B}$
22 k	10 k	0.1			0 V ____ V p-p	
22 k	10 k	0.05			0 V ____ V p-p	
22 k	10 k	0.2			0 V ____ V p-p	
22 k	2.2 k	0.1			0 V ____ V p-p	
10 k	10 k	0.1			0 V ____ V p-p	

(b)

Figure 12–11 555 astable (clock) multivibrator: (a) circuit diagram; (b) data table.

shown in Figure 12-11a. The output frequency (f_{out}) and the duty cycle (DC) can be calculated using the values of R_A, R_B, and C_1. If a diode (D_1) is placed across R_B, the output voltage waveform becomes symmetrical and has a 50% duty cycle. Capacitor C_1 charges up through D_1 and R_A and discharges through R_B.

Materials Needed:

1 +5-V power supply
1 Standard or digital voltmeter
1 Oscilloscope
1 555 timer IC
1 2.2-kΩ resistor at 0.5 W (R_B)
2 10-kΩ resistors at 0.5 W (R_A, R_B)
1 22-kΩ resistor at 0.5 W (R_A, R_B)
1 0.01-μF capacitor at 25 WV dc (C_2)
1 0.05-μF capacitor at 25 WV dc (C_1)
1 0.1-μF capacitor at 25 WV dc (C_1)
1 0.2-μF capacitor at 25 WV dc (C_1)
1 1N4002 diode or equivalent
1 Breadboard for constructing circuit

Procedure:

1. Construct the circuit shown in Figure 12-11a using the values of R_A, R_B, and C_1 as given in the first line of the data table of Figure 12-11b. (Do not connect the diode D_1 at this time.)
2. Calculate f_{out} and record in the data table.
3. Using the oscilloscope at the output (pin 3), measure f_{out} and record in the data table.
4. Draw the representative output voltage waveform in the data table. (Indicate peak-to-peak voltage.)
5. Using the values of R_A and R_B, calculate the duty cycle and record it in the data table.
6. Repeat steps 2 through 5 for the remaining values of R_A, R_B, and C_1 given in the data table.
7. Using R_A = 10 kΩ, R_B = 10 kΩ, and C_1 = 0.1μF, add diode D_1 across R_B as shown in Figure 12-11a. Since $R_A = R_B$ and the diode is in place, the output frequency can be calculated by the formula

$$f_{out} = \frac{1.49}{2R_AC}$$

Calculate the f_{out} and record here: ______.

8. Measure f_{out} with the oscilloscope and note that the duty cycle is 50%.

Fill-in Questions:

1. The f_{out} of a 555 astable multivibrator is determined by __________, ______, and __________.

2. If the resistance or capacitance is increased, f_{out} __________.

3. If the resistance or capacitance is decreased, f_{out} __________.

4. Without the use of a diode, the output voltage waveform will be __________ and the duty cycle will be __________ than 50%.

5. With the use of a diode across R_B and $R_A = R_B$, the output voltage waveform will be __________ and the duty cycle will be ______ %.

6. With the use of a diode across R_B and the proper values of R_A and R_B, the duty cycle can range from ______ to ______ %.

EXPERIMENT 4. 555 MISSING-PULSE DETECTOR

Objective:

To demonstrate how a 555 timer can monitor a series of negative-going pulses and detect if one of the pulses is missing.

Introduction:

Figure 12-12a shows a 555 missing-pulse detector circuit. Components R_B, R_L, and Q_1 are not part of the actual circuit, but only serve to invert the positive-going pulses of the signal generator. Switch S_1 is used to interrupt the input trigger pulses to see the 555 output indication of missing pulses. The input trigger pulses are applied to the base of Q_2 and the trigger terminal (pin 2) of the 555 timer simultaneously. When a negative-going pulse is present, the 555 is triggered into operation and its output goes high; however, Q_2 is on for half a cycle, which in effect shorts out C_1 for this period of time. During the next half-cycle, the input trigger pulse is high, Q_2 is off, and capacitor C_1 begins to charge up to $+V_{cc}$ through R_A. Components R_A and C_1 are selected to set the time constant a little longer than the time between pulses so that, before the voltage charge across C_1 reaches the upper threshold voltage, another input trigger pulse is present, which discharges C_1 and keeps the output of the 555 timer high. When a pulse is missing, C_1 is allowed to charge up to the threshold voltage, which resets the 555 timer, causing the output to go low, and LED_1 gives an indication of a missing pulse. The relationship of the circuit voltage waveforms is seen in Figure 12-12b.

Materials Needed:

- *1* +5-V power supply
- *1* Standard or digital voltmeter
- *1* Oscilloscope (dual trace preferred)
- *1* Square-wave generator
- *1* 555 timer IC
- *1* 2N2222 transistor (Q_1)
- *1* 2N3906 transistor (Q_2)
- *1* LED at $V_F \approx 2$ V (LED_1)
- *1* 220-Ω resistor at 0.5 W (R_S)
- *1* 2.2-kΩ resistor at 0.5 W (R_L)
- *1* 5.6-kΩ resistor at 0.5 W (R_A)
- *1* 10-kΩ resistor at 0.5 W (R_B)
- *1* 0.01-μF capacitor at 25 WV dc (C_2)
- *1* 0.1-μF capacitor at 25 WV dc (C_1)
- *1* SPDT switch
- *1* Breadboard for constructing circuit

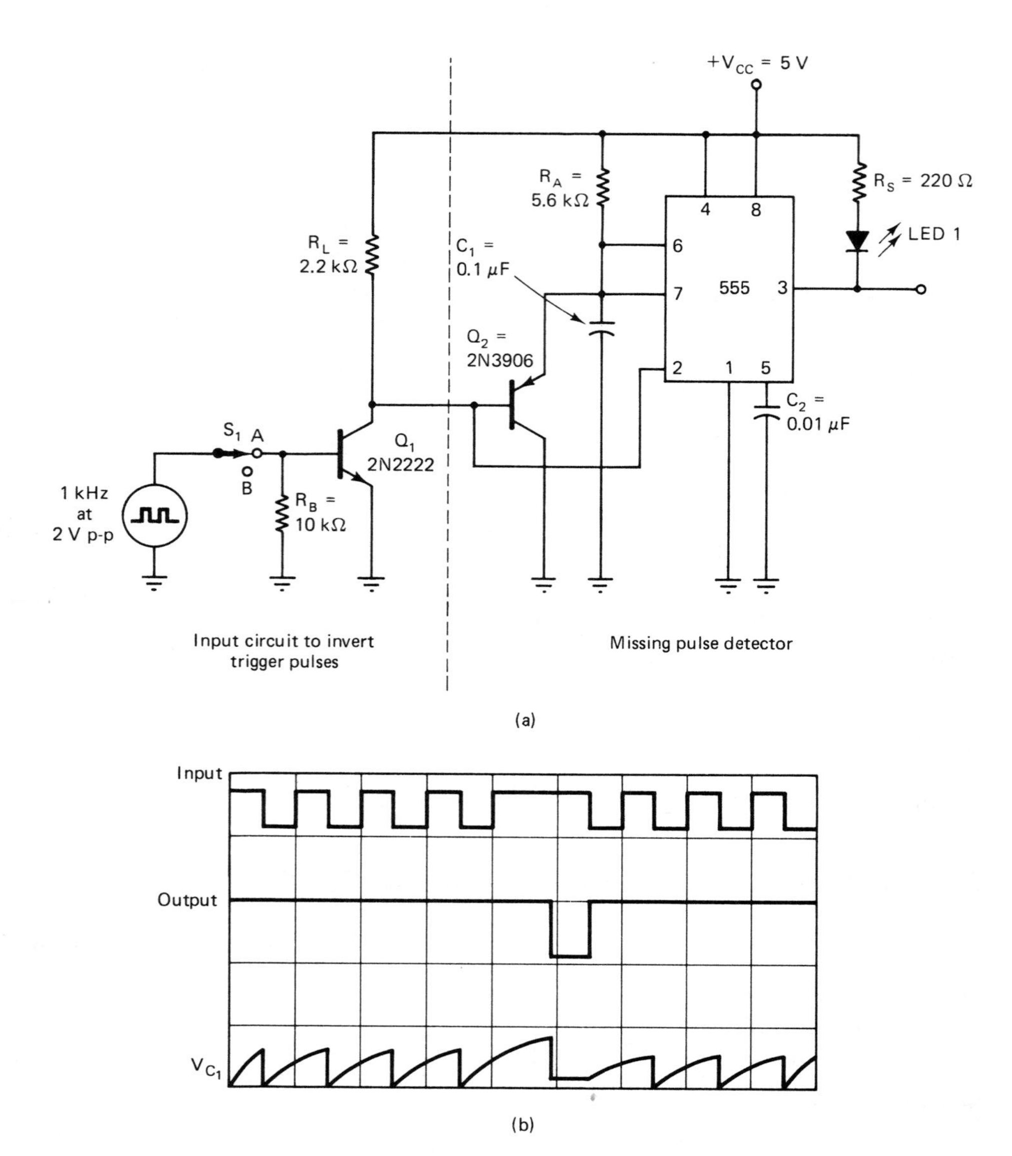

Figure 12–12 555 missing pulse detector: (a) circuit diagram; (b) comparison of voltage waveforms.

Procedure:

1. Construct the circuit shown in Figure 12-12a.
2. Using the oscilloscope, set the square-wave generator at 1 kHz with an amplitude of 2 V p-p.
3. Calculate the period of time for a half-cycle using the formula

$$t_{1/2} = \frac{1/f}{2}$$

Record here: ______

4. Calculate the time to reach the threshold voltage using the formula

$$t = 1.1R_AC_1$$

Record here: ______

5. Set up the oscilloscope for dc and place it at the output (pin 3) of the 555 timer.
6. Momentarily open and close S_1 and compare the oscilloscope display with the voltage waveforms shown in Figure 12-12b. It will be impossible to only omit a single pulse; however, the out-

put voltage will go low and LED_1 will turn on.

7. Place the oscilloscope across C_1 and repeat step 6. Compare this waveform to the one shown in Figure 12-12b.

Fill-in Questions:

1. When input trigger pulses are detected by the missing pulse detector, the output of the 555 timer is ____________ .
2. When a missing pulse is detected by the missing pulse detector, the output of the 555 timer goes ____________ .
3. For the 555 missing pulse detector to operate correctly, the charge time of R_A and C_1 should be slightly ____________ than the time between input trigger pulses.

EXPERIMENT 5. 555 PULSE-WIDTH MODULATION

Objective:

To demonstrate how the pulse width of an output voltage waveform from a 555 timer can be varied by applying voltage to the control voltage terminal.

Introduction:

In this experiment, the 555 timer is wired as a square-wave astable multivibrator. In the first part, a dc voltage is applied to the control voltage terminal (pin 5), which changes the threshold voltage level, thereby causing the pulse width to change at the output. As the dc voltage increases, the width of the output pulse increases. This circuit arrangement could also be used to fine adjust a specific frequency of the oscillator. In the second part, an ac voltage is applied to the control voltage terminal. This voltage not only varies the width of the pulse, but causes its position to change, similar to frequency modulation.

Materials Needed:

1 +15-V power supply
1 Standard or digital voltmeter
1 Oscilloscope (dual trace preferred)
1 Sine-wave generator
1 555 timer IC
1 1N4002 diode or equivalent (D_1)
2 1-kΩ resistors at 0.5 W (R_A, R_B)
2 10-kΩ resistors at 0.5 W (R_A, R_B)
1 10-kΩ linear potentiometer
2 0.1-μF capacitors at 25 WV dc (C_1, C_2)
1 Breadboard for constructing circuit

Procedure:

1. Construct the circuit shown in Figure 12-13a.
2. Connect the oscilloscope to the output terminal (pin 3) of the 555 timer. Also, connect a wire from this point to the external trigger or sync input of the oscilloscope. Set the time-sweep switch of the oscilloscope to external. This keeps the display stationary when the pulse width is varied.
3. Set the wiper of R_c to ground.
4. Adjust R_c to set V_{mod} at +2 V, as indicated in the first line of the data table of Figure 12-13b.
5. Set the time base of the oscilloscope as indicated in the data table.
6. Measure the time the pulse is high (t_{high}) for one cycle and record in the data table.
7. Measure the time the pulse is low (t_{low}) for one cycle and record in the data table.
8. Sketch an approximate representation of the output voltage waveform in the data table.
9. Repeat steps 5 through 8 for the remaining values of V_{mod} given in the data table.
10. Construct the circuit shown in Figure 12-13c.
11. Connect the oscilloscope as given in step 2, and set the time/cm switch for 0.2 ms/cm or 0.5 ms/cm.
12. Adjust the sine-wave generator for 300 Hz at 7 V p-p.

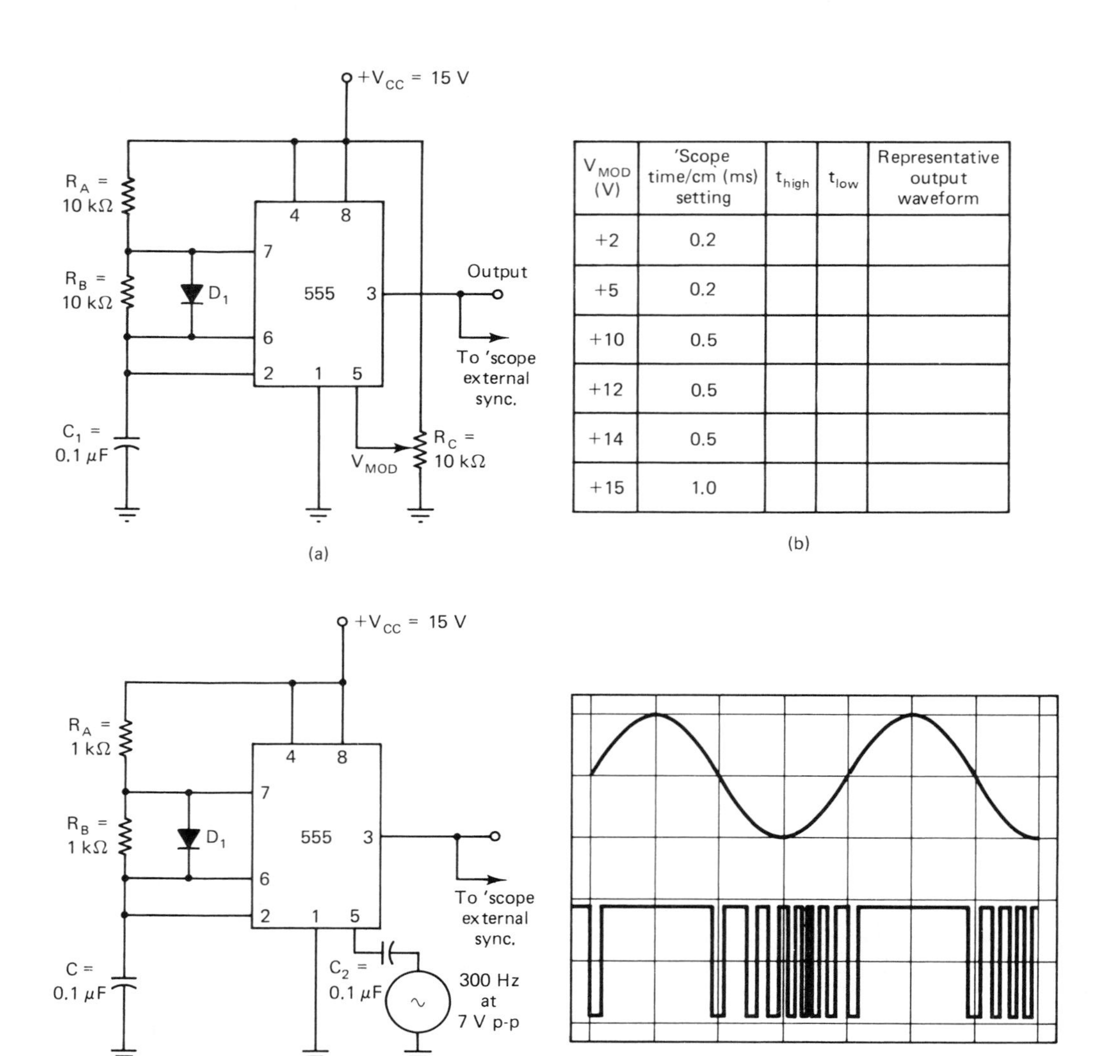

V_{MOD} (V)	'Scope time/cm (ms) setting	t_{high}	t_{low}	Representative output waveform
+2	0.2			
+5	0.2			
+10	0.5			
+12	0.5			
+14	0.5			
+15	1.0			

Figure 12–13 (a) pulse-width modulation; (b) data table; (c) pulse position; (d) waveform graph.

13. Compare the oscilloscope display with the waveform graph shown in Figure 12-13d. Notice that on the positive peaks of the sine wave the output pulse is wider, because the threshold voltage is higher. The negative peak of the sine wave causes the threshold voltage to be less, and the 555 timer has time to complete more cycles.

Fill-in Questions:

1. When the voltage on the control voltage terminal of a 555 timer becomes more positive, the threshold voltage is ______________ and the output pulses are ______________ .

2. When the voltage on the control voltage terminal of a 555 timer becomes less positive, the threshold voltage is ______________ and the output pulses become ______________ .

3. The control voltage terminal of a 555 timer can serve as a ______________ frequency adjust when the circuit is an astable multivibrator.

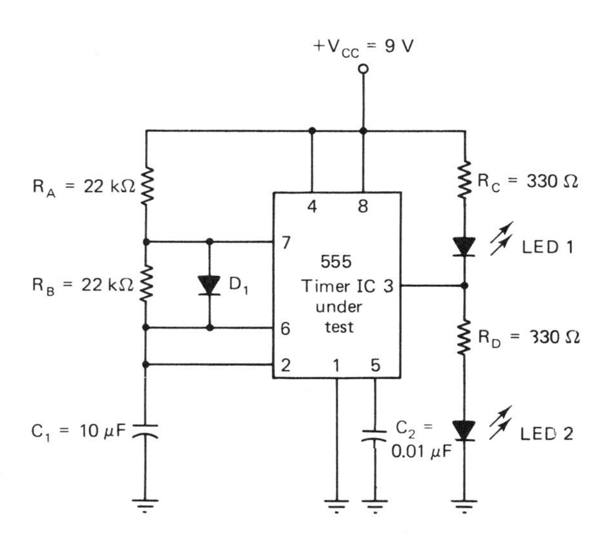

Figure 12–14 555 timer IC tester.

EXPERIMENT 6. 555 TIMER IC TESTER

Objective:

To show a simple circuit for testing the 555 timer IC.

Introduction:

The circuit shown in Figure 12-14 is a 555 astable multivibrator. Its output frequency is about 3 Hz. When the output is low, LED_1 lights, and when the output is high, LED_2 lights. This is only a functional test indicating that the output can turn on and off, which is usually sufficient to prove that the 555 timer under test is functioning. It does not indicate that some of the components within the IC might have changed value, which could affect its precision operation (only an exact replacement in an actual circuit with another 555 timer could determine this). This tester could be mounted on a small perforated board using an IC socket and a 9-V radio battery.

Materials Needed:

1 +9-V power supply
1 555 timer IC (for testing)
2 LEDs at V_F = 2 V (LED_1, LED_2)
1 1N4002 diode or equivalent (D_1)
2 330-Ω resistors at 0.5 W (R_C, R_D)
2 22-kΩ resistors at 0.5 W (R_A, R_B)
1 10-μF capacitor at 25 WV dc (C_1)
1 0.01-μF capacitor at 25 WV dc (C_2)
1 Breadboard for constructing circuit

Procedure:

1. Construct the circuit shown in Figure 12-14.
2. Observe the action of the two LEDs.

Fill-in Questions:

1. When LED_1 is on, the output of the 555 timer is ______________ .

2. When LED_2 is on, the output of the 555 timer is ______________ .

SECTION 12-5 555 TIMER BASIC TROUBLESHOOTING APPLICATION: 555 TIMER ASTABLE MULTIVIBRATOR

Construct the circuit shown below. Open or short the components as listed and record the voltages in the proper place in the table. The abbreviations will help indicate the voltage conditions associated with each problem. All voltages are referenced to ground.

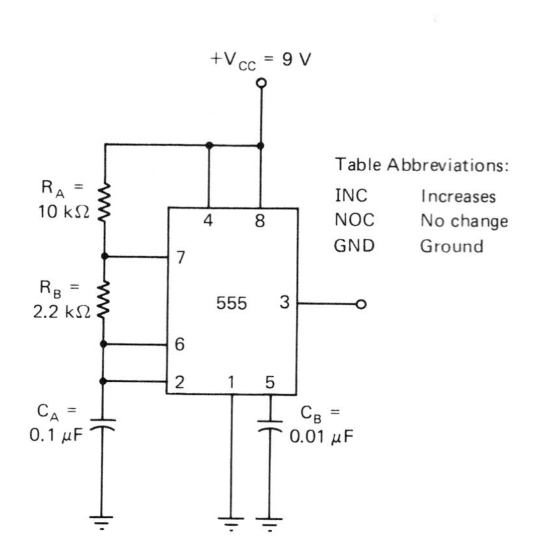

Condition	V_{out} (dc)	V_{out} (signal)	Comments
Normal			All voltages are proper
R_A open	INC		No charge path for C_A; output saturates
R_B open	INC		No charge path for C_A; output saturates
C_A open	INC		Wiring capacitance causes high frequency (≈ 200 kHz)
C_A shorted	INC		No charging; output saturates
C_B open	NOC		No apparent trouble; check for noise or 60-Hz modulation
C_B shorted	GND		2/3 $+V_{CC}$ reference voltage shorted to ground

Figure 12–E.1

SECTION 12-6 MORE EXERCISES

Perform the following exercises before beginning Section 12-7.

1. Match the terminal function of a 555 timer in column A with its proper description given in column B.
Pin

1. Ground ____

2. Trigger ____

3. Output ____

4. Reset ____

5. Control voltage ____

6. Threshold ____

7. Discharge ____

8. $+V_{CC}$ ____

a. Consists of a totem-pole circuit arrangement.
b. The output goes low when the voltage on this terminal reaches two-thirds of $+V_{CC}$.
c. Initiates the timing cycle.
d. The highest potential power supply connection.
e. Used for pulse-width modulation.
f. The lowest potential power supply connection.
g. Provides a discharge path for the capacitor.
h. Connected directly to flip-flop that controls output.

2. Draw the 555 one-shot multivibrator circuit from Figure 12-10a and calculate the width of the output pulse for the circuit given the various values of R and C. Use the formula $T = 1.1RC$.

a. $R = 1\ k\Omega$
$C = 0.01\ \mu F$

$T =$ ______ ms

b. $R = 3.3\ k\Omega$
$C = 0.02\ \mu F$

$T =$ ______ ms

c. $R = 10\ k\Omega$
$C = 0.01\ \mu F$

$T =$ ______ ms

d. $R = 56\ k\Omega$
$C = 0.033\ \mu F$

$T =$ ______ ms

3. Draw the 555 astable multivibrator from Figure 12-11a (without the diode) and calculate the output frequency, the time the output is low, the time the output is high, the total time for one cycle, and the duty cycle for the various values given. Use the formulas $f = 1.49/(R_A + 2R_B)C$, $t_{low} = 0.693(R_B)C$, $t_{high} = 0.693(R_A + R_B)C$, $T_{(total)} = t_{low} + t_{high}$ or $0.693(R_A + 2R_B)C$, and $DC = R_B/R_A + 2R_B$.

a. $R_A = 1\ k\Omega$
$R_B = 1\ k\Omega$
$C = 0.1\ \mu F$

$f =$ ______ Hz

$t_{low} =$ ______

$t_{high} =$ ______

$T =$ ______

DC = ______

b. $R_A = 10\ k\Omega$
$R_B = 2.2\ k\Omega$
$C = 0.01\ \mu F$

$f =$ ______ Hz

$t_{low} =$ ______

$t_{high} =$ ______

$T =$ ______

DC = ______

c. $R_A = 56\ k\Omega$
$R_B = 33\ k\Omega$
$C = 0.02\ \mu F$

$f =$ ______ Hz

$t_{low} =$ ______

$t_{high} =$ ______

$T =$ ______

DC = ______

d. $R_A = 33\ k\Omega$
$R_B = 56\ k\Omega$
$C = 0.02\ \mu F$

$f =$ ______ Hz

$t_{low} =$ ______

$t_{high} =$ ______

$T =$ ______

DC = ______

4. List the correct 555 timer circuit for each input and output signal shown below.

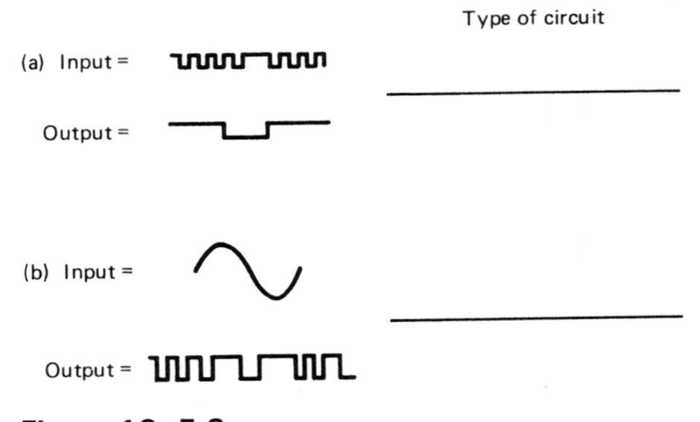

Figure 12–E.2

SECTION 12-7 555 TIMER INSTANT REVIEW

The 555 timer IC contains a group of circuits used for precision timing in time-delay circuits and square-edge oscillators, with the minimum of external components.

SECTION 12-8 SELF-CHECKING QUIZZES

12-8a 555 TIMER: TRUE–FALSE QUIZ

Place a T for true or an F for false to the left of each statement.

_____ **1.** The 555 timer can be used as an amplifier.

_____ **2.** Voltage comparators are a necessary part of the 555 timer.

_____ **3.** The reset terminal is connected directly to the flip-flop inside the 555 timer.

_____ **4.** A positive-going pulse to the trigger terminal of a 555 timer initiates a timing cycle.

_____ **5.** The output of a 555 timer can be high ($\approx +V_{cc}$) or low (≈ 0 V), but not between these limits.

_____ **6.** The discharge terminal permits an external capacitor to discharge through a transistor inside the 555 timer.

_____ **7.** The three equal resistors inside the 555 timer bias one comparator at two-thirds of $+V_{cc}$ and the other comparator at one-third of $+V_{cc}$.

_____ **8.** The reset terminal of a 555 timer is used to discharge the external capacitor when the circuit is an astable multivibrator.

_____ **9.** The control voltage terminal can be used to bypass noise and power supply variations to ground.

_____ **10.** When the voltage rises above two-thirds of $+V_{cc}$ on the threshold terminal of a 555 timer, the output goes high.

12-8b 555 TIMER: MULTIPLE-CHOICE QUIZ

Circle the correct answer for each question.

1. In a 555 monostable multivibrator $R_A = 68$ kΩ and $C_1 = 0.1$ μF; therefore, the timing cycle equals:

a. 6.8 ms **b.** 7.48 s

c. 7.48 ms **d.** 0.134 ms

Questions 2 and 3 refer to Figure 12-A.

2. The f_{out} is about:

a. 7 kHz

b. 8 kHz

c. 10.6 kHz

d. none of the above

3. The duty cycle is about:

a. 30%

b. 68%

c. 24%

d. none of the above

4. The control voltage terminal on a 555 timer can be used to:

a. bypass noise and power supply variation to ground

b. fine-adjust the f_{out} of an astable multivibrator

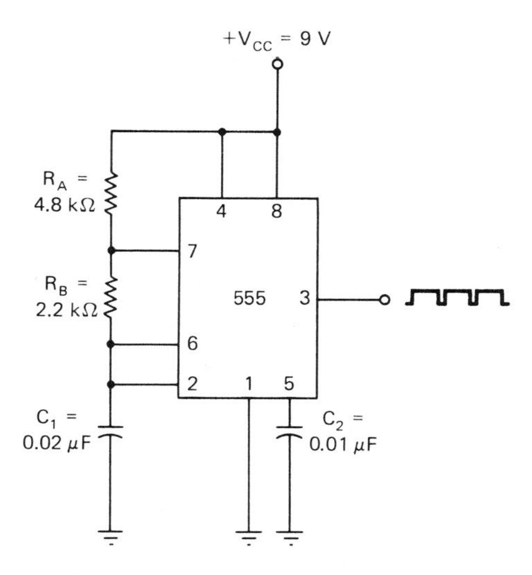

Figure 12–A

c. modulate the output pulse width

d. all of the above

e. none of the above

5. Which one of the following circuits is *not* suitable for the 555 timer?

a. astable multivibrator

b. voltage follower

c. missing-pulse detector

d. pulse-width modulator

e. monostable multivibrator

ANSWERS TO EXPERIMENTS AND QUIZZES

Experiment 1. **(1)** negative-going, two **(2)** low

Experiment 2. **(1)** R_A, C_1 **(2)** interfere **(3)** reshape a pulse, stretch a pulse, bounceless switch

Experiment 3. **(1)** R_A, R_B, C_1 **(2)** decreases **(3)** increases **(4)** unsymmetrical, less **(5)** symmetrical, 50 **(6)** 5, 95

Experiment 4. **(1)** high **(2)** low **(3)** longer

Experiment 5. **(1)** higher, wider **(2)** less, narrower **(3)** fine

Experiment 6. **(1)** low **(2)** high

True–False. **(1)** F **(2)** T **(3)** T **(4)** F **(5)** T **(6)** T **(7)** T **(8)** F **(9)** T **(10)** F

Multiple Choice. **(1)** c **(2)** b **(3)** c **(4)** d **(5)** b

Basic Electronics Final Examination

Circle the most correct answer for each question.

1. The majority carriers in *P*-type semiconductor material are:

   a. holes. **b.** electrons.

   c. ions. **d.** neutrons.

2. Valence electrons:

   a. are found closest to the nucleus of an atom.

   b. indicate a balanced condition of the atom.

   c. can easily become free electrons.

   d. none of the above.

3. The conduction diagram for a conductor:

   a. has a large forbidden region.

   b. has a moderate forbidden region.

   c. has a narrow forbidden region.

   d. does not have a forbidden region.

4. Valence electrons shared by atoms are called:

   a. electron-pair combinations.

   b. covalent bonding.

   c. ionization.

   d. all of the above.

5. The process of adding impurities to silicon or germanium to form *N*-type and *P*-type semiconductor material is called:

   a. crystallizing. **b.** doping.

   c. bonding. **d.** mixing.

6. Silicon doped with donor atom material produces:

   a. *n*-type material. **b.** *p*-type material.

   c. ionization. **d.** neutrons.

7. The *P–N* junction shown in Figure T-1 is:

   a. forward biased.

   b. reverse biased.

   c. neutral biased.

   d. none of the above.

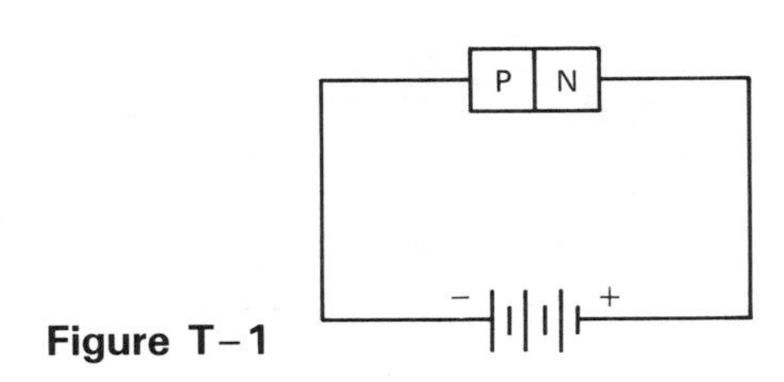

Figure T–1

8. Which of the following is *not* true of a forward-biased diode that is in saturation?

   a. Holes flow toward the junction.

   b. Electrons flow toward the junction.

   c. Its forward resistance is infinite.

   d. The voltage drop across it is about 0.7 V.

9. The band on one end of a diode indicates its:

   a. voltage rating. b. temperature rating.

   c. cathode. d. anode.

10. The ohmmeter shown in Figure T-2 should read:

    a. low ohms.

    b. very high ohms.

    c. infinite ohms.

    d. none of the above.

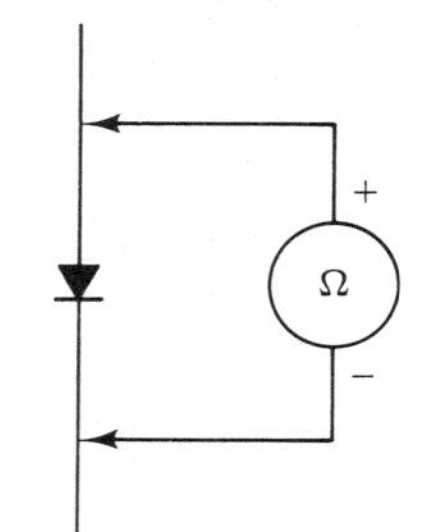

Figure T–2

11. If a diode has −2.5 V on its anode and −3.2 V on its cathode, it is:

    a. conducting.

    b. not conducting.

    c. cut off.

    d. cannot be determined.

12. When a diode is reverse biased, it will have:

    a. low resistance.

    b. a voltage drop of 0.7 V.

    c. maximum power loss.

    d. a leakage current.

13. The voltmeter in Figure T-3 should read:

    a. + 0.7 V. b. + 1.4 V.

    c. + 2.1 V. d. + 10 V.

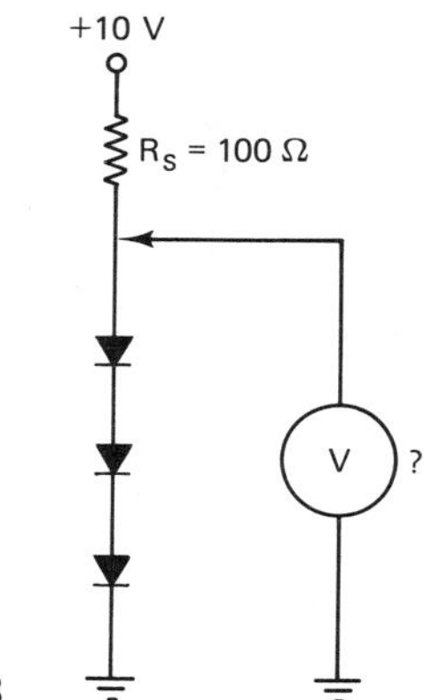

Figure T–3

14. The output voltage waveform for the circuit shown in Figure T-4 is: a. b. c. d.

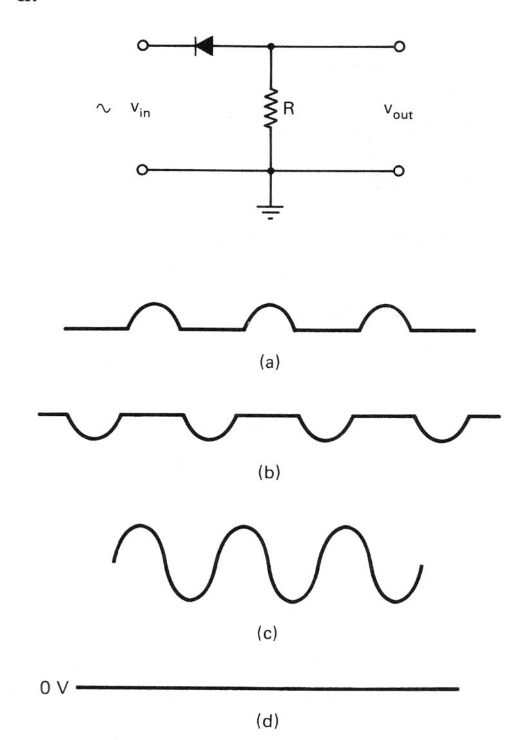

Figure T–4

15. If the supply voltage to a zener-diode regulator circuit increases, the output voltage (V_Z) will:

    a. increase.

    b. remain fairly constant.

    c. decrease.

    d. none of the above.

16. If the load current decreases in a zener-diode regulator circuit, the zener current:

a. remains the same.

b. increases.

c. decreases.

d. none of the above.

17. Referring to Figure T-5, the value of R_S is:

a. 300. b. 115.

c. 75. d. 60.

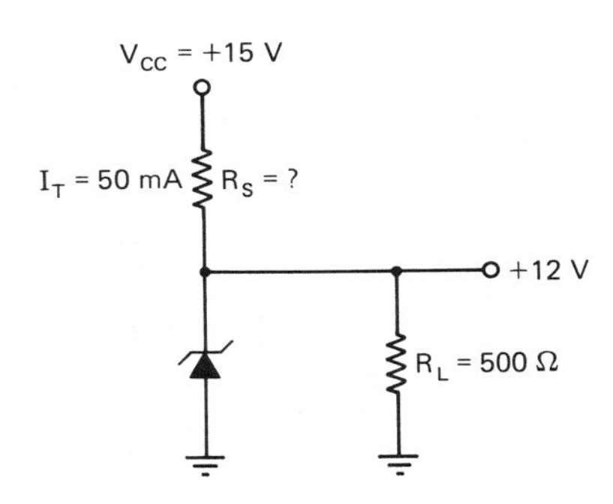

Figure T–5

18. Referring to Figure T-5, the value of I_Z is:

a. 50 mA. b. 26 mA.

c. 24 mA. d. 20 mA.

19. Referring to Figure T-6, V_{out} equals: a. b. c. d.

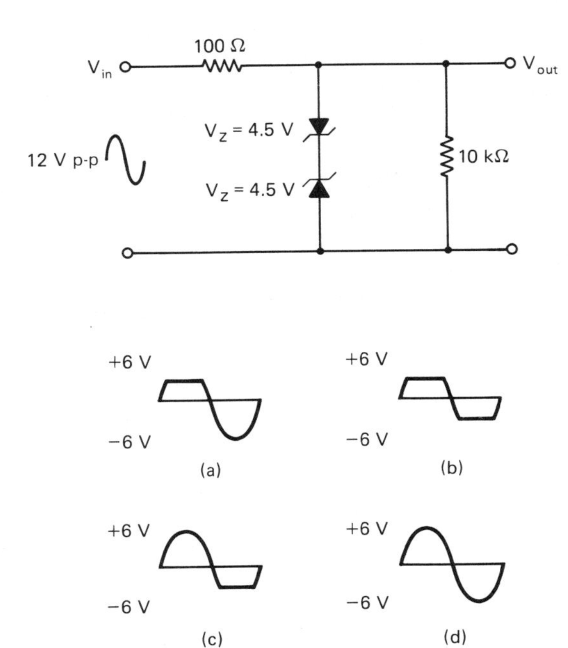

Figure T–6

20. To operate properly, an *NPN* transistor must have junctions:

a. EB forward biased and CB reverse biased.

b. EB reverse biased and CB forward biased.

c. EB forward biased and CB forward biased.

d. EB reverse biased and CB reverse biased.

21. If a bipolar transistor with a beta of 150 has an I_B of 30 μA, its I_C will be:

a. 2.5 mA. b. 3.75 mA.

c. 4.5 mA. d. 5.0 mA.

22. When I_B increases in a bipolar transistor:

a. I_C remains the same.

b. I_C decreases.

c. I_C increases.

d. V_E decreases.

23. A transistor is similar to:

a. an inductor.

b. a capacitor.

c. a variable resistor.

d. a transformer.

24. Referring to Figure T-7, the I_C of the circuit will be about:

a. 1.5 mA. b. 3.0 mA.

c. 4.3 mA. d. 5.45 mA.

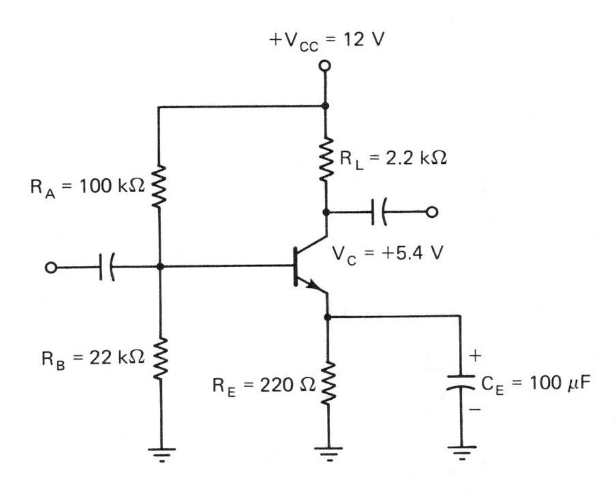

Figure T–7

25. Referring to Figure T-7, V_B will be about:

a. 1.9 V. **b.** 2.1 mA.
c. 0.6 V. **d.** 1.3 V.

26. Referring to Figure T-7, when the input voltage goes positive, the output voltage:

a. goes positive. **b.** goes negative.
c. remains fixed. **d.** none of the above.

27. The reason for the capacitor in the emitter circuit is that it:

a. provides positive feedback.
b. reduces the dc current flow in the transistor.
c. provides negative feedback for frequency control.
d. reduces negative feedback and provides stability.

28. The term I_{CO} refers to leakage current between:

a. emitter–base.
b. collector–base.
c. emitter–collector.
d. all of the above.

29. Measurements are made on a class A amplifier. With a V_{CC} of +12 V, the voltage readings shown in Figure T-8 indicate that the transistor is:

a. cut off. **b.** saturated.
c. operating normally. **d.** none of the above.

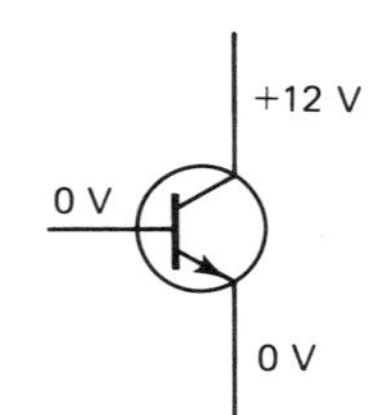

Figure T–8

30. Measurements are made on another class A amplifier. With a V_{CC} of +12 V, the voltage readings shown in Figure T-9 indicate that the transistor is:

a. cut off. **b.** saturated.
c. operating normally. **d.** none of the above.

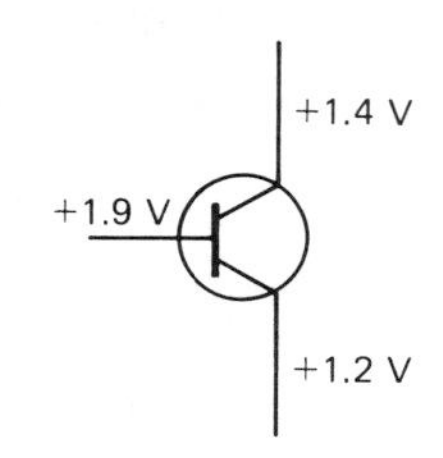

Figure T–9

31. The proper relationship for current through a bipolar transistor is:

a. $I_C = I_E + I_B$. **b.** $I_B = I_C + I_E$.
c. $I_C = I_E - I_B$. **d.** $I_E = I_C - I_B$.

32. The circuit that has the most voltage gain is the:

a. common-base circuit.
b. common-emitter circuit.
c. common-collector circuit.
d. emitter-follower circuit.

33. Beta is expressed as β =:

a. I_C/I_B. **b.** I_B/I_E.
c. I_C/I_E. **d.** I_E/I_B.

34. A common-emitter amplifier has an input voltage of 0.05 V p-p and an output voltage of 3 V p-p. The voltage gain of the amplifier is:

a. 1.5. **b.** 30.
c. 60. **d.** none of the above.

35. A transistor circuit with a voltage gain of less than 1 is the:

a. common-collector amplifier.
b. common-base amplifier.
c. common-emitter amplifier.
d. none of the above.

36. The stability of a transistor can be affected by:

a. sound. **b.** light.
c. heat. **d.** all of the above.

37. A *PNP* transistor has −9 V on its collector, −0.6 V on its base, and 0 V on its emitter. The transistor is:

a. properly biased.

b. improperly biased.

c. partially biased.

d. none of the above.

38. The ohmmeter is Figure T-10 indicates that the transistor is:

a. normal. b. shorted.

c. open. d. none of the above.

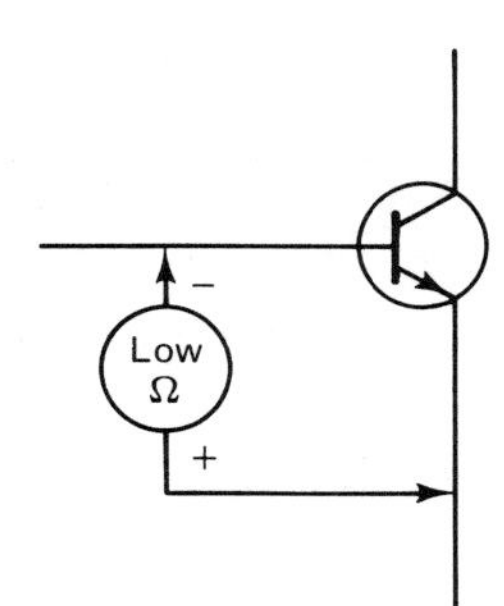

Figure T−10

39. A bipolar transistor is a:

a. voltage-operated device.

b. current-operated device.

c. resistance-operated device.

d. all of the above.

40. The current flowing through a JFET is controlled by:

a. V_{GS}. b. V_{DS}.

c. the gate current. d. answers a and c.

41. I_{DSS}:

a. occurs when $V_{GS} = 0$ V.

b. is called saturation current.

c. occurs when the gate and source are shorted.

d. all of the above.

42. The JFET is a:

a. bipolar device.

b. unipolar device.

c. resistance device.

d. none of the above.

43. A JFET operates in the:

a. depletion region.

b. enhancement region.

c. both regions.

d. none of the above.

44. The JFET is a:

a. normally on device.

b. normally off device.

c. both a and b.

d. none of the above.

45. The ohmmeter shown in Figure T-11 indicates the JFET is:

a. normal. b. open.

c. shorted. d. difficult to determine.

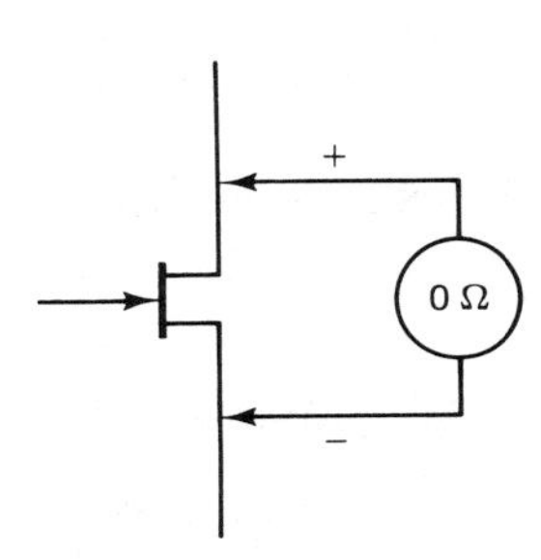

Figure T−11

46. Referring to Figure T-12, the bias for this circuit is developed by:

a. R_D. b. R_G.

c. R_S. d. C_S.

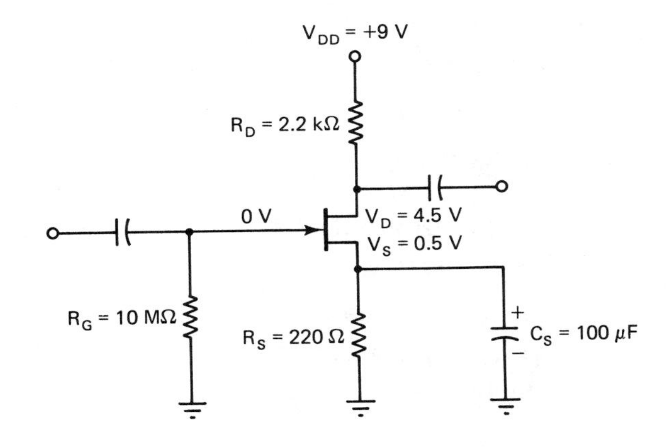

Figure T−12

47. Referring to Figure T-12, when the input signal goes negative:

a. V_D goes negative.

b. V_S goes positive.

c. V_G remains at 0 V.

d. I_D decreases.

48. Referring to Figure T-12, if $V_D = +9$ V, the probable cause could be:

a. R_D open. b. R_G shorted.

c. R_S shorted. d. R_S open.

49. The main advantage of FETs over bipolar transistors is:

a. less operating voltage required.

b. less amplitude distortion.

c. a higher input resistance.

d. easier to solder.

50. The insulating part of a MOSFET gate is the:

a. *n*-type region.

b. *p*-type region.

c. silicon dioxide.

d. metal gate.

51. A depletion-type MOSFET can be biased:

a. with positive voltage.

b. with negative voltage.

c. at zero volts.

d. all of the above.

52. An enhancement-type MOSFET is a:

a. normally on device.

b. normally off device.

c. current-operated device.

d. answers b and c.

53. Care should be used in handling MOSFETs because:

a. they break easily when dropped.

b. they should be soldered with a soldering gun only.

c. static charges can rupture the metal oxide gate.

d. all of the above.

54. The meter in Figure T-13 indicates that the MOSFET is:

a. normal. b. needs repairing.

c. open. d. shorted.

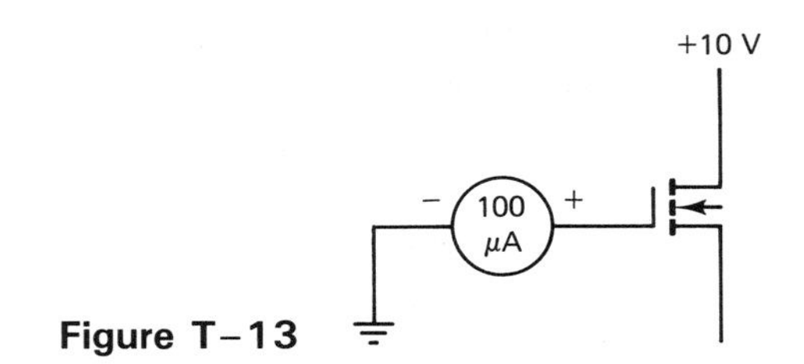

Figure T-13

55. When a MOSFET is in the pinch-off region:

a. no current flows through it.

b. it cannot be used as an amplifier.

c. it is in the normal operating region.

d. answers a and b.

56. Referring to Figure T-14, V_G is:

a. 0 V. b. 3.3 V.

c. 5.5 V. d. 6.3 V.

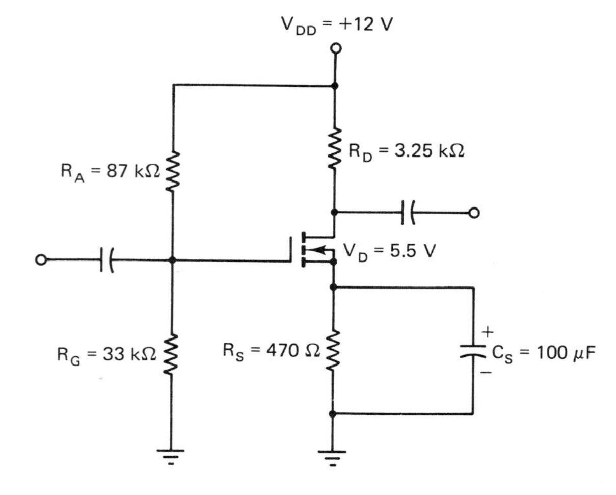

Figure T-14

57. Referring to Figure T-14, I_D is:

a. 2 mA. b. 3.3 mA.

c. 5.5 mA. d. 11.7 mA.

58. Referring to Figure T-14, if $V_D = +1.9$ V, $V_G = +12$ V, and $V_S = +1.5$ V, the probable cause could be:

a. R_D open. b. R_S open.

c. R_A open. d. R_G open.

59. Referring to Figure T-14, if $V_D = +12$ V, $V_G = 0$ V, and $V_S = 0$ V, the probable cause could be:

a. R_A open.
b. C_S shorted.
c. R_D open.
d. R_G open.

60. A UJT turns on (fires) when V_E exceeds:

a. V_{B1}.
b. V_P.
c. V_{B2}.
d. V_v.

61. When fired, the current through a UJT increases because:

a. the capacitor charges.
b. V_E is greater than V_{B2}.
c. the resistance between E and B_1 decreases.
d. the resistance between E and B_2 decreases.

62. Referring to Figure T-15, the output frequency is primarily dependent on:

a. R_E and C_E with a frequency of about 30 kHz.
b. R_2 and C_E with a frequency of about 330 Hz.
c. R_1 and C_E with a frequency of about 30 kHz.
d. R_E and C_E with a frequency of about 3.3 kHz.

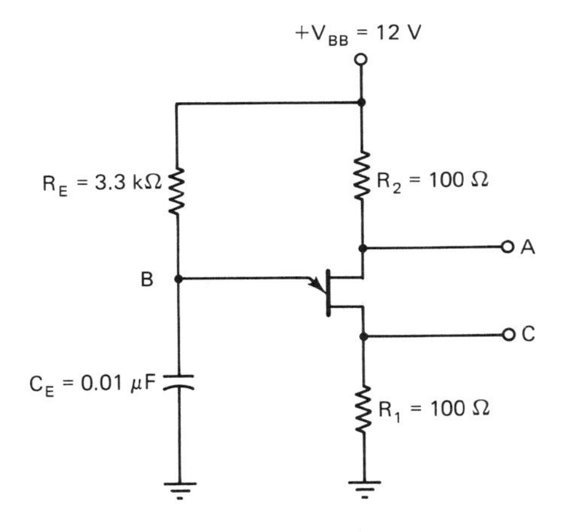

Figure T-15

63. Referring to Figure T-15, the voltage waveform at point C would resemble a:

a. negative-going pulse train.
b. positive-going pulse train.
c. sawtooth pulse train.
d. none of the above.

64. Referring to Figure T-15, the voltage waveform at point B would resemble a:

a. ramp-voltage pulse train.
b. negative-going pulse train.
c. positive-going pulse train.
d. sawtooth pulse train.

65. A PUT:

a. is a thyristor.
b. can be turned on with a negative-going pulse applied at the gate.
c. can be turned on if the anode voltage is made more positive than the gate voltage.
d. all of the above.

66. The PUT is programmable depending on the:

a. exact manufacturer's identification number.
b. values of the gate resistors selected.
c. amount of current required by the circuit.
d. answers b and c.

67. An SCR can be triggered on by:

a. exceeding its breakover voltage.
b. limiting its holding current.
c. a positive pulse to the gate.
d. answers a and c.

68. Once triggered on, an SCR can be turned off in a dc circuit when:

a. a negative voltage is applied to the gate.
b. the positive voltage is removed from the gate.
c. a positive voltage is again applied to the gate.
d. the holding current falls below the minimum point.

69. When an SCR is used with ac voltage and is triggered into operation, it conducts:

a. on neither alternation of the cycle because it is used only with dc circuits.

b. for only the positive alternation of the cycle.

c. for only the negative alternation of the cycle.

d. for the entire cycle.

70. Referring to Figure T-16, the trigger time capable with this circuit is approximately:

a. 0°–90°. **b.** 0°–180°.

c. 90°–270°. **d.** 0°–360°.

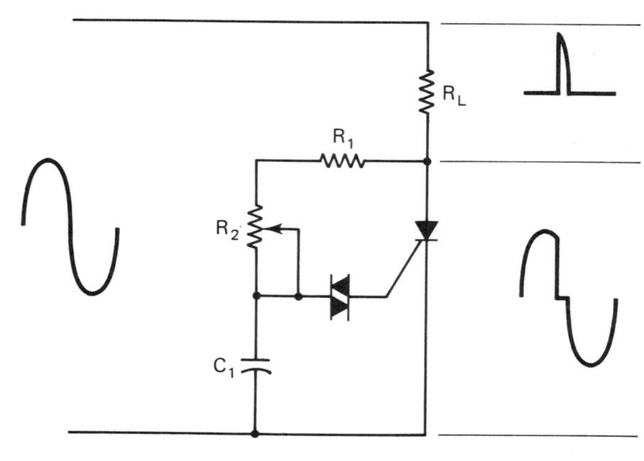

Figure T-16

71. Referring to Figure T-16, and the voltage waveforms, the SCR is triggered and begins conducting:

a. before 45°.

b. between 45° and 90°.

c. after 90°.

d. none of the above.

72. The circuit shown in Figure T-17 is an overvoltage protection circuit. The voltage at the gate of the SCR is 0.45 V and the SCR is not conducting. If the input voltage developed a 10-V spike momentarily, the reaction of the circuit would be:

a. nothing and the SCR would remain off.

b. that the SCR turns on and the reset switch opens.

c. that the SCR turns on, absorbs the voltage spike, and then turns off.

d. that the SCR turns on, shunting current away from the load, and remains in this condition until the reset switch is pressed.

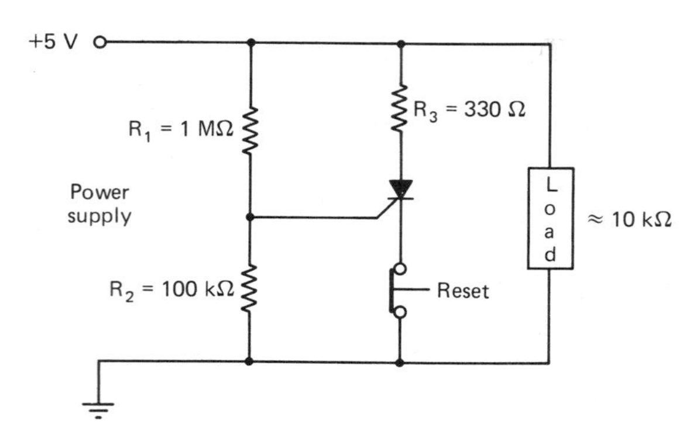

Figure T-17

73. The DIAC is similar to:

a. two zener diodes connected back to back.

b. two SCRs connected face to face.

c. an SCR and a zener diode connected together.

d. none of the above.

74. The DIAC can be used to:

a. trigger other thyristors.

b. limit voltage peaks.

c. neither answer a nor b.

d. answers a and b.

75. Thyristors are primarily used as:

a. impedance-matching devices.

b. solid-state switches.

c. amplifiers.

d. all of the above.

76. A TRIAC:

a. is like a DIAC, but has a gate lead.

b. can conduct in both directions.

c. must be triggered by a positive-going pulse.

d. answers a and b.

77. Referring to Figure T-18, a possible voltage waveform that might be across a TRIAC is: a. b. c. d.

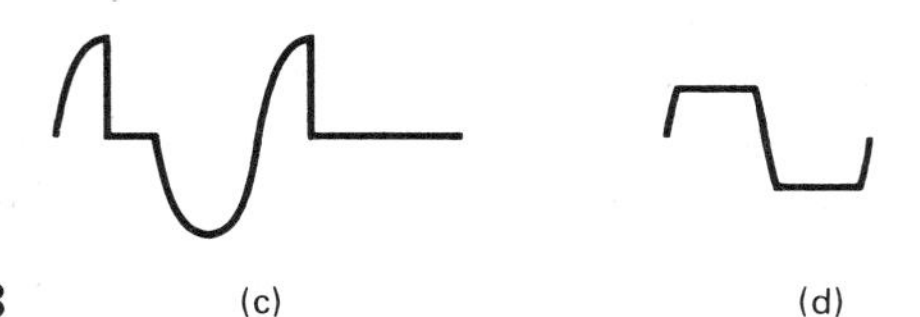

Figure T-18

78. The thyristor that *can not* be tested with an ohmmeter is the:

a. TRIAC. **b.** SCR.
c. DIAC. **d.** PUT.

79. Particles of light are called:

a. photons. **b.** ions.
c. neutrons. **d.** litrons.

80. Which one of the following devices is *not* a photodetector?

a. phototransistor **b.** LASCR
c. LED **d.** photoresistor

81. Referring to Figure T-19, the value of R_S is:

a. 100. **b.** 200.
c. 500. **d.** 600.

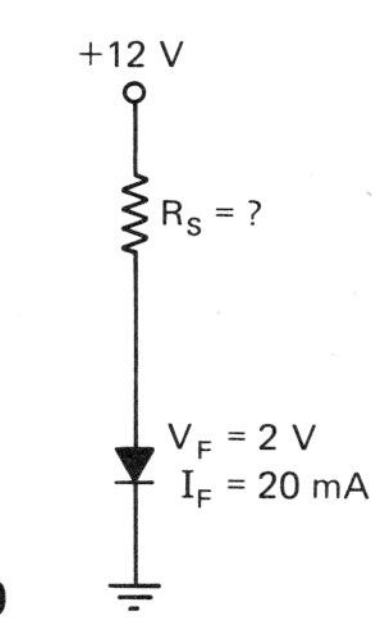

Figure T-19

82. Each solar cell produces 0.25 V with 50 μA of current under full sunlight. There are there parallel branches of these cells with four cells in each branch. The total voltage and current capability is:

a. 0.25 V at 200 μA.
b. 1 V at 150 μA.
c. 0.75 V at 200 μA.
d. 1.5 V at 100 μA.

83. Which of the following devices *can not* be tested with an ohmmeter?

a. photoresistor **b.** phototransistor
c. LASCR **d.** none of the above

84. Referring to Figure T-20, when light strikes the phototransistor, the:

a. LED turns off, the LASCR turns on, and L_1 goes on.
b. LED turns on, the LASCR turns on, and L_1 goes on.
c. LED turns on, the LASCR turns off, and L_1 goes off.
d. LED turns off, the LASCR turns on, and L_1 goes off.

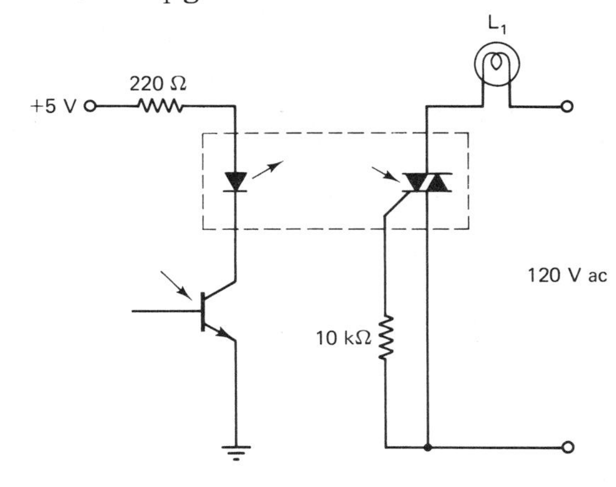

Figure T-20

85. A standard op amp has:

a. low input impedance, high gain, and high output impedance.
b. low input impedance, high gain, and low output impedance.
c. high input impedance, high gain, and low output impedance.
d. high input impedance, low gain, and high output impedance.

86. Referring to Figure T-21, when V_{in} = +5.5 V, V_{out} is:

a. −1.375 V. **b.** +9.1 V.
c. +13.5 V. **d.** none of the above.

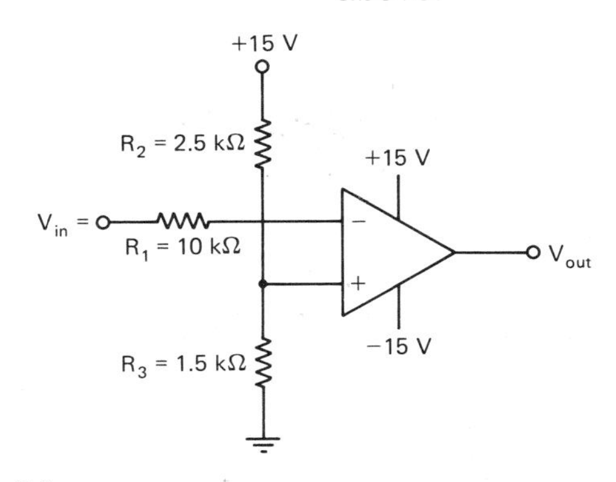

Figure T-21

87. Referring to Figure T-21, when V_{in} = + 6.2 V, V_{out} is:

a. +1.55 V. **b.** −13.5 V.
c. −8.25 V. **d.** none of the above.

88. If a signal of the same amplitude and phase angle is applied to both inputs of an op amp simultaneously and the output is nearly zero, the op amp is said to have a:

a. high common-mode rejection ratio.
b. very low gain.
c. high slew rate.
d. narrow bandwidth.

89. Error voltages at the output of an op amp can be corrected by:

a. offset nulling.
b. increasing the slew rate.
c. decreasing the common-mode rejection ratio.
d. all of the above.

90. The type of feedback generally applied to an op amp to control gain and stability is:

a. positive feedback.
b. regenerative feedback.
c. positive and regenerative feedback.
d. negative feedback.

91. Referring to Figure T-22, V_{out} is:

a. +2 V. **b.** −4 V.
c. +4 V. **d.** −8 V.

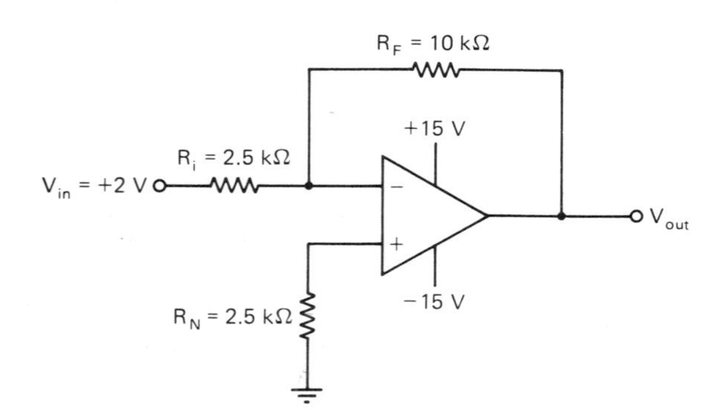

Figure T–22

92. Referring to Figure T-23, V_{out} is:

a. +2 V. **b.** −2 V.
c. +8 V. **d.** −8 V.

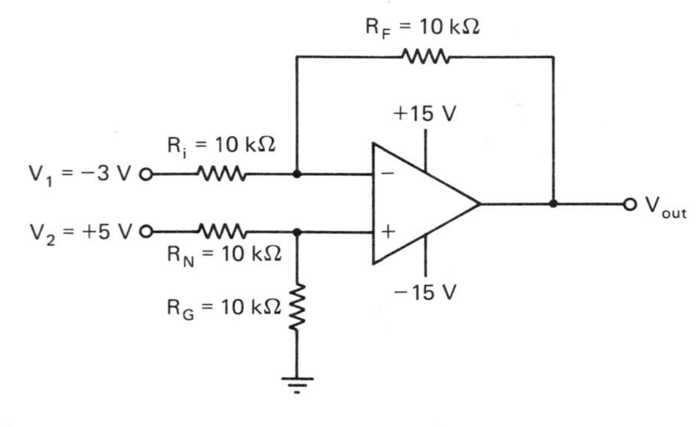

Figure T–23

93. Referring to Figure T-24, V_{out} is:

a. +4 V. **b.** −4 V.
c. +8 V. **d.** −5 V.

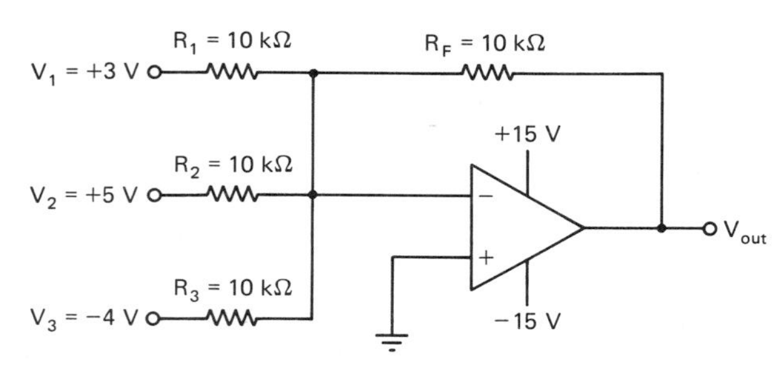

Figure T–24

94. The circuit shown in Figure T-25 is a:

a. low-pass filter.
b. high-pass filter.
c. bandpass filter.
d. notch filter.

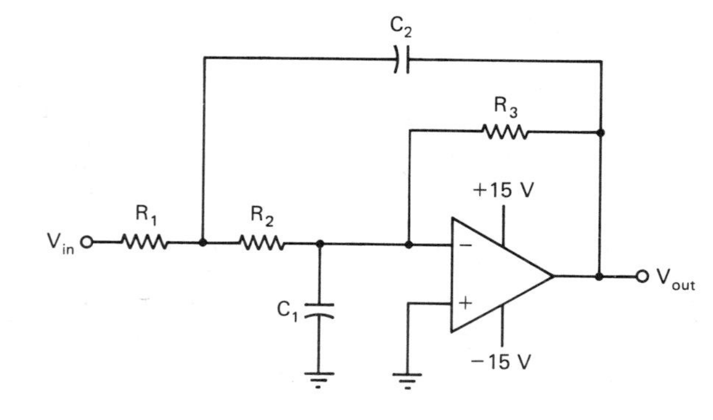

Figure T–25

95. Some of the internal circuits of a 555 timer IC are:

a. op-amp comparator.

b. flip-flop.

c. discharge transistor.

d. all of the above.

96. The internal reference voltages for the 555 timer IC are developed by a:

a. capacitor.

b. flip-flop.

c. voltage comparator.

d. three-resistor voltage divider.

97. When the discharge transistor inside the 555 timer is conducting, the output of the IC is:

a. high. **b.** low.

c. between high and low. **d.** none of the above.

98. The 555 timer is *not* used as:

a. an oscillator.

b. a one-shot multivibrator.

c. an audio amplifier.

d. a missing pulse detector.

99. Referring to Figure T-26, when Q_{24} is conducting:

a. LED_1 is on. **b.** LED_2 is on.

c. both LEDS are on. **d.** both LEDS are off.

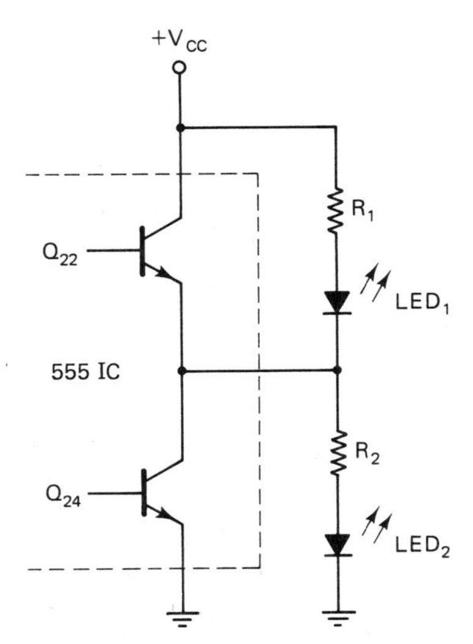

Figure T-26

100. Referring to Figure T-27, the components that determine the output frequency are:

a. R_A and C_2. **b.** R_A, R_B, and C_2.

c. R_A, R_B, and C_1. **d.** none of the above.

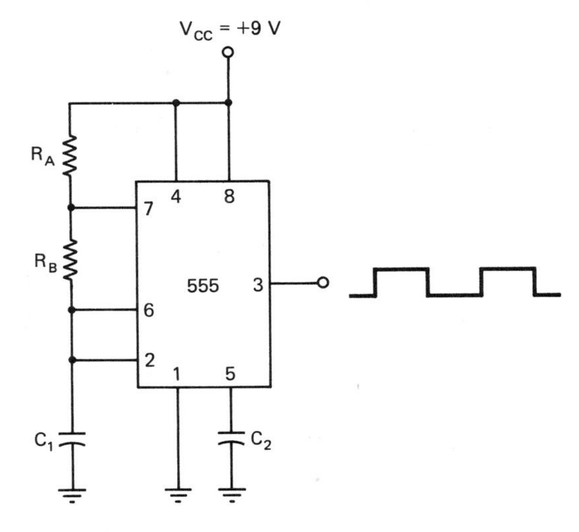

Figure T-27

Appendix I

Component Parts List for Experiments

Active devices	*Quantity*	*Part number*
Diode	4	1N4001 or equivalent
Zener diode	2	1N5231 with V_z = 5.1 V or equivalent
Zener diode	1	1N960 with V_z = 9.1 V or equivalent
NPN transistor	1	2N2222 or equivalent
Power transistor	1	2N3055 or equivalent
PNP transistor	1	2N3906 or equivalent
N-channel JFET	1	2N3823 or equivalent
N-channel MOSFET	1	40673 or equivalent
V-MOSFET	1	VN10KM or VN67AF or equivalent
UJT	1	2N2646 or equivalent
PUT	1	2N6027 or equivalent
SCR	1	C106Y1 or MCR 106-2 (Motorola)
DIAC	1	R.S. 276-1050 (Radio Shack) or equivalent with V_{BO} = 40 V
TRIAC	1	2N5754 or equivalent
LED	2	With a $V_F \approx 2$ V
Photoresistor	1	R.S. 276-116 (Radio Shack) or equivalent
Photodetectors	–	Photodiode, phototransistor, photofet, LASCR, Photodarlington transistor, solar cell
Photocouplers	–	LED/phototransistor, LED/darlington, LED/photoresistor, LED/LASCR
OP-amp IC	1	741 or equivalent
Precision timer IC	1	555 or equivalent

Switches	*Quantity*
SPST	1
SPDT	2
DPST	2
TPST	1

Miscellaneous	*Quantity*
8-Ω loudspeaker	1
12-V relay at less than 1 A	1
6.3-V center-tapped transformer	1
12-V center-tapped transformer	1
24-V center-tapped transformer	1
1.5-V battery	1

Resistors (all values at 0.5 W)	*Quantity*
10 Ω	1
100 Ω	2
220 Ω	1
330 Ω	2
470 Ω	1
500 Ω	1
1 kΩ	2
1.2 kΩ	1
2.2 kΩ	1
3.3 kΩ	1
4.7 kΩ	1
5.6 kΩ	1
6.8 kΩ	1
10 kΩ	9
22 kΩ	2
33 kΩ	1
47 kΩ	1
56 kΩ	1
68 kΩ	1
86 kΩ	1
100 kΩ	1
220 kΩ	1
470 kΩ	1
500 kΩ	1
1 MΩ	1
4.7 MΩ	1
10 MΩ	1

Potentiometers (linear taper)	*Quantity*
100 Ω	1
500 Ω	1
10 kΩ	2
50 kΩ	1
100 kΩ	1
1 MΩ	1

Capacitors	*Quantity*
0.01 μF at 25 WV dc	2
0.02 μF at 25 WV dc	1
0.05 μF at 25 WV dc	1
0.1 μF at 25 WV dc	2
0.2 μF at 25 WV dc	1
1 μF at 25 WV dc	2
10 μF at 25 WV dc	1
5 μF at 50 WV dc	1
100 μF at 12 WV dc	1
220 μF at 15 WV dc	1
470 μF at 50 WV dc	2

Appendix II

Interpreting and Understanding Data Sheets

Sometimes a technician needs to know the operating parameters and limitations of a particular active component or device to repair a piece of equipment. In other cases, a technician may be working on the development of a particular device or system and needs to be familiar with the specifications of a component to set up a proper circuit for testing or modification. In either situation, a technician should be familiar with specifications and data sheets.

Every electronic component created will have data sheets with specifications for that component supplied by the manufacturer. Several companies have compiled manuals of the specifications of all the components and placed them into tabular form. Figure A-1 shows how two bipolar transistors may appear. The 2N3565 transistor is generally used as a small-signal voltage amplifier, and the 2N3055 is used as an output power device to drive speakers, relays, or some other device. The first column shows the transistor identification number. The type column indicates *N* if it is *NPN* type or *P* if it is a *PNP* type. The Ma column shows if the basic material is silicon (Si) or germanium (Ge). Absolute maximum ratings are given to protect the particular device. If these ratings are exceeded, the circuit performance or service life of the device is usually impaired. Breakdown voltages refer to the reverse voltage applied to one selected junction while normally the other junction is open circuited. These voltages are given as:

BV_{ceo} = DC breakdown voltage, collector to emitter, with the base open

BV_{cbo} = DC breakdown voltage, collector to base, with the emitter open

BV_{ebo} = DC breakdown voltage, emitter to base, with the collector open

The average maximum continuous total power dissipation, P_T, of a transistor is equal to $V_c \times I_c$ for a given temperature, which is usually 25 °C. The total allowable collector current $I_{c\,max}$ may also be given.

The maximum leakage current, I_{cbo} or I_{co}, is from the collector to the base with the emitter open. A large value of I_{cbo}, as compared to the rating, usually indicates a partially shorted transistor.

The interval of time taken for electrons to travel from the emitter to the collector is termed transit time. Transit time places a limit on the highest frequency than can be amplified by a transistor. The reaction of the transit time acts like a small capacitance in parallel with the base and emitter. At low frequencies, this capacitance has little or no effect on the gain of the transistor. However, at higher frequencies, the capacitance has a shunting effect, which reduces the transistor's gain. The frequency at which the gain, alpha for a common-base circuit, falls to 70% is termed the alpha-cutoff frequency, f_{ab}.

The gain or amplification factor of a transistor is called beta, for a common-emitter circuit, and is represented as h_{fe}. Beta can vary somewhat under different circuit operating conditions, and a minimum and maximum range is usually indicated with specific voltages, current, and frequency.

Package information may include the type of case and a drawing reference. Sometimes a picture of the case is shown and the drawings indicate the physical dimensions of the package.

This appendix also shows some data sheets of integrated circuits. These sheets may include device description, an equivalent schematic diagram, special features, absolute maximum ratings, pin configurations, DC electrical characteristics, typical performance characteristic graphs, and some applications. Studying data sheets can reveal many interesting facts about a device and also increase your knowledge of the operating principles of the device.

The following IC data sheets are reprinted with permission granted by Signetics Corporation, a subsidiary of U.S. Philips Corp., 811 E. Arque Ave., Sunnyvale, Calif. 94086.

Figure A–1 Example of specifications and diagrams used in data sheets for discrete devices.

Transistor number	Type	Ma	ABS Max ratings at 25°C				Max I_{CBO}	f_{ab}	h_{fe}		Package	
			BV_{CEO} (V)	BV_{CBO} (V)	BV_{EBO} (V)	P_T (W)	At max V_{CB} (A)	(H)	V_{CE} = 5 V f = 1 kH I_C = 1 mA		Case type	DWG
									Min	Max		
2N3565	N	Si	25	30	6	200 m	0.05 μ	40 M	120	750	TO-92	D-1
2N3055	N	Si	70	100	7	115	5.0 m	10 k	20	100	TO-3 TO-202	D-2 D-3

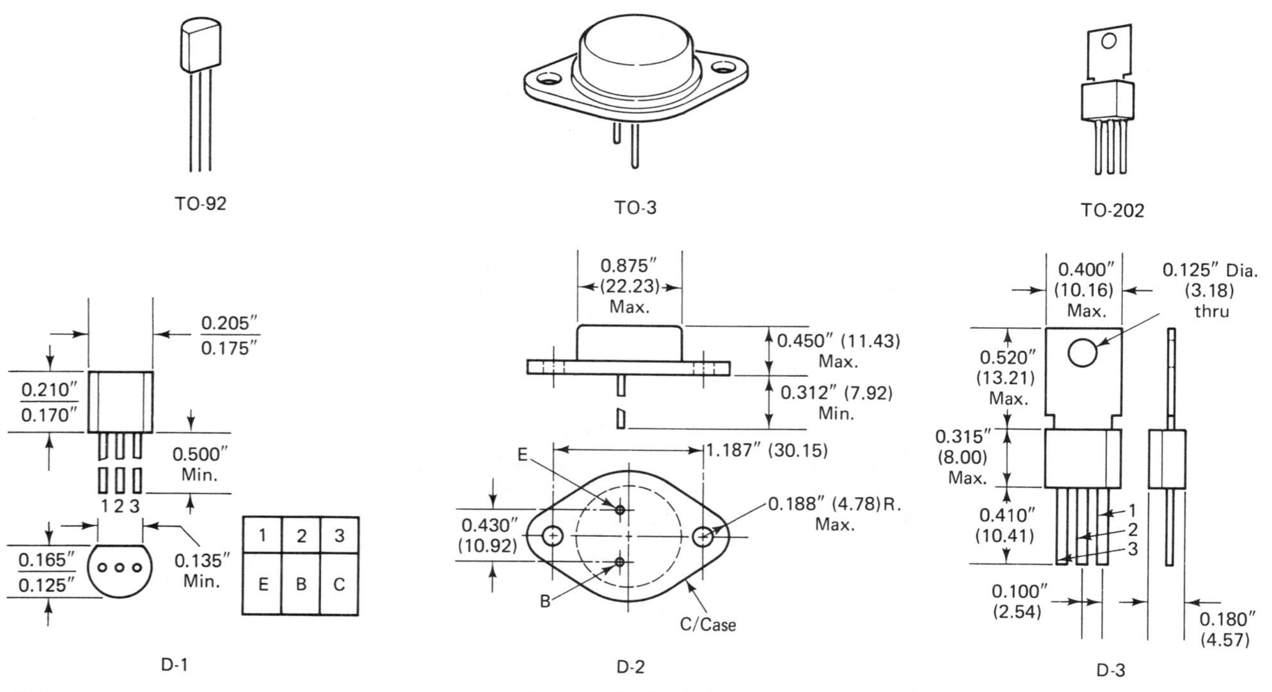

μA741/741C-N,FE
SA741C-N
μA741C-D

DESCRIPTION

The μA741 is a high performance operational amplifier with high open loop gain, internal compensation, high common mode range and exceptional temperature stability. The μA741 is short-circuit protected and allows for nulling of offset voltage.

FEATURES

- **Internal frequency compensation**
- **Short circuit protection**
- **Excellent temperature stability**
- **High input voltage range**

PIN CONFIGURATION

D,N,FE PACKAGE

OFFSET NULL 1 · INV INPUT 2 · NON INV INPUT 3 · V- 4 · OFFSET NULL 5 · OUTPUT 6 · V+ 7 · NC 8

ORDER PART NO.

μA741N μA741FE
μA741CN μA741CFE
SA741CN
μA741CD

EQUIVALENT SCHEMATIC

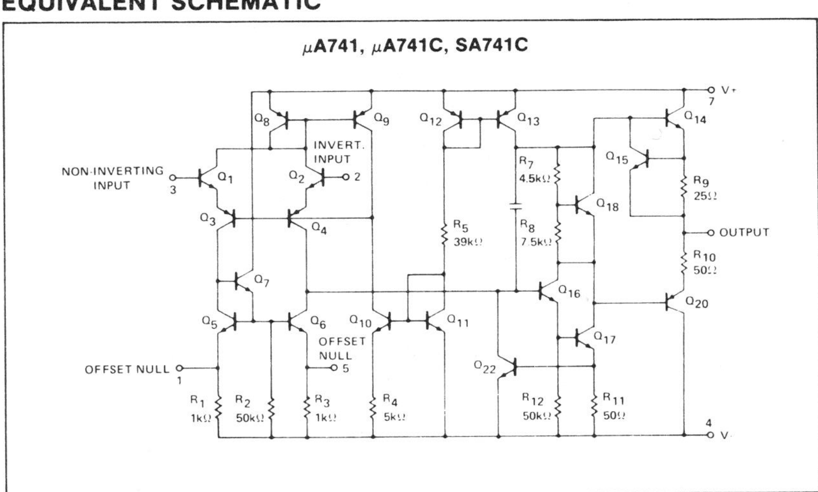

ABSOLUTE MAXIMUM RATINGS

PARAMETER	RATING	UNIT
Supply voltage		
μA741C	±18	V
μA741	±22	V
Internal power dissipation		
N package	500	mW
FE package	1000	mW
Differential input voltage	±30	V
Input voltage[1]	±15	V
Output short-circuit duration	Continuous	
Operating temperature range		
μA741C	0 to +70	°C
SA741C	-40 to +85	°C
μA741	-55 to +125	°C
Storage temperature range	-65 to +150	°C
Lead temperature (soldering 60sec)	300	°C

NOTE
1. For supply voltages less than ±15V, the absolute maximum input voltage is equal to the supply voltage.

µA741/741C-N,FE
SA741C-N
µA741C-D

DC ELECTRICAL CHARACTERISTICS $T_A = 25°C$, $V_S = \pm15V$, unless otherwise specified.

PARAMETER		TEST CONDITIONS	µA741			µA741C			UNIT
			Min	Typ	Max	Min	Typ	Max	
V_{OS}	Offset voltage	$R_S = 10k\Omega$		1.0	5.0		2.0	6.0	mV
		$R_S = 10k\Omega$, over temp.		1.0	6.0			7.5	mV
I_{OS}	Offset current			20	200		20	200	nA
		Over temp.						300	nA
		$T_A = +125°C$		7.0	200				nA
		$T_A = -55°C$		20	500				nA
I_{BIAS}	Input bias current			80	500		80	500	nA
		Over temp.						800	nA
		$T_A = +125°C$		30	500				nA
		$T_A = -55°C$		300	1500				nA
V_{OUT}	Output voltage swing	$R_L = 10k\Omega$	±12	±14		±12	±14		V
		$R_L = 2k\Omega$, over temp.	±10	±13		±10	±13		V
A_{VOL}	Large signal voltage gain	$R_L = 2k\Omega$, $V_O = \pm10V$	50	200		20	200		V/mV
		$R_L = 2k\Omega$, $V_O = \pm10V$, over temp.	25			15			V/mV
	Offset voltage adjustment range			±30			±30		mV
P_{SRR}	Supply voltage rejection ratio	$R_S \leq 10k\Omega$					10	150	µV/V
		$R_S \leq 10k\Omega$, over temp.		10	150				µV/V
CMRR	Common mode rejection ratio								dB
		Over temp.	70	90					dB
I_{CC}	Supply current			1.4	2.8		1.4	2.8	mA
		$T_A = +125°C$		1.5	2.5				mA
		$T_A = -55°C$		2.0	3.3				mA
V_{IN}	Input voltage range	(µA741, over temp.)	±12	±13		±12	±13		V
R_{IN}	Input resistance		0.3	2.0		0.3	2.0		MΩ
P_d	Power consumption			50	85		50	85	mW
		$T_A = +125°C$		45	75				mW
		$T_A = -55°C$		45	100				mW
R_{OUT}	Output resistance			75			75		Ω
I_{SC}	Output short-circuit current			25			25		mA

μA741/741C-N,FE
SA741C-N
μA741C-D

DC ELECTRICAL CHARACTERISTICS (Cont'd) $T_A = 25°C$, $V_S = \pm15V$, unless otherwise specified.

	PARAMETER	TEST CONDITIONS	SA741C			UNIT
			Min	Typ	Max	
V_{OS}	Offset voltage	$R_S = 10k\Omega$		2.0	6.0	mV
		$R_S = 10k\Omega$, over temp.			7.5	mV
I_{OS}	Offset current			20	200	nA
		Over temp.			500	nA
I_{BIAS}	Input bias current			80	500	nA
		Over temp.			1500	nA
V_{OUT}	Output voltage swing	$R_L = 10k\Omega$	±12	±14		V
		$R_L = 2k\Omega$, over temp.	±10	±13		V
A_{VOL}	Large signal voltage gain	$R_L = 2k\Omega$, $V_O = \pm10V$	20	200		V/mV
		$R_L = 2k\Omega$, $V_O = \pm10V$, over temp.	15			V/mV
	Offset voltage adjustment range			±30		mV
P_{SRR}	Supply voltage rejection ratio	$R_S \leq 10k\Omega$		10	150	μV/V
CMRR	Common mode rejection ratio					dB
I_{CC}	Supply current			1.4	2.8	mA
V_{IN}	Input voltage range	(μA741, over temp.)	±12	±13		V
R_{IN}	Input resistance		0.3	2.0		MΩ
P_d	Power consumption			50	85	mW
R_{OUT}	Output resistance			75		Ω
I_{SC}	Output short-circuit current			25		mA

AC ELECTRICAL CHARACTERISTICS $T_A = 25°C$, $V_S = \pm15V$, unless otherwise specified.

PARAMETER	TEST CONDITIONS	μA741, μA741C			UNIT
		Min	Typ	Max	
Parallel input resistance	Open loop, f = 20Hz				MΩ
Parallel input capacitance	Open loop, f = 20Hz		1.4		pF
Unity gain crossover frequency	Open loop		1.0		MHz
Transient response unity gain	$V_{IN} = 20mV$, $R_L = 2k\Omega$, $C_L \leq 100pf$				
Rise time			0.3		μs
Overshoot			5.0		%
Slew rate	$C \leq 100pf$, $R_L \geq 2k$, $V_{IN} = \pm10V$		0.5		V/μs

TYPICAL PERFORMANCE CHARACTERISTICS

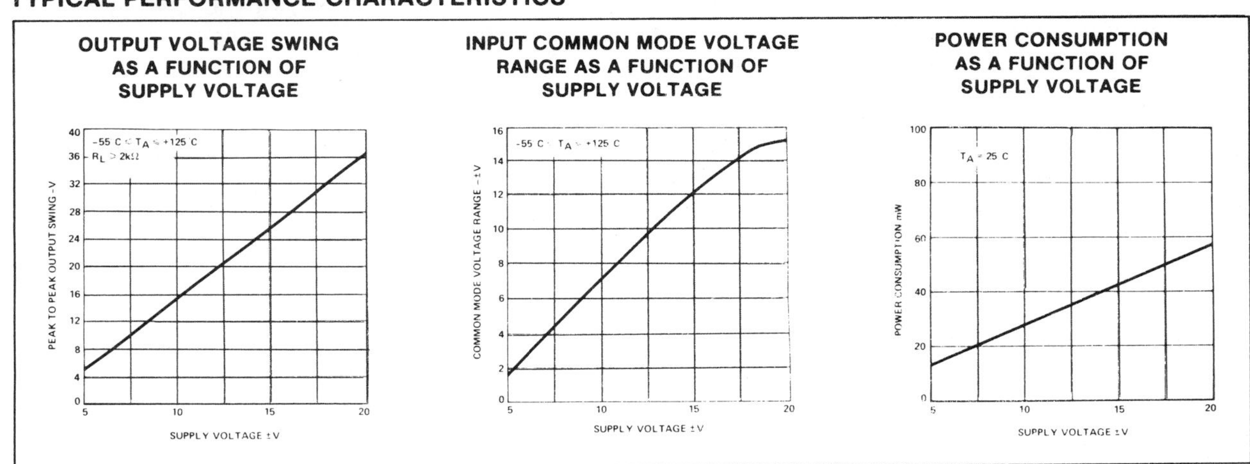

TYPICAL PERFORMANCE CHARACTERISTICS (Cont'd)

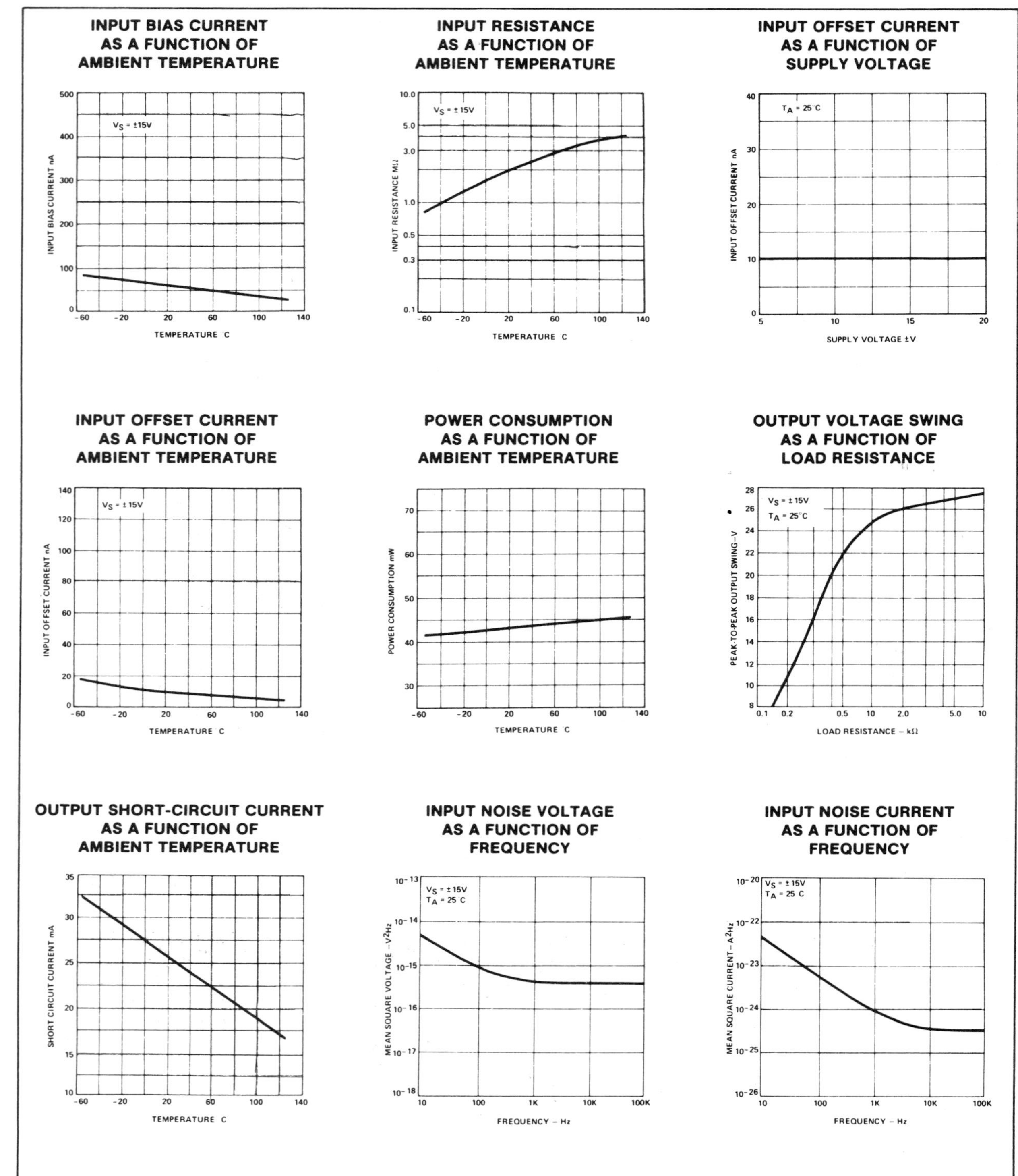

TYPICAL PERFORMANCE CHARACTERISTICS (Cont'd)

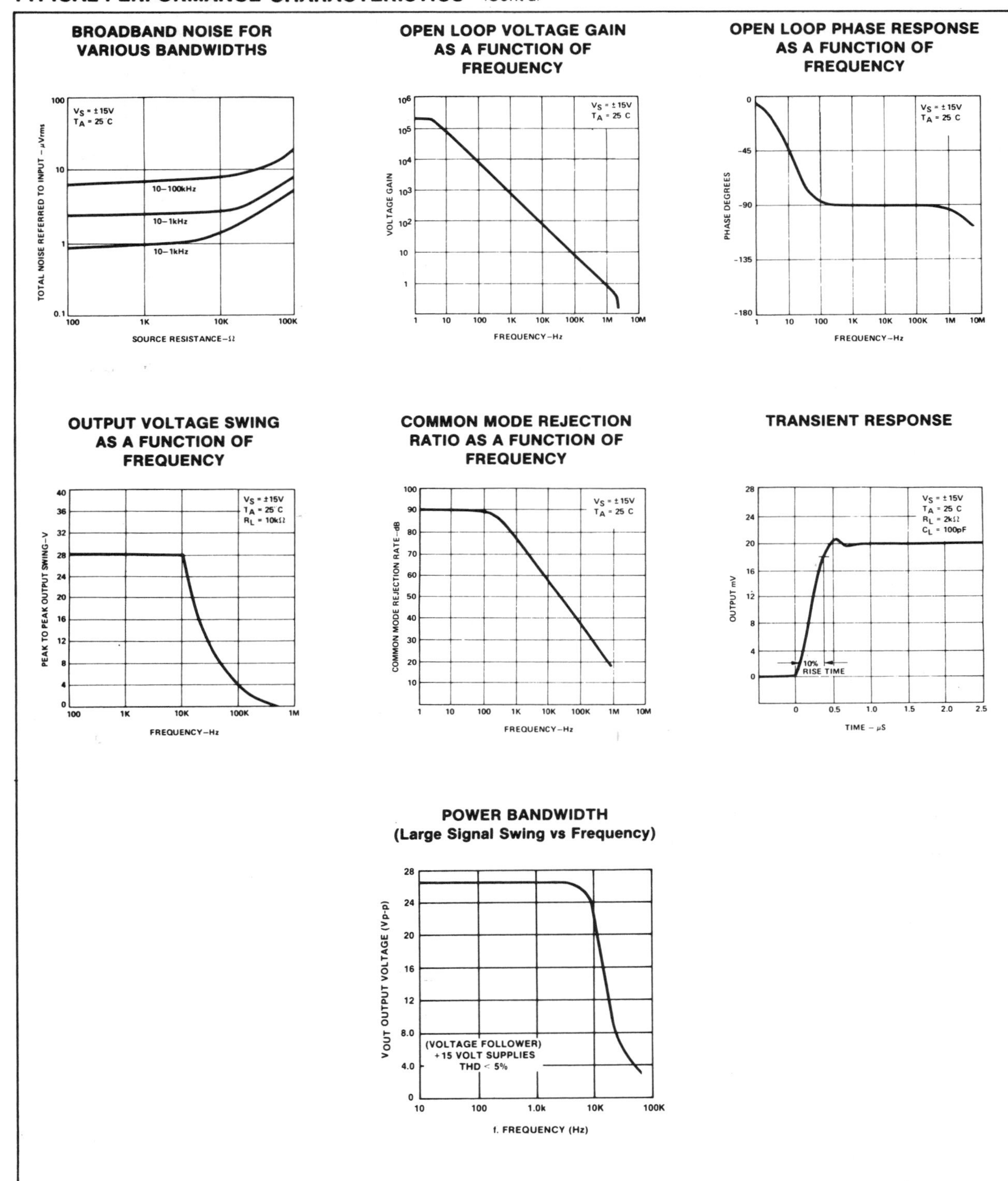

NE/SE5514-F,N

DESCRIPTION

The NE/SE5514 family of Quad Operational Amplifiers sets new standards in Bipolar Quad Amplifier Performance. The amplifiers feature low input bias current and low offset voltages. Pin-out is identical to LM324/LM348 which facilitates direct product substitution for improved system performance. Output characteristics are similar to a μA741 with improved slew and drive capability.

FEATURES

- **Low input bias current: <±3nA**
- **Low input offset current: <±3nA**
- **Low input offset voltage: <1mV**
- **Low supply current: 1.5mA/Amp**
- **1 V/μsec slew rate**
- **High input impedance: 100MΩ**
- **High common mode impedance: 10GΩ**
- **Internal compensation for unity gain**

APPLICATIONS

- **AC amplifiers**
- **RC active filters**
- **Transducer amplifiers**
- **DC gain block**
- **Instrumentation amplifier**

PIN CONFIGURATION

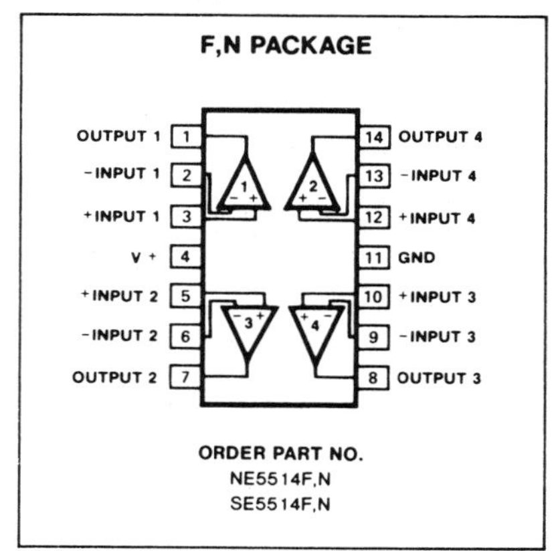

ABSOLUTE MAXIMUM RATINGS

	PARAMETER	RATING	UNIT
V_{CC}	Supply voltage	±16	V
V_{DIFF}	Differential input voltage	32	V
V_{IN}	Input voltage	0 to 32	V
	Output short to ground	Continuous	
TS	Storage temperature range	−65 to +150	°C
T_{SOLD}	Lead soldering temperature	300	°C
T_A	Operating temperature range		
	NE5514	0 to 70	°C
	SE5514	−55 to +125	°C

EQUIVALENT SCHEMATIC

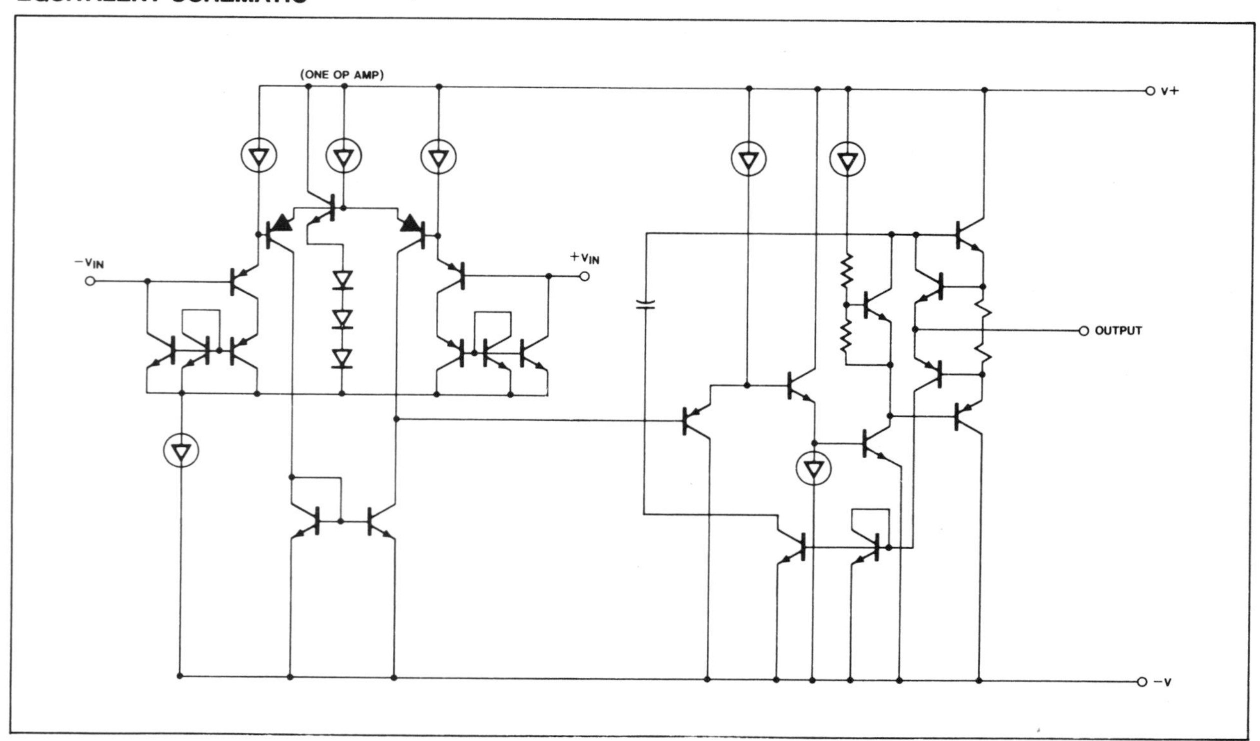

NE / SE5514-F,N

ELECTRICAL CHARACTERISTICS $V_{CC} = \pm 15V$, F.R. = −55°C to +125°C (SE) 0°C to 70°C (NE)

	PARAMETER	TEST CONDITIONS	SE5514			NE5514			UNIT
			Min	Typ	Max	Min	Typ	Max	
V_{OS}	Input offset voltage	$R_S = 100\Omega$, $T_A = +25°C$, T_A = F.R.		0.7 1	2 3		1 1.5	5 6	mV
I_{OS}	Input offset current	$R_S = 100k\Omega$, $T_A = +25°C$, T_A = F.R.		3 4	10 20		6 8	20 30	nA
I_B	Input bias current	$R_S = 100k\Omega$, $T_A = +25°C$, T_A = F.R.		3 4	10 20		6 8	20 30	nA
R_{IN}	Input resistance differential	$T_A = 25°C$		100			100		MΩ
V_{CM}	Input common mode range	$T_A = 25°C$, T_A = F.R.	± 13.5 ± 13.5	± 13.7 ± 13.2		± 13.5 ± 13	± 13.7 ± 13.2		V
CMRR	Input common-mode rejection ratio	$V_{CC} = \pm 15V$, $V_{IN} = \pm 13.5V$ (RM), $T_A = 25°C$, $V_{IN} = \pm 13V$ (F.R.), T_A = F.R.	70	100		70	100		dB
AVOL GAIN	Large-signal voltage gain	$R_L = 2k\Omega$, $T_A = 25°C$ $V_C = \pm 10V$, T_A = F.R.	50 25	200		50 25			V/mV
S.R.	Slew rate	$T_A = 25°C$	0.6	1		0.6	1		V/µs
GBW	Small-signal unity gain bandwidth	$T_A = 25°C$		1			1		MHz
θ_M	Phase margin	$T_A = 25°C$		45			45		Degr
V_{OUT}	Output voltage swing	$R_L = 2k\Omega$, $T_A = 25°C$ T_A = F.R.	± 13 ± 12.5	± 13.5 ± 13		± 13 ± 12.5	± 13.5 ± 13		V
V_{OUT}	Output voltage swing	$R_L = 600\Omega$*, $T_A = 25°C$ T_A = F.R.	± 10 ± 8	± 11.5 ± 9		± 10 ± 8	± 11.5 ± 9		V
I_{CC}	Power supply current	R_L = Open, $T_A = 25°C$ T_A = F.R.		6 7	10 12		6 7	10 12	mA
PSRR	Power supply rejection ratio	$T_A = 25°C$, T_A = F.R.	80 80	110 100		80 80	110 100		dB
AA	Amplifier to amplifier coupling	f = 1kHz to 20kHz, $T_A = 25°C$		−120			−120		dB
HD	Total harmonic distortion	f = 10kHz, $T_A = 25°C$ V_O = 7VRMS		0.01			0.01		%
V_{INN}	Input-noise voltage	f = 1kHz, $T_A = 25°C$		30			30		$nV/\sqrt{Hz}$

NOTE

*For operation at elevated temperature, N package must be derated based on a thermal resistance of 95°C/W junction to ambient.

FOUR QUADRANT PHOTO-CONDUCTIVE DETECTOR AMPLIFIER

When operating a photo diode in the photoconductive mode (reverse biased) very small currents in the micro ampere range must be sensed in the photo active operating region. Dark currents in the nano amperes are common. Generally, for this reason, J-FET input preamps are used to prevent interaction and accuracy degradation due to input bias currents.

The 5514 has sufficiently low input bias current (6na) to allow its use under these circuit constraints as shown in a possible design used to sense four quadrant motion of a light source. By proper summing of the signals from the X and Y axes, four quadrant output may be fed to an X-Y plotter, oscilloscope or computer for simulation. (See figure 1).

The wide input common mode voltage range of the device allows a +10 volt supply to be used to drive the signal bridge giving high sensitivity and improved signal to noise. Obviously, input balancing is critical to achieving common mode signal rejection in addition to adequate shielding of the sensor leads. The sensor head itself must be shielded and the shield grounded to signal common to avoid unwanted noise pick up from power line and other local noise sources. Amplifier response may be shaped to aid in noise reduction by more complex filter configurations. If possible the 5514 should be located in close proximity to the sensor head.

System balance may be done under dark field conditions if adequate photo detector tracking results. However, for high accuracy systems a bipolar balance adjust added to the non-inverting output stage is more desirable. With this latter method the signal bridge is balanced for a null output under uniform light field conditions using the bridge balance pot as shown. D.C. offset is then adjusted using the balance pot on the output amplifier under dark field conditions.

FOUR QUADRANT PHOTO DETECTOR

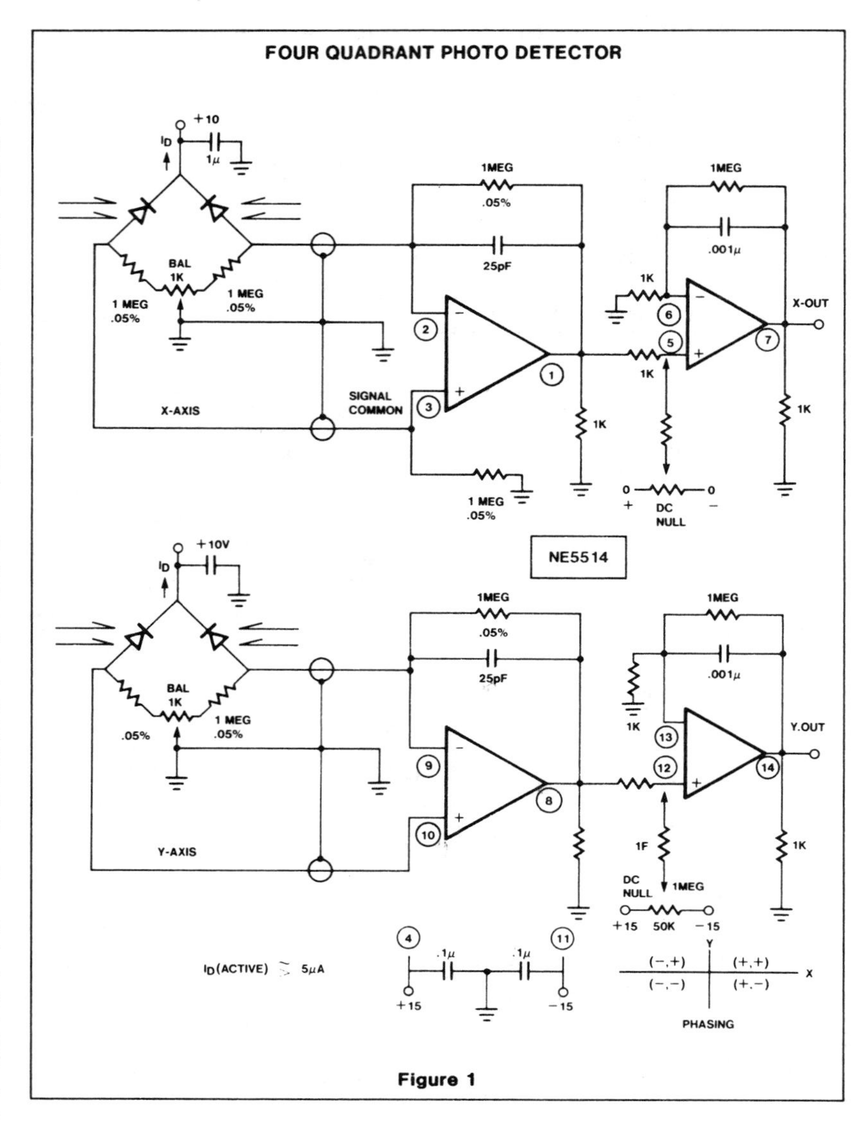

Figure 1

MULTI-TONE BANDPASS FILTER FOR PLL TONE DECODER

In the design of a multiple tone signaling system, particularly where signals are transmitted over long lines, noise and adjacent channel interference may be a significant barrier to reliable communications.

By the use of narrow band active pre-filters to attain selectivity and gain, the effective signal to noise ratio is greatly improved. The NE/SE5514 is easily adapted to such filter configurations due to its inherent stability. In addition its very high input impedance drastically reduces loading on the passive networks and allows for increased "Q" and large value resistors.

The circuit in Figure 2 demonstrates multiple feedback filters operating at four of the standard signaling frequencies. More channels may be added to increase the capacity of the system.

Test results obtained from this filter configuration were as follows:

Wide band signal to noise	63dB
Gain (Mid band)	30dB
Q (effective)	≈ 30
Output	OdBM (.775v_{rms})

Note that the amplifiers are operated from a single +12 volt supply and are biased to half V_{CC} by a simple resistive divider at point B which connects to all non-inverting inputs.

4-STATION 0–50° TEMPERATURE SENSOR

By using an NPN transistor as a temperature sensing element, the NE5514 forms the basis for a multi-station temperature sensor as shown in Figure 3. The principle used is fundamental to the current-voltage relationship of a forward biased junction. The current flow across the base-emitter junction is determined by absolute temperature in the following way:

$$I_E = -(I_C + I_B)$$

and $I_E \propto I_S \exp(V_{BE}/V_T)$; $V_T = \frac{kt}{q}$

therefore, $V_{BE} \propto V_T \ln I_E/I_S$

Where I_E is the forward current and I_S is the saturation current inherent in the junction, I_E must be high enough such that the I_S variation with temperature is small relative to I_E ($I_E >> I_S$). I_S is typically .05 pA, therefore, setting I_E to 1 or 2 μA gives the desired condition.

Diode D_1 serves to substantially reduce error due to power supply variation by giving a fixed voltage reference. To calibrate the sensor adjust R_4 for "O" volts output from the NE5514 at 0°C. Adjust R_6 tracking resistor for a scale factor of 100 millivolts per °C output.

Only the transistor need be placed in the temperature controlled environment. Figure 4 shows the addition of an A/D converter and display to give a digital thermometer.

NE/SE5514
MFB BANDPASS FILTER
FOR
MULTI-CHANNEL TONE DECODER

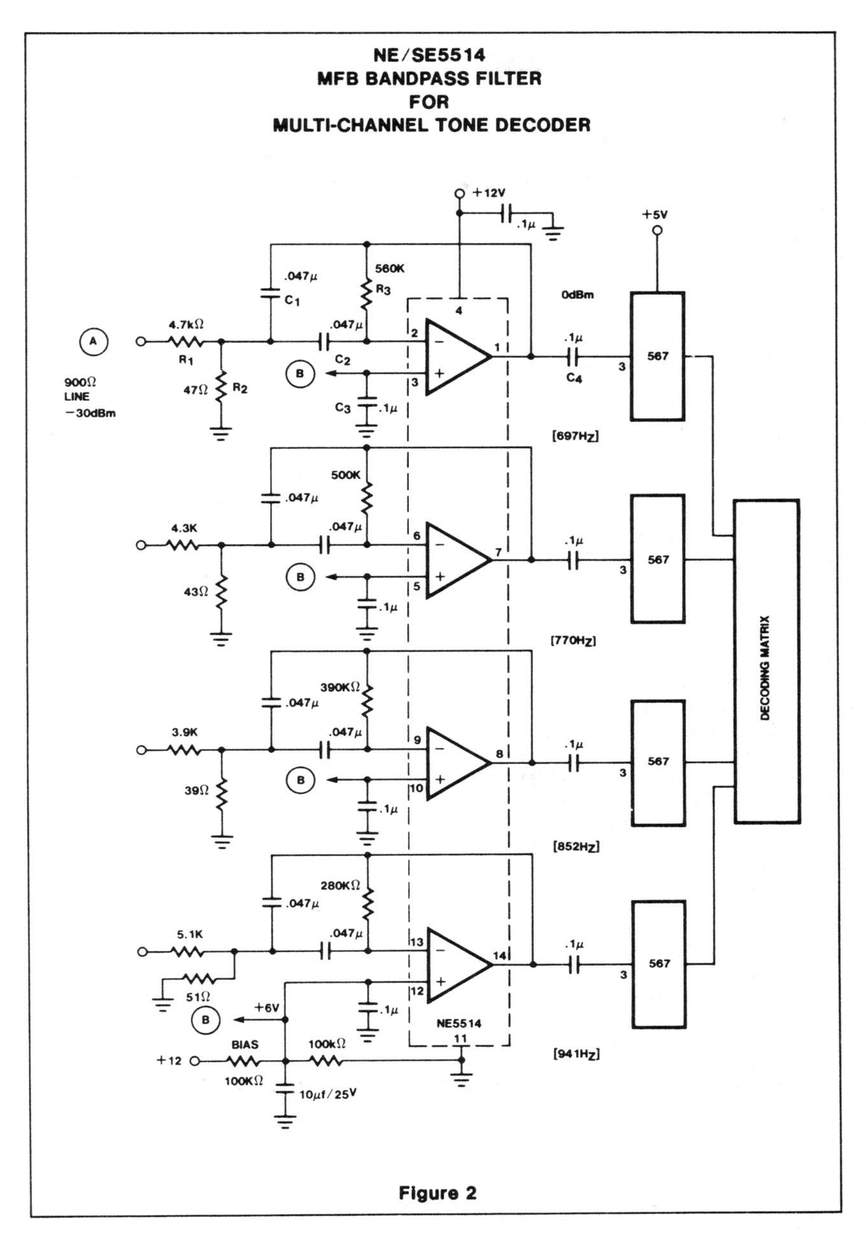

Figure 2

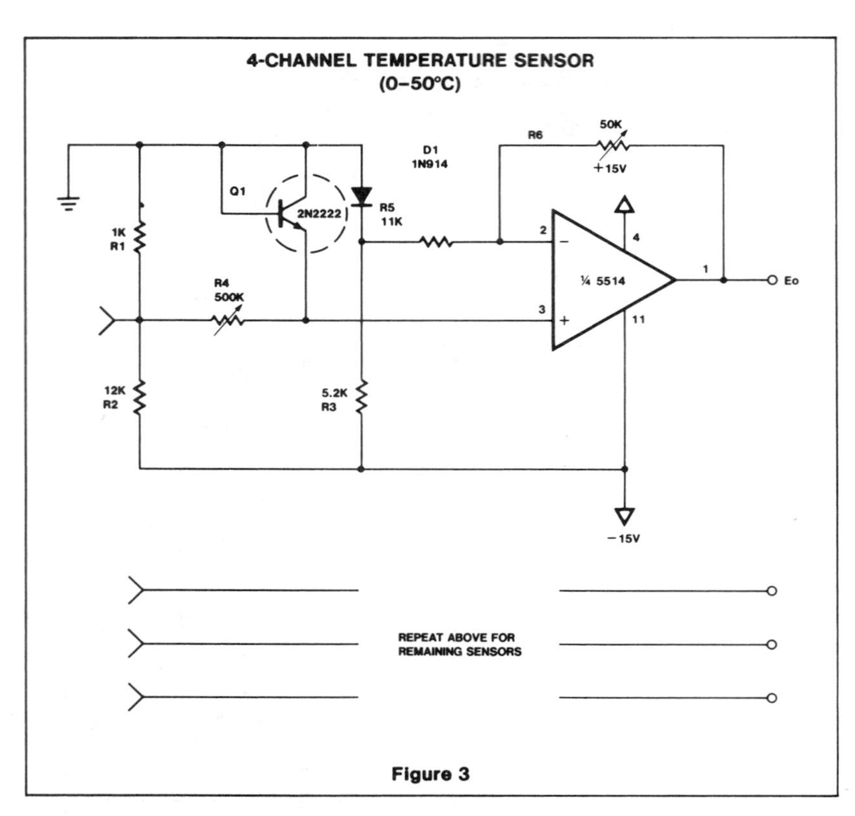

Figure 3

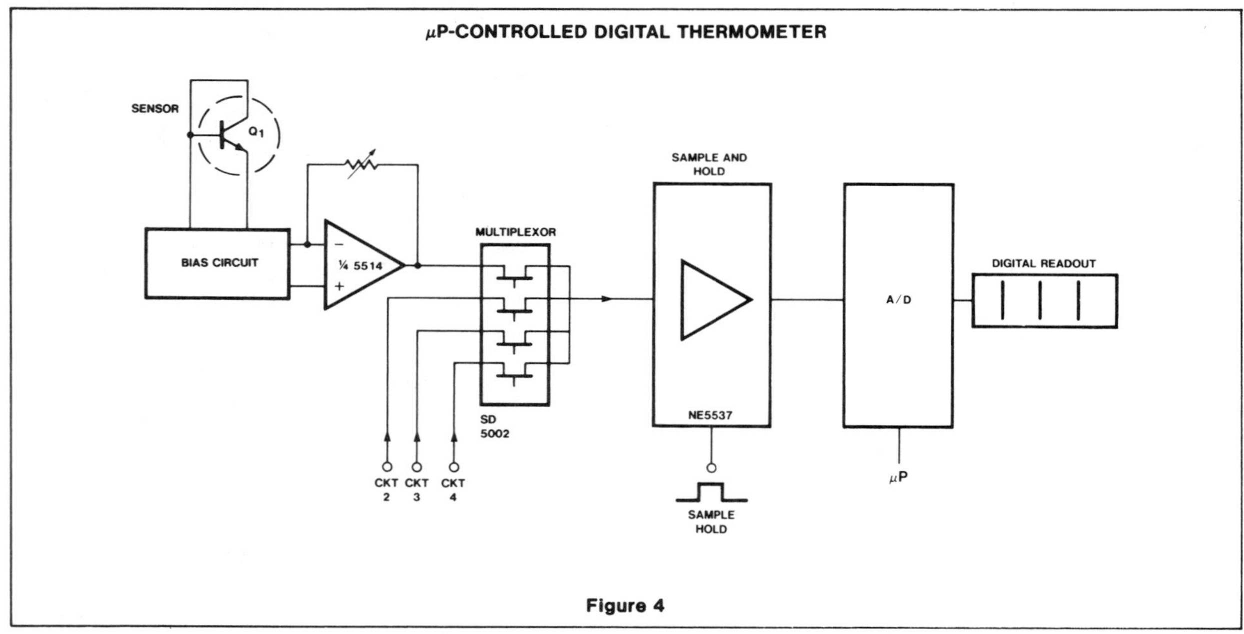

Figure 4

NE/SE555/SE555C-F,H,N,FE

DESCRIPTION

The 555 monolithic timing circuit is a highly stable controller capable of producing accurate time delays, or oscillation. In the time delay mode of operation, the time is precisely controlled by one external resistor and capacitor. For a stable operation as an oscillator, the free running frequency and the duty cycle are both accurately controlled with two external resistors and one capacitor. The circuit may be triggered and reset on falling waveforms, and the output structure can source or sink up to 200mA.

FEATURES

- **Turn off time less than 2μs**
- **Maximum operating frequency greater than 500kHz**
- **Timing from microseconds to hours**
- **Operates in both astable and monostable modes**
- **High output current**
- **Adjustable duty cycle**

PIN CONFIGURATIONS

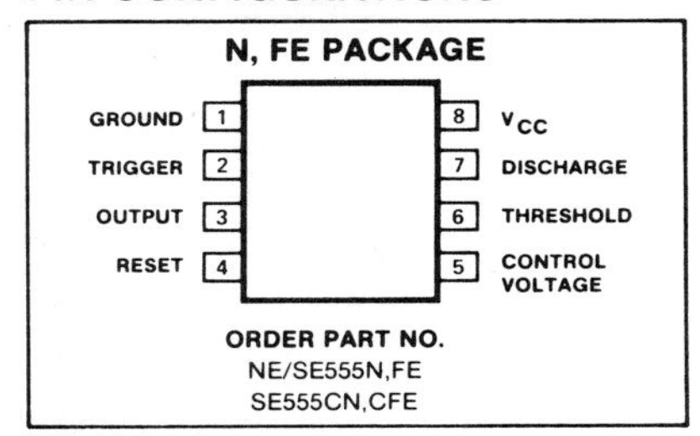

- **TTL compatible**
- **Temperature stability of 0.005% per °C**

APPLICATIONS

- **Precision timing**
- **Pulse generation**
- **Sequential timing**
- **Time delay generation**
- **Pulse width modulation**
- **Pulse position modulation**
- **Missing pulse detector**

F PACKAGE

Pin		Pin	
GND	1	14	VCC
NC	2	13	NC
TRIGGER	3	12	DISCHARGE
OUTPUT	4	11	NC
NC	5	10	THRESHOLD
RESET	6	9	NC
NC	7	8	CONTROL VOLTAGE

ORDER PART NO.
NE/SE555F
SE555CF

H PACKAGE*

VCC 8, GROUND 1, TRIGGER 2, OUTPUT 3, RESET 4, CONTROL VOLTAGE 5, THRESHOLD 6, DISCHARGE 7

ORDER PART NO.
NE/SE555H
SE555CH

*Metal cans (H) not recommended for new designs

ABSOLUTE MAXIMUM RATINGS

PARAMETER	RATING	UNIT
Supply voltage		
SE555	+18	V
NE555, SE555C,	+16	V
Power dissipation	600	mW
Operating temperature range		
NE555	0 to +70	°C
SE555, SE555C	-55 to +125	°C
Storage temperature range	-65 to +150	°C
Lead temperature (soldering, 60sec)	300	°C

EQUIVALENT SCHEMATIC

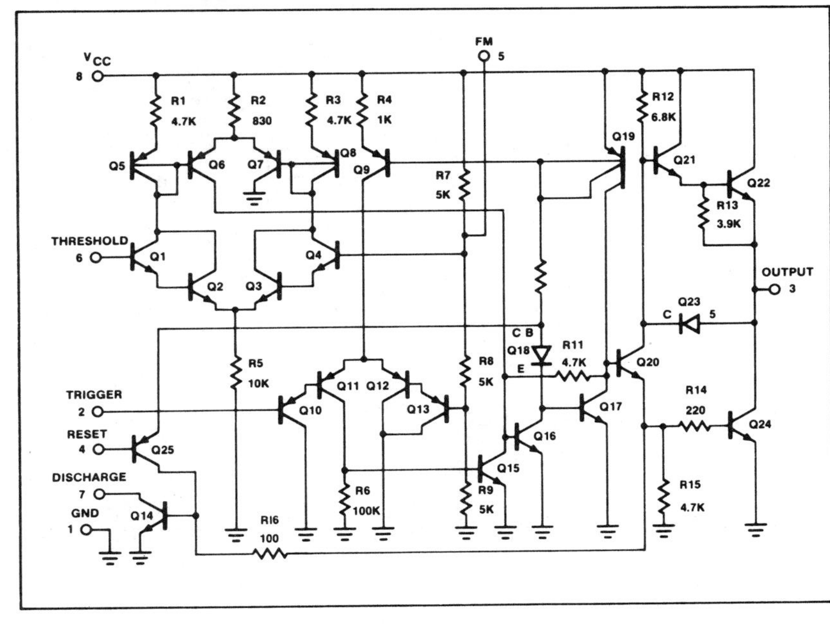

BLOCK DIAGRAM

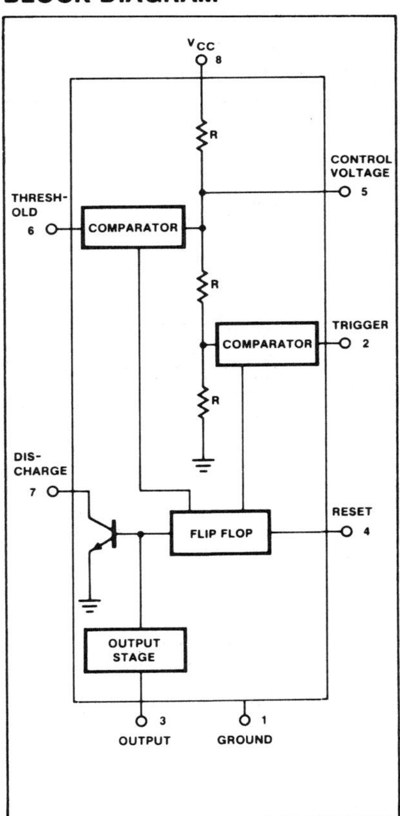

DC ELECTRICAL CHARACTERISTICS T_A = 25°C, V_{CC} = +5V to +15 unless otherwise specified.

PARAMETER	TEST CONDITIONS	SE555			NE555/SE555C			UNIT
		Min	Typ	Max	Min	Typ	Max	
Supply voltage		4.5		18	4.5		16	V
Supply current (low state)[1]	V_{CC} = 5V R_L = ∞		3	5		3	6	mA
	V_{CC} = 15V R_L = ∞		10	12		10	15	mA
Timing error (monostable)	R_A = 2KΩ to 100KΩ							
Initial accuracy[2]	C = 0.1μF		0.5	2.0		1.0	3.0	%
Drift with temperature			30	100		50		ppm/°C
Drift with supply voltage			0.05	0.2		0.1	0.5	%/V
Timing error (astable)	R_A, R_B = 1kΩ to 100kΩ							
Initial accuracy[2]	C = 0.1μF		1.5			2.25		%
Drift with temperature	V_{CC} = 15V		90			150		ppm/°C
Drift with supply voltage			0.15			0.3		%/V
Control voltage level	V_{CC} = 15V	9.6	10.0	10.4	9.0	10.0	11.0	V
	V_{CC} = 5V	2.9	3.33	3.8	2.6	3.33	4.0	V
Threshold voltage	V_{CC} = 15V	9.4	10.0	10.6	8.8	10.0	11.2	V
	V_{CC} = 5V	2.7	3.33	4.0	2.4	3.33	4.2	V
Threshold current[3]			0.1	0.25		0.1	0.25	μA
Trigger voltage	V_{CC} = 15V	4.8	5.0	5.2	4.5	5.0	5.6	V
	V_{CC} = 5V	1.45	1.67	1.9	1.1	1.67	2.2	V
Trigger current	V_{TRIG} = 0V		0.5	0.9		0.5	2.0	μA
Reset voltage[4]		0.4	0.7	1.0	0.4	0.7	1.0	V
Reset current			0.1	0.4		0.1	0.4	mA
Reset current	V_{RESET} = 0V		0.4	1.0		0.4	1.5	mA
Output voltage (low)	V_{CC} = 15V							
	I_{SINK} = 10mA		0.1	0.15		0.1	0.25	V
	I_{SINK} = 50mA		0.4	0.5		0.4	0.75	V
	I_{SINK} = 100mA		2.0	2.2		2.0	2.5	V
	I_{SINK} = 200mA		2.5			2.5		V
	V_{CC} = 5V							
	I_{SINK} = 8mA		0.1	0.25		0.3	0.4	V
	I_{SINK} = 5mA		0.05	0.2		0.25	0.35	V
Output voltage (high)	V_{CC} = 15V							
	I_{SOURCE} = 200mA		12.5			12.5		V
	I_{SOURCE} = 100mA	13.0	13.3		12.75	13.3		V
	V_{CC} = 5V							
	I_{SOURCE} = 100mA	3.0	3.3		2.75	3.3		V
Turn off time[5]	V_{RESET} = V_{CC}		0.5	2.0		0.5		μs
Rise time of output			100	200		100	300	ns
Fall time of output			100	200		100	300	ns
Discharge leakage current			20	100		20	100	na

NOTES

1. Supply current when output high typically 1mA less.
2. Tested at V_{CC} = 5V and V_{CC} = 15V.
3. This will determine the maximum value of $R_A + R_B$, for 15V operation, the max total R = 10 megohm, and for 5V operation, the max total R = 3.4 megohm.
4. Specified with trigger input high.
5. Time measured from a positive going input pulse from 0 to 0.8 x V_{CC} into the threshold to the drop from high to low of the output. Trigger is tied to threshold.

TYPICAL PERFORMANCE CHARACTERISTICS

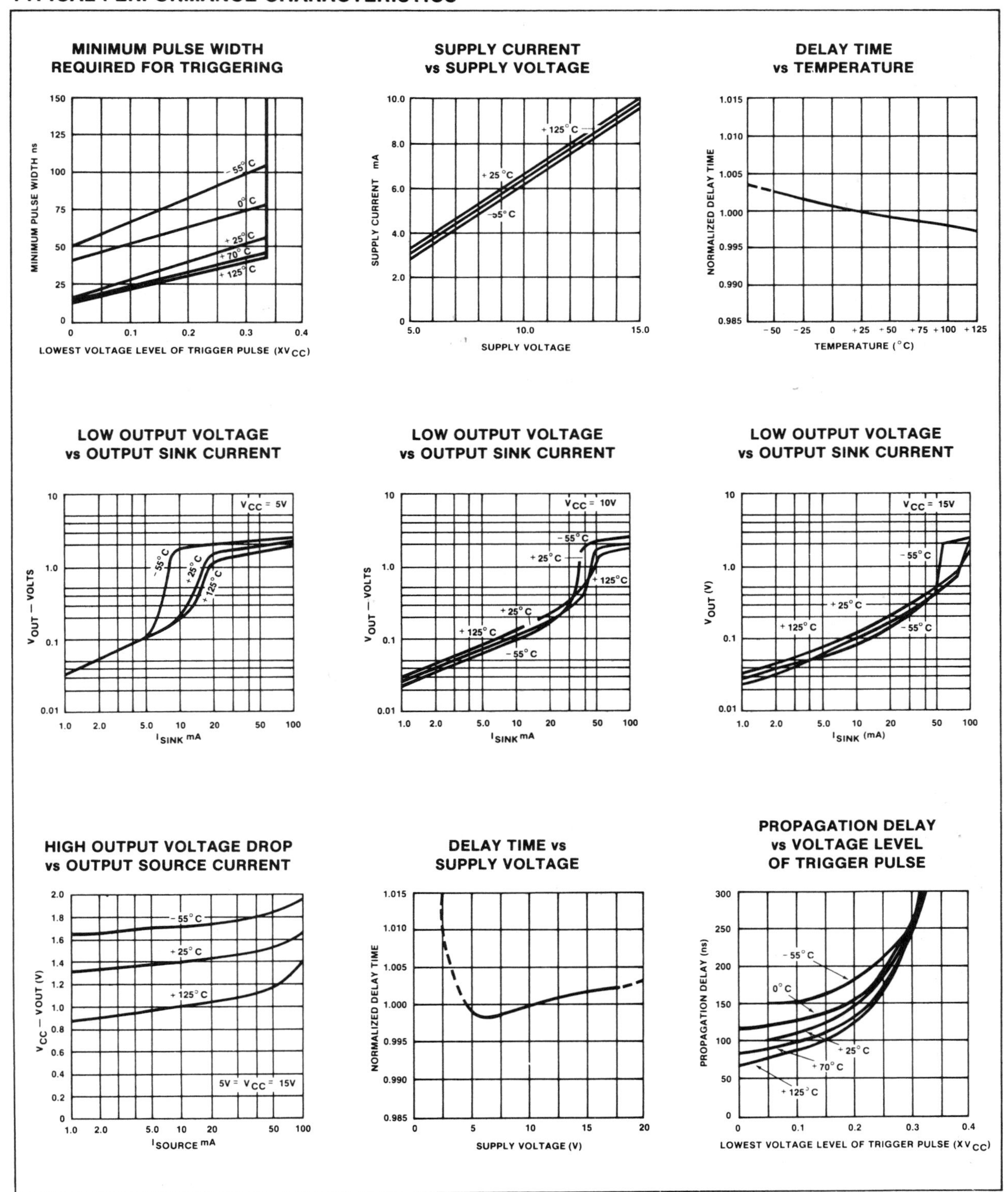

Index

V

Z